Constructions and Combinatorial
Problems in Design of Experiments

A WILEY PUBLICATION IN MATHEMATICAL STATISTICS

Constructions and Combinatorial Problems in Design of Experiments

DAMARAJU RAGHAVARAO

Professor and Head
Department of Mathematics and Statistics
Punjab Agricultural University
Ludhiana, India

John Wiley & Sons, Inc.
New York · London · Sydney · Toronto

Library of Congress Catalogue Card Number: 78-148508

ISBN 0-471-70485-7

Printed in the United States of America.

10 9 8 7 6 5 4 3 2 1

Dedicated to the memory of my parents

Lakshminarasamma and Venkatrayudu

Preface

The design of experiments consists of two parts: (a) the analysis and (b) the constructions and combinatorial problems. There are excellent textbooks on the analysis of the design of experiments. Kempthorne's book *The Design and Analysis of Experiments* meets the requirements of mathematical statisticians, and the book by Cochran and Cox, *Experimental Designs*, is useful to experimental workers. Though these and other books on the design of experiments and combinatorial mathematics cover portions of the constructions and combinatorial problems, a full-length treatment of the subject has not so far appeared in book form. The present work aims at this.

Courses and research in the combinatorial theory of the design of experiments are on the increase, and the literature is available only in scattered papers in journals. This volume will serve as a suitable text in its area. It is hoped that it will also be useful as a secondary and primary reference for statisticians and mathematicians doing research on the design of experiments.

The material is covered in 17 full-length chapters and two appendices. Chapters 1, 2, and 3 cover orthogonal latin squares and their generalizations, Chapters 4 through 11 cover incomplete block designs, Chapter 12 covers nonexistence theorems on incomplete block designs, Chapters 13 through 16 cover factorial experiments, and Chapter 17 covers weighing designs. In Chapter 1 results on complete sets of mutually orthogonal latin squares are discussed while results on noncomplete sets are postponed to Chapter 3. Orthogonal arrays, which are generalizations of orthogonal latin squares, are covered in Chapter 2. In Chapters 2 and 3 we make certain applications of balanced incomplete block (BIB) designs and partially balanced incomplete block (PBIB) designs, and though cross references to these concepts are made, persons reading this subject matter for the first time are advised to read Chapters 2 and 3 after mastering Chapters 5 and 8.

Chapter 4 lays the foundation for a series of chapters on incomplete block designs. BIB and PBIB designs are considered in Chapters 5 and 8, respectively. To keep the volume of this book within reasonable limits, only two-associate-class PBIB designs are considered in full details, with some mention of the higher-associate-class designs being made. Results on systems of distinct representatives, tactical configuration, partial geometry, and symmetrical unequal-block arrangements are dealt with in Chapters 6, 7, 9, and 11, respectively. Dualizing known incomplete block designs to get new incomplete block designs is a useful technique and is considered in Chapter 10. Though designs satisfy the ordinary parametric conditions, yet they may not exist. Necessary conditions for the existence of designs using the Hasse–Minkowski invariant are presented in Chapter 12.

Chapters 13 and 14 consider the confounding aspect of symmetrical and asymmetrical factorial experiments, respectively, whereas Chapters 15 and 16 deal with fractional replications and rotatable designs. Chapter 17 considers the various aspects of weighing designs.

Each chapter contains a list of references, and in most cases a bibliography as well to help the reader survey the literature in a particular area.

Since this book is meant for both statisticians and mathematicians, to assist them two appendices are added. Appendix A presents the mathematical background for statisticians, and Appendix B the statistical background for mathematicians.

D. RAGHAVARAO

Ludhiana, India
November 1970

Acknowledgments

It gives me great pleasure to express my deep gratitude to my sister and brother-in-law, Mrs. U. Seethalakshmi and Mr. U. V. Subbarao, who have given me strong support in shaping my career after the death of my parents. Excepting for their interest in me, I would not have been what I am now. I am also very thankful to my wife, Mrs. D. V. Rathnam, and my children D. Venkata Lakshmi and D. Venkatrayudu for allowing me to work long hours at home without demanding much of my attention.

I am grateful to Dr. S. S. Shrikhande, Director of the Centre for Advanced Training and Research in Mathematics, Bombay, in whose company I started appreciating combinatorial problems of the design of experiments, which is the genesis of the present work.

I also record my thanks to Dr. M. S. Randhawa and Dr. A. S. Kahlon, Vice-Chancellor and Dean, College of Basic Sciences and Humanities, respectively of the Punjab Agricultural University for permitting this book to be published by John Wiley and Sons, Inc.

Finally I express my gratitude to Dr. S. S. Shrikhande, Dr. Esther Seiden, Dr. B. V. Shah, Dr. C. R. Nair, Mr. Suresh K. Tharthare, and the Editors of the *Annals of Mathematical Statistics, Journal of the American Statistical Association, Canadian Journal of Mathematics*, and *Biometrics* for allowing me to reproduce some material from their publications.

D. R.

Contents

Constructions and Combinatorial

Problems in Design of Experiments

CHAPTER 1

Complete Sets of Mutually Orthogonal
Latin Squares

1.1. LATIN SQUARES AND ORTHOGONAL LATIN SQUARES

Latin square designs are normally used in experiments to remove the heterogeneity of experimental material in two directions. These designs require that the number of replications equal the number of treatments or varieties. The nomenclature of the older literature used the terms "treatments" or "varieties" rather than the currently used terms "elements" or "symbols," and "blocks" were used for "sets."

Definition 1.1.1. A latin square arrangement is an arrangement of s symbols in s^2 cells arranged in s rows and s columns, such that every symbol occurs once in each row and in each column. The term s is known as the order of the latin square.

If the symbols are taken as A, B, C, D, a latin square arrangement of order 4 is as follows:

$$
\begin{array}{cccc}
A & B & C & D \\
B & C & D & A \\
C & D & A & B \\
D & A & B & C
\end{array}
$$

(1.1.1)

A latin square is said to be in the standard form if the symbols in the first row and first column are in natural order, and it is said to be in the semi-standard form if the symbols of the first row are in natural order. Some authors denote both of these concepts by the term "standard form." However, there is a need to distinguish between these two concepts. The standard form is used for randomizing the latin-square designs, and the semistandard form is needed for studying the properties of the orthogonal latin squares.

1

We now introduce the concept of orthogonal latin squares.

Definition 1.1.2. If in two latin squares of the same order, when super-imposed on one another, every ordered pair of symbols occurs exactly once, the two latin squares are said to be orthogonal. If the symbols of one latin square are denoted by Latin letters and the symbols of the other are denoted by Greek letters, the pair of orthogonal latin squares is also called a graeco-latin square.

Definition 1.1.3. If in a set of latin squares every pair is orthogonal, the set is called a set of mutually orthogonal latin squares (MOLS). It is also called a hypergraecolatin square.

As an example of the graecolatin square, consider the following two latin squares of order 4:

(1.1.2)

$$
\begin{array}{cccc}
A & B & C & D \\
B & A & D & C \\
C & D & A & B \\
D & C & B & A
\end{array}
\qquad
\begin{array}{cccc}
\alpha & \gamma & \delta & \beta \\
\beta & \delta & \gamma & \alpha \\
\gamma & \alpha & \beta & \delta \\
\delta & \beta & \alpha & \gamma
\end{array}
$$

When these are superimposed on one another, we get

(1.1.3)

$$
\begin{array}{cccc}
A\alpha & B\gamma & C\delta & D\beta \\
B\beta & A\delta & D\gamma & C\alpha \\
C\gamma & D\alpha & A\beta & B\delta \\
D\delta & C\beta & B\alpha & A\gamma
\end{array}
$$

We can verify that in (1.1.3) every pair of ordered Latin and Greek symbols occurs exactly once, and hence the two latin squares under consideration constitute a graecolatin square.

In the next section we show that the cardinality of the set of MOLS is bounded above by $s - 1$. A set of $s - 1$ MOLS is known as a complete set of MOLS. Complete sets of MOLS of order s exist when s is a prime or a prime power, and we study a method for their construction in Section 1.3. It may be noted that there may be more than one complete set of MOLS of order s. Much of the work of this chapter is due to Bose (1938).

1.2. UPPER BOUND FOR THE NUMBER OF ORTHOGONAL LATIN SQUARES

Consider a set of MOLS of order s. Without loss of generality, we assume that each latin square of the set is in semistandard form. Let us number the rows and columns from $0, 1, \cdots, s - 1$. Consider the symbol occurring in

the (i, j)th cell at the ith row and jth column of each of the latin squares in
this set $(i \neq 0)$. Clearly the symbol j cannot occur in any latin square in this
position, because it occurs in the 0th row of this column in each latin square.
Furthermore, the symbols are distinct, for otherwise the condition of pairwise
orthogonality of the MOLS will be violated. Since it is possible to have at
most $s - 1$ distinct symbols different from j, the number of latin squares in
this set cannot exceed $s - 1$. We state this result in the form of the following
theorem:

Theorem 1.2.1. The total number of MOLS of order s is at most $s - 1$.

1.3. CONSTRUCTION OF COMPLETE SETS OF MOLS

In this section we describe a method of constructing a complete set of
$s - 1$ MOLS of order s, where s is a prime or a prime power.

Let s be a prime or a prime power and let x be a primitive root of $GF(s)$,
the Galois field of order s. Let the elements of $GF(s)$ be chosen as $\alpha_0 = 0$,
$\alpha_1 = 1$, $\alpha_2 = x, \cdots, \alpha_{s-1} = x^{s-2}$. Let us construct the latin square L_i whose
(r, t)th position is filled by the element

(1.3.1)
$$\alpha_r + \alpha_i \alpha_t \qquad i = 1, 2, \cdots, s - 1;$$
$$r, t = 0, 1, 2, \cdots, s - 1.$$

We now verify that L_i is a latin square, and latin squares L_i and L_j $(i \neq j)$
are orthogonal. In L_i in the rth row $(r = 0, 1, \cdots, s - 1)$ each of the s distinct
elements of $GF(s)$ occurs exactly once. In fact, when t varies from
$0, 1, \cdots, s - 1$, $\alpha_i \alpha_t$ will take each value of $GF(s)$, and hence $\alpha_r + \alpha_i \alpha_t$ will
take each value of $GF(s)$. In the same manner we can see that each element of
$GF(s)$ will occur exactly once in the tth column $(t = 0, 1, \cdots, s - 1)$. Finally,
let us consider two latin squares L_i and L_j as well as their (r, t)th cells. In
L_i the element is $\alpha_r + \alpha_i \alpha_t = \alpha_x$ (say) and in L_j it is $\alpha_r + \alpha_j \alpha_t = \alpha_y$ (say).
Given α_x, α_y, α_i, and α_j, we can solve the above equations for α_r and α_t:

(1.3.2)
$$\alpha_r = \frac{\alpha_i \alpha_y - \alpha_j \alpha_x}{\alpha_i - \alpha_j},$$

$$\alpha_t = \frac{\alpha_x - \alpha_y}{\alpha_i - \alpha_j}.$$

Since these are unique, the ordered pair (α_x, α_y) will occur on the superimposition of L_j on L_i in the (r, t)th cell as determined by (1.3.2). Thus every
ordered pair occurs exactly once.

The method indicated above can be considerably simplified by considering some additional properties of the symbols in the latin squares. The element in the $(r, t + 1)$th cell of L_i is

$$
\begin{aligned}
\alpha_r + \alpha_i \alpha_{t+1} &= x^{r-1} + x^{i-1} x^t \\
&= x^{r-1} + x^i x^{t-1} \\
&= \alpha_r + \alpha_{i+1} \alpha_t,
\end{aligned}
$$

(1.3.3)

which is the element in the (r, t)th cell of L_{i+1}. Hence the $(t + 1)$th column of L_i is the same as the tth column of L_{i+1}. The element in the $(r, 1)$th cell of L_i is

$$
\begin{aligned}
\alpha_r + \alpha_i \alpha_1 &= x^{r-1} + x^{i-1} x^0 \\
&= x^{r-1} + x^{i-1} x^{s-1} \\
&= x^{r-1} + x^i x^{s-2} \\
&= \alpha_r + \alpha_{i+1} \alpha_{s-1},
\end{aligned}
$$

(1.3.4)

which is the element in the $(r, s - 1)$th cell of L_{i+1}. Hence the first column of L_i is same as the $(s - 1)$th column of L_{i+1}.

Therefore, if L_1 is constructed, $L_2, \cdots, L_{s-1}$ can be constructed by cyclically permuting columns such that the $(t + 1)$th column of L_i is the same as the tth column of L_{i+1} $(1 \le t \le s - 2)$ and the first column of L_i is the same as the $(s - 1)$th column of L_{i+1}. The latin square L_1 can easily be seen to be the addition table of the elements of $GF(s)$.

We use this method to construct a complete set of MOLS of order 4. The elements of $GF(2^2)$ are $\alpha_0 = 0$, $\alpha_1 = 1$, $\alpha_2 = x$, $\alpha_3 = x + 1$, and the latin square L_1, which is the addition table of the elements of $GF(2^2)$, is

(1.3.5)
$$
L_1 = \begin{matrix}
0 & 1 & x & 1+x \\
1 & 0 & 1+x & x \\
x & 1+x & 0 & 1 \\
1+x & x & 1 & 0
\end{matrix}
$$

The other two latin squares generated by the method of this section are

(1.3.6)
$$
L_2 = \begin{matrix}
0 & x & 1+x & 1 \\
1 & 1+x & x & 0 \\
x & 0 & 1 & 1+x \\
1+x & 1 & 0 & x
\end{matrix}
$$

$$(1.3.7) \qquad L_3 = \begin{matrix} 0 & 1+x & 1 & x \\ 1 & x & 0 & 1+x \\ x & 1 & 1+x & 0 \\ 1+x & 0 & x & 1 \end{matrix}$$

The three latin squares L_1, L_2, L_3 of (1.3.5), (1.3.6), and (1.3.7) constitute a complete set of MOLS of order 4.

1.4. CONNECTION BETWEEN COMPLETE SETS OF MOLS AND PG(2, s)

We now show that there is a $1:1$ relation between a complete set of MOLS and the finite projective geometry PG(2, s). In PG(2, s) there are $s^2 + s + 1$ points and the same number of lines. Furthermore, on each line there are $s + 1$ points, and through each point there pass $s + 1$ lines. A single line passes through every pair of points, and every pair of lines intersects at a single point.

Let l be any line of PG(2, s) and X_R, X_C, X_1, $\cdots$, X_{s-1} the points on l. The $s(s + 1)$ lines other than l can be separated into $(s + 1)$ sets of s each, such that any two lines of the same set intersect at some point X on l. If we omit the line l and the points on it and designate the remaining points and lines as finite points and lines, the set of lines intersecting at X becomes a pencil of finite lines, denoted by $[X]$. This separates the finite lines into pencils of lines $[X_R]$, $[X_C]$, $[X_1]$, $[X_2]$, $\cdots$, $[X_{s-1}]$. We number the lines of each pencil in any arbitrary manner by the integers $0, 1, \cdots, s-1$. We can associate the lines of $[X_R]$ and $[X_C]$ with the rows and columns of an $s \times s$ square so that the point of intersection of line r of $[X_R]$ and line t of $[X_C]$ represents the (r, t)th cell of the square. Corresponding to the pencil $[X_i]$, $i = 1, 2, \cdots, s - 1$, we form a square by putting in the (r, t)th cell the number corresponding to that line of $[X_i]$ which passes through the point corresponding to the cell (r, t). It is easy to verify that L_i is a latin square. Through the point corresponding to the (r, t)th cell there passes exactly one line, say u, of $[X_i]$ and exactly one line, say v, of $[X_j]$, $i, j = 1, 2, \cdots, s - 1$. Hence L_i and L_j are orthogonal latin squares $(i \neq j)$.

Conversely, suppose we are given a complete set of MOLS $L_1, L_2, \cdots, L_{s-1}$ of order s in integers $0, 1, \cdots, s - 1$. Write down two squares L_R (L_C) of order s in which the symbol i appears in all cells of the row (column) numbered i ($i = 0, 1, \cdots, s - 1$). Write down the s^2 points (i, j) ($i, j = 0, 1, \cdots, s - 1$). Join the points with the same first coordinate by a line. We thus get s lines, which have no common point. Let them intersect at a new point X_R. Similarly draw the s lines joining the points with the same second coordinate and let them intersect at a new point X_C. Take the latin square L_i and let the s^2 cells of this latin square be identified with the above mentioned s^2 points; join the points corresponding to the same integer in L_i and let the s lines thus formed intersect at a new point X_i $(i = 1, 2, \cdots, s - 1)$. Finally join $X_R, X_C, X_1, \cdots, X_{s-1}$ by a line. We thus get PG(2, s).

1.5. NONEXISTENCE OF COMPLETE SETS OF MOLS

Since the existence of a complete set of MOLS is equivalent to the existence of PG(2, s), we now consider the condition for the existence of PG(2, s). These results were first obtained by Bruck and Ryser (1949) and form a particular case of the results of Chowla and Ryser (1950) and Shrikhande (1950) given in two different forms. It may be added that Shrikhande and Raghavarao (1964) showed that Chowla and Ryser's fundamental result is equivalent to that of Shrikhande; but Chowla and Ryser have not completely deduced all cases in their theorem from their basic result.

Let $m = s^2 + s + 1$. With the PG(2, s) we associate an mth order square matrix $N = (n_{ij})$, where $n_{ij} = 1$ or 0, depending on whether the ith point occurs on the jth line or not $(i, j = 1, 2, \cdots, m)$. We call N the incidence matrix of PG(2, s). The concept of incidence matrices will be very useful in combinatorial problems of the design of experiments, and the reader's attention is invited in this connection especially to Chapters 4, 5, 7, 8, 10, and 12. From the properties of PG(2, s) we observe that

$$(1.5.1) \qquad NN' = sI_m + E_{m, m},$$

where I_m is the identity matrix of order m and $E_{m, m}$ is a $m \times m$ matrix with $+1$ everywhere. Clearly $|NN'| \neq 0$; hence N is a nonsingular matrix and thus

$$(1.5.2) \qquad NN' \sim I_m,$$

where $\sim$ denotes that the matrices on the left and right sides of the symbol are rationally congruent. Using the properties of the Hasse–Minkowski invariant given in Appendix A,

$$(1.5.3) \qquad C_p(NN') = \left(s, (-1)^{s(s+1)/2} m\right)_p (-1, -1)_p = (-1, -1)_p$$

for all primes p, where $(a, b)_p$ is the Hilbert norm residue symbol. Thus we have the following theorem:

Theorem 1.5.1. A necessary condition for the existence of PG(2, s) is that

$$(1.5.4) \qquad \left(s, (-1)^{s(s+1)/2} m\right)_p = 1$$

for all primes p.

From the above theorem we deduce the following:

Corollary 1.5.1.1. If $s \equiv 1$ or 2 (mod 4) and the square free part of s contains a prime congruent to 3 (mod 4), then there does not exist PG(2, s).

Proof. First we show that

$$
\begin{aligned}
\left(s,(-1)^{s(s+1)/2}\left(s^2+s+1\right)\right)_p &= \left(s,(-1)^{s(s+1)/2}\right)_p\left(s, s^2+s+1\right)_p \\
&= \left(s,(-1)^{s(s+1)/2}\right)_p\left(s,(s+1)^2-s\right)_p \\
&= \left(s,(-1)^{s(s+1)/2}\right)_p\left((s+1)^2,-s\right)_p \\
&= \left(s,(-1)^{s(s+1)/2}\right)_p.
\end{aligned}
$$

(1.5.5)

When $s \equiv 1$ or 2 (mod 4) and the square free part of s contains a prime congruent to 3 (mod 4), then $(s,-1)_p = -1$. In that case the necessary condition will be violated and PG(2, s) does not exist.

From the corollary we can see that PG(2, s) does not exist for $s = 6, 14, 21, 22$, etc.

1.6. EMBEDDING TO A COMPLETE SET OF MOLS

We now consider the problem of extending a given set of MOLS to a complete set of MOLS. We prove the following theorem:

Theorem 1.6.1. Any set of $s - 2$ MOLS of order s can be extended to a complete set of $s - 1$ MOLS.

Proof. Let the symbols of each latin square be the integers $0, 1, 2, \cdots, s - 1$. Without loss of generality take the set $L_1, L_2, \cdots, L_{s-2}$ of MOLS in semi-standard form. Form a new square L containing the initial row in the natural order and put into the cell (r, t) for $r \neq 0$ the integer, different from t, that does not occur in this cell in any L_i $(i = 1, 2, \cdots, s - 2)$. It can be easily verified that L is a latin square, orthogonal to each of the latin squares $L_1, L_2, \cdots, L_{s-2}$.

We state the following result and leave its proof as an exercise to the reader after he completes reading Section 8.9:

Theorem 1.6.2. If $s \neq 4$, any set of $s - 3$ MOLS of order s can be uniquely extended to a complete set of $s - 1$ MOLS.

Another embedding result, based on the work of Bruck (1963), is given in Theorem 9.5.2.

In conclusion it is to be noted that the construction of a complete set of MOLS of order s is known in the literature when s is a prime or a prime power. However, sets of MOLS of any order except 6 exist, and we describe them in Chapter 3.

REFERENCES

1. Bose, R. C. (1938). On the application of properties of Galois fields to the problem of construction of hypergraecolatin squares. *Sankhyā*, **3**, 328–338.
2. Bruck, R. H. (1963). Finite nets II. Uniqueness and imbedding. *Pacific J. Math.*, **13**, 421–457.
3. Bruck, R. H., and Ryser, H. J. (1949). The nonexistence of certain finite projective planes. *Can. J. Math.*, **1**, 88–93.
4. Chowla, S., and Ryser, H. J. (1950). Combinatorial problems. *Can. J. Math.*, **2**, 93–99.
5. Shrikhande, S. S. (1950). The impossibility of certain symmetrical balanced incomplete block designs. *Ann. Math. Stat.*, **21**, 106–111.
6. Shrikhande, S. S., and Raghavarao, D. (1964). A note on the nonexistence of symmetric balanced incomplete block designs. *Sankhyā*, **A26**, 91–92.

BIBLIOGRAPHY

Albert, A. A., and Sandler, R. *An Introduction to Finite Projective Planes*. Holt, Rinehart and Winston, New York, 1968.

Dembowski, P. *Finite Geometries*, Springer Verlag, 1968.

Shrikhande, S. S. A note on mutually orthogonal latin squares. *Sankhyā*, **A23**, 115–116 (1961).

Shrikhande, S. S. Combinatorial problems. Lecture Notes No. 1, Research and Training School, Indian Statistical Institute (1961).

CHAPTER 2

Orthogonal Arrays

2.1. INTRODUCTION AND DEFINITION

The concept of latin squares and orthogonal latin squares was generalized by Kishen (1942) to latin cubes and hypercubes and orthogonal latin cubes and hypercubes. Kishen defined latin cubes and orthogonal latin cubes of the first order as follows:

Definition 2.1.1. An arrangement of s symbols in s squares of s rows and s columns is said to be a latin cube of the first order of side s if each symbol occurs exactly s times in each square, in the ith rows of all squares ($i = 1, 2, \cdots, s$), and the jth columns of all squares ($j = 1, 2, \cdots, s$). Two latin cubes of the first order are said to be orthogonal if on their superimposition every symbol of the first cube occurs exactly s times with each symbol of the second cube.

The following arrangement is an example of a $4 \times 4 \times 4$ latin cube of the first order in Latin letters and Greek letters, and the two cubes are orthogonal:

	Square 1				Square 2		
$A\alpha$	$B\gamma$	$C\delta$	$D\beta$	$A\alpha$	$B\gamma$	$C\delta$	$D\beta$
$B\beta$	$A\delta$	$D\gamma$	$C\alpha$	$B\beta$	$A\delta$	$D\gamma$	$C\alpha$
$C\gamma$	$D\alpha$	$A\beta$	$B\delta$	$C\gamma$	$D\alpha$	$A\beta$	$B\delta$
$D\delta$	$C\beta$	$B\alpha$	$A\gamma$	$D\delta$	$C\beta$	$B\alpha$	$A\gamma$

	Square 3				Square 4		
$A\alpha$	$B\gamma$	$C\delta$	$D\beta$	$A\alpha$	$B\gamma$	$C\delta$	$D\beta$
$B\beta$	$A\delta$	$D\gamma$	$C\alpha$	$B\beta$	$A\delta$	$D\gamma$	$C\alpha$
$C\gamma$	$D\alpha$	$A\beta$	$B\delta$	$C\gamma$	$D\alpha$	$A\beta$	$B\delta$
$D\delta$	$C\beta$	$B\alpha$	$A\gamma$	$D\delta$	$C\beta$	$B\alpha$	$A\gamma$

9

Kishen (1949) showed that the maximal number of mutually orthogonal $s \times s \times s$ cubes of the first order is $s^2 + s - 2$ and gave methods of constructing them when s is a prime or a prime power. He also made the following definition:

Definition 2.1.2. A second-order latin cube of side s is an arrangement of s^2 symbols in s squares of s rows and s columns such that each symbol occurs exactly once in each square, in the ith rows of all squares ($i = 1, 2, \cdots, s$), and the jth columns of all squares ($j = 1, 2, \cdots, s$).

The arrangement

Square 1			Square 2			Square 3		
A	E	I	D	H	C	G	B	F
B	F	G	E	I	A	H	C	D
C	D	H	F	G	B	I	A	E

can be seen to be a second-order latin cube of side s. The construction of such cubes when s is a prime or a prime power was given by Kishen (1949). Another method of construction for any s was given by Saxena (1960).

The concept of latin hypercubes and orthogonal latin hypercubes was also introduced by Kishen (1942, 1949).

Rao (1946) generalized the concepts further to obtain hypercubes of strength d that are equivalent to latin and orthogonal latin cubes and hypercubes of the first order when $d = 2$. The present notion of orthogonal arrays is also due to Rao (1947), and hypercubes of strength d form a subclass of orthogonal arrays.

Orthogonal arrays are defined as follows:

Definition 2.1.3. A $k \times N$ matrix A with entries from a set of s (≥ 2) elements is called an orthogonal array of size N, k constraints, s levels, strength t, and index λ if any $t \times N$ submatrix of A contains all possible $t \times 1$ column vectors with the same frequency λ. Such an array is denoted by (N, k, s, t); N is also called the number of assemblies.

In view of this definition, trivially we must have $N = \lambda s^t$. When $\lambda = s^{t-d}$, the orthogonal array reduces to the hypercube of strength d. In fact Rao's definition of hypercubes of strength d is as follows:

Definition 2.1.4. Let there be m factors, each of which occurs at s levels. Of the s^m possible combinations, a subset of s^t combinations can be called an $[m, s, t]$ array, which is said to be of strength d if all combinations of any

d of the m factors occur s^{t-d} times; such an arrangement is denoted by $[m, s, t, d]$.

Comparing Definitions 2.1.3 and 2.1.4, we note that a hypercube $[m, s, t, d]$ of strength d is an orthogonal array (s^t, m, s, d) of index s^{t-d}.

The rows of orthogonal arrays with $s = 2$ and $t = 2$ in elements $+1$ and -1, are orthogonal. For such an array of index λ there will be $\lambda s^2 = 4\lambda$ assemblies, and the maximum number of constraints will be shown in Theorem 2.2.4 to be $4\lambda - 1$. Let A be one such array. If we form

$$(2.1.1) \qquad\qquad H_{4\lambda} = \begin{pmatrix} E_{1, 4\lambda} \\ A \end{pmatrix},$$

then

$$(2.1.2) \qquad\qquad H_{4\lambda} H'_{4\lambda} = 4\lambda I_{4\lambda}.$$

The terms $H_{4\lambda}$ are known in mathematical literature as Hadamard matrices, whose construction was systematically studied for the first time by Paley (1933). They play a very useful role in combinatorial mathematics. We introduce them as optimum weighing designs in Chapter 17 and study them there in detail. We can easily observe that, given $H_{4\lambda}$, we can reverse the steps of the above method of construction and get an orthogonal array $(4\lambda, 4\lambda - 1, 2, 2)$ of index λ.

Also, orthogonal arrays of strength 2 have close relationships with generalized Hadamard matrices, group-divisible designs of Section 8.5, and affine resolvable balanced incomplete block designs of Section 5.4, and we refer the reader to the original paper of Shrikhande (1964).

We can easily verify that the existence of orthogonal arrays of index unity and $t = 2$ is equivalent to the existence of $k - 2$ MOLS of order s. Orthogonal arrays (N, k, s, t) may not exist for an arbitrary parameter values. An important problem in this direction is to construct the arrays with the maximum possible number of constraints.

2.2. MAXIMUM NUMBER OF CONSTRAINTS

Since the existence of an orthogonal array $(N, k, s, 2)$ of index unity is equivalent to the existence of $k - 2$ MOLS of order s, in view of the preceding chapter $k \leq s + 1$ and the equality is attained when s is a prime or a prime power.

Using the property that orthogonal arrays can be used for fractionally replicated designs to measure up to all t factor interactions, Rao (1947) proved the following:

Theorem 2.2.1. In an orthogonal array (N, k, s, t) of index λ

$$(2.2.1) \qquad N - 1 \geq \binom{k}{1}(s - 1) + \cdots + \binom{k}{u}(s - 1)^u \qquad \text{if } t = 2u$$

and

$$(2.2.2) \quad N - 1 \geq \binom{k}{1}(s - 1) + \cdots + \binom{k}{u}(s - 1)^u + \binom{k-1}{u}(s - 1)^{u+1}$$

$$\text{if } t = 2u + 1.$$

Bush (1952b) substantially improved Rao's bounds (2.2.1) and (2.2.2) for orthogonal arrays of index unity. He proved the following:

Theorem 2.2.2. For an orthogonal array (s^t, k, s, t) of index unity, if $t \leq s$, then

$$(2.2.3) \qquad\qquad k \leq s + t - 1 \qquad \text{if } s \text{ is even}$$

and

$$(2.2.4) \qquad\qquad k \leq s + t - 2 \qquad \text{if } s \text{ is odd and } t \geq 3.$$

The proof of the theorem is based on the following lemma:

Lemma 2.2.1. For an orthogonal array $(s^3, k, s, 3)$ of index unity and $s \geq 3$

$$(2.2.5) \qquad\qquad k \leq s + 2 \qquad \text{if } s \text{ is even}$$

and

$$(2.2.6) \qquad\qquad k \leq s + 1 \qquad \text{if } s \text{ is odd.}$$

Proof. Clearly $s + 3$ constraints are impossible. Suppose $(s^3, s + 2, s, 3)$ can be constructed. Then each symbol occurs s^2 times in each row, and every pair of symbols occurs together s times in any two specified rows. Consider the symbols of the first column and let θ be any symbol occurring in that column. With the symbols of the first column we can form $s + 1$ pairs with θ as one of the symbols in each pair. We have to form the pairs in the other columns containing the selected symbols $(s + 1)(s - 1) = s^2 - 1$ times, and this enumeration exhausts the replications of the selected element. Hence any two columns of the array must have either common symbols in two rows or no common symbol in any row. In the latter case the columns are said to be disjoint. Since we can form $(s + 2)(s + 1)/2$ pairs of symbols from the symbols of the first columns, it is possible to have disjoint columns as $(s + 2)(s + 1)(s - 1)/2 < (s^3 - 1)$. Let us consider two disjoint columns and $s - 1$ more columns, with the first two rows identical with the first two rows of the second column, as below:

$$s - 1 \text{ columns}$$

$$
\begin{array}{cc@{\qquad}cccc}
\theta_1 & \phi_1 & \phi_1 & \phi_1 & \cdots & \phi_1 \\
\theta_2 & \phi_2 & \phi_2 & \phi_2 & \cdots & \phi_2 \\
\cdot & \cdot \\
\cdot & \cdot \\
\cdot & \cdot \\
\theta_{s+2} & \phi_{s+2}
\end{array}
$$

Here $\theta_i \neq \phi_i$, $i = 1, 2, \cdots, s + 2$. Since every triplet occurs exactly once, the symbols $\theta_3, \cdots, \theta_{s+2}$ must occur exactly once in the 3rd, 4th, $\cdots$, $(s + 2)$th rows in columns 3 through $s + 1$. But, since two columns have either two symbols or none in common, we get a contradiction when s is odd. Thus the lemma is established.

Proof of Theorem 2.2.2. From an orthogonal array $(s^t, k + 1, s, t)$, by writing down the portion that has the same symbol in the first row and omitting the first row, we get an orthogonal array $(s^{t-1}, k, s, t - 1)$. Thus increasing the strength by unity results in the increase of constraints by at most unity. For any strength t we have increased the strength by $t - 3$ from strength 3, the number of constraints given by Lemma 2.2.1 will be increased at most by $t - 3$, and hence the theorem.

We shall demonstrate in Section 2.4 that the bounds can actually be attained.

The following bound on the number of constraints for orthogonal arrays of index unity was also given by Bush (1952b):

Theorem 2.2.3. In an orthogonal array (s^t, k, s, t) of index unity, if $s \leq t$, then

(2.2.27) $$k \leq t + 1.$$

Proof. Let, if possible, $t + 2$ constraints be constructed. We distinguish three cases: (a) $s = 2$, (b) $s = 3$, and (c) $s > 3$, and treat these cases separately.

Case a. Let us divide the $t + 2$ rows into two parts A and B, with three rows and $t - 1$ rows, respectively. Without loss of generality, let us assume that the first column contains the symbol 0 everywhere. Clearly there exists another column that has the symbol 0 in the last $t - 1$ rows; let this column, which must have 1 in the first three rows, be the second column. There will also be two more columns with 1 in the first row of B and 0 in the other $t - 2$ rows of B; let them be the third and fourth columns. These four columns can then be represented as follows:

$$
\begin{array}{cccc}
 & 0 & 1 & * & * \\
A & 0 & 1 & * & * \\
 & 0 & 1 & * & * \\
\hline
 & 0 & 0 & 1 & 1 \\
 & 0 & 0 & 0 & 0 \\
t-1 \text{ rows} & . & . & . & . \\
B & . & . & . & . \\
 & . & . & . & . \\
 & 0 & 0 & 0 & 0 \\
\end{array}
$$

where the spaces marked by asterisks are to be suitably filled by 0 or 1. The third and fourth columns have $t - 2$ elements in common with each of the first two columns, and they can have at most one symbol in common with the first two columns in the first three rows—an impossible construction. Hence the result in this case.

Case b. We again divide the array into two parts A and B, with three and $t - 1$ rows, respectively. By carefully enumerating the possibilities, we shall have the 15 columns shown below:

$$
\begin{array}{ccccccccccccccc}
0 & 1 & 2 & 0 & * & * & 0 & * & * & 0 & * & * & 0 & * & * \\
0 & 1 & 2 & * & 0 & * & * & 0 & * & * & 0 & * & * & 0 & * \\
0 & 1 & 2 & * & * & 0 & * & * & 0 & * & * & 0 & * & * & 0 \\
\hline
0 & 0 & 0 & 1 & 1 & 1 & 2 & 2 & 2 & 0 & 0 & 0 & 0 & 0 & 0 \\
0 & 0 & 0 & 0 & 0 & 0 & 0 & 0 & 0 & 1 & 1 & 1 & 2 & 2 & 2 \\
. & . & . & . & . & . & . & . & . & . & . & . & . & . & . \\
. & . & . & . & . & . & . & . & . & . & . & . & . & . & . \\
. & . & . & . & . & . & . & . & . & . & . & . & . & . & . \\
0 & 0 & 0 & 0 & 0 & 0 & 0 & 0 & 0 & 0 & 0 & 0 & 0 & 0 & 0 \\
\end{array}
$$

with A spanning the top three rows and $t-1$ rows / B the lower portion.

where the spaces marked by asterisks are to be suitably filled by 1 or 2. Let us consider the four columns with 0 in the first row, omitting the first column, and let us try to fill the positions of the asterisks with 1 or 2. These columns have at least $t - 2$ common symbols, and if the same pair is used to fill two of the four columns, there will be two identical t-plets, a contradiction of the unity index. Thus we fill these positions with the four ordered pairs from 1, 2. Then eventually two of these columns will produce a repeated t-plet, one identical with that from the second column and the other identical with that from the third column, which is again a contradiction. Hence $k \leq t + 1$ in this case.

Case c. Again dividing the array into two parts A and B, with three and $t - 1$ rows, respectively, and enumerating the possibilities, we have $s + s(s - 1)(t - 1)$ columns, given by array for case c. The spaces marked by asterisks are to be filled by the elements $1, 2, \cdots, s - 1$. Let the first s columns constitute the set F. Consider the $(s - 3)(s - 1)(t - 1)$ columns which do not contain the symbol 0 in the A portion excluding the set F. In these columns let θ be the symbol that occurs most frequently in the first row of A. Then, since every symbol occurs exactly once in each of the s blocks of columns, θ occurs at least $(s - 3)(t - 1)$ times in these columns. The element θ will also occur in some column of F in the first row of A. Suppose that ϕ, φ are the symbols occurring in the second and third rows of that column in F. We can easily verify that no symbol in the second row of A can be ϕ and the symbol in the third row can be φ if the column, not belonging to F, contains θ in the first row of A.

We can form $(s - 2)^2$ ordered pairs of symbols by using symbols other than 0 and θ. Distinct pairs of symbols, excepting the ϕ, φ pair, are to be used to fill the second and third rows of A where θ occurs in the first row, in columns not in F; failing this, the index of unity will be violated. Hence $(s - 2)^2 \geq (s - 2) + (s - 3)(t - 1)$, which on simplification yields $(3 - s)(t - s + 1) \geq 0$, a contradiction. Thus the theorem is established in this case also, and the proof is complete.

Array for case c

	0****	0**** ***	0****	0*******	0****	***0*******	0****
A	0****	*0*** ***	*0***	*0******	*0***	****0******	*0***
	0****	**0** ***	**0**	**0*****	**0**	*****0*****	**0**

$0\cdots0$ $1\cdots1\cdots s-1\cdots s-1$ $0\cdots0\cdots$ 0 $\cdots$ 0 $\cdots0\cdots0\cdots$ 0 $\cdots$ 0
$0\cdots0$ $0\cdots0\cdots$ 0 $\cdots$ 0 $1\cdots1\cdots s-1\cdots s-1\cdots0\cdots0\cdots$ 0 $\cdots$ 0

$t - 1$
rows
B

$0\cdots0$ $0\cdots0\cdots$ 0 $\cdots$ 0 $0\cdots0\cdots$ 0 $\cdots$ 0 $\cdots1\cdots1\cdots s-1\cdots s-1$

$\quad s \qquad s \qquad\qquad s \qquad\quad s \qquad s \qquad\qquad s \qquad s \qquad\qquad s$

Plackett and Burman (1943–1944) considered optimum multifactorial experiments, which are essentially orthogonal arrays of strength 2. They established the following theorem:

Theorem 2.2.4. For an orthogonal array $(\lambda s^2, k, s, 2)$ we have the inequality

(2.2.8)
$$k \leqq \left[\frac{\lambda s^2 - 1}{s - 1} \right],$$

where $[x]$ stands for the greatest integer of x.

Proof. Two columns of an array are said to have i coincidences if the same symbols occur in i rows of these columns. For an orthogonal array $(\lambda s^t, k, s, t)$ let the first column have n_i columns with i coincidences $(i = 0, 1, \cdots, k)$. Then

(2.2.9)
$$\sum_{i=0}^{k} n_i = \lambda s^t - 1$$

and

(2.2.10)
$$\sum_{i=0}^{k} \binom{i}{h} n_i = \binom{k}{h} (\lambda s^{t-h} - 1) \qquad 1 \leq h \leq t.$$

Formula (2.2.9) is obvious. Formula (2.2.10) can be established by counting in two ways the possible $h \times 1$ column vectors formed by columns other than the first one by considering subarrays with h rows. The h rows can be selected in $\binom{k}{h}$ ways, and each subarray has $(\lambda s^{t-h} - 1)$ columns with the same column as the first one. Hence the possible $h \times 1$ vectors in the first way of enumeration is $\binom{k}{h} (\lambda s^{t-h} - 1)$. Again, any column that has i coincidences with the first contributes nothing or $\binom{i}{h}$ to this number, depending on whether $i < h$ or $i \geqq h$; hence this number, in the second way of enumeration, is $\sum_{i=0}^{k} \binom{i}{h} n_i$, and by equating these two numbers, we get (2.2.10). In case $t = 2$, by writing (2.2.9) and (2.2.10) and defining

(2.2.11)
$$f(x) = \sum_{i=0}^{k} (i - x)(i - 1 - x) n_i,$$

we see that

$$0 \leq f(x) = \sum_{i=0}^{k} i(i-1)n_i - 2x \sum_{i=0}^{k} in_i + x(x+1) \sum_{i=0}^{k} n_i$$

$$(2.2.12) \qquad = \lambda(k(k-1) - 2kxs + x(x+1)s^2)$$

$$- (k(k-1) - 2kx + x(x+1))$$

and hence

$$(2.2.13) \qquad \lambda \geq \frac{k(k-1) - 2kx + x(x+1)}{k(k-1) - 2kxs + x(x+1)s^2}.$$

Letting $\alpha = k - 1 - xs$ and simplifying (2.2.13), we get

$$(2.2.14) \qquad \frac{\lambda s^2 - 1}{s-1} \geq k \left(\frac{1 + \alpha(s-\alpha)}{D} \right),$$

where

$$(2.2.15) \qquad \begin{aligned} D &= (s-1)(k - \alpha - 1) + \alpha(\alpha+1) \\ &= k(s-1) - (\alpha+1)(s-1-\alpha). \end{aligned}$$

Let

$$(2.2.16) \qquad \lambda - 1 = a(s-1) + b \qquad 0 \leq b < s - 1; a \geq 0.$$

Then

$$(2.2.17) \qquad \frac{\lambda s^2 - 1}{s-1} = \lambda s + \lambda + a + \frac{b}{s-1}.$$

Suppose there exists an array with $k = \lambda s + \lambda + a + 1$ constraints. Then $k - 1 = s(\lambda + a) + b + 1$. Choosing $x = \lambda + a$ and simplifying, we see that

$$(2.2.18) \qquad \frac{\alpha(s-\alpha)}{D} > 0,$$

and from (2.2.14) and (2.2.17) we get

$$(2.2.19) \qquad \frac{b}{s-1} > 1,$$

a contradiction. Hence $k \leq \lambda s + \lambda + a$, which is the required result.

The above proof is due to Bose and Bush (1952). The bound on k given in (2.2.8) is the same as Rao's bounds when $t = 2$. The bound (2.2.8) was shown by Bose and Bush (1952) to be improved to the following:

Theorem 2.2.5. If $\lambda - 1$ is not divisible by $s - 1$, then for the orthogonal array $(\lambda s^2, k, s, 2)$ we have

(2.2.20)
$$k \leq \left[\frac{\lambda s^2 - 1}{s - 1}\right] - [\theta] - 1,$$

where

(2.2.21)
$$\theta = \frac{\sqrt{1 + 4s(s - 1 - b)} - (2s - 2b - 1)}{2},$$

b being the remainder on dividing $\lambda - 1$ by $s - 1$.

Proof. Since $\lambda - 1$ is not divisible by $s - 1$, we have

(2.2.22)
$$\lambda - 1 = a(s - 1) + b, \qquad 0 < b < s - 1.$$

Let

(2.2.23)
$$k = \lambda s + \lambda + a - n,$$

and then

(2.2.24)
$$k - 1 = s(\lambda + a) + b - n.$$

Choosing x as in the preceding theorem, we can show by using the second value of D of (2.2.15), after some simplification, that

(2.2.25)
$$(b - n)(b + 1 - n) - s(b - 2n) > 0.$$

Hence if n is an integer $(b > n \geq 0)$ for which (2.2.25) is violated, then $k = \lambda s + \lambda + a - n$ and all higher values are impossible. By replacing the inequality in (2.2.25) by equality, we can see that the quadratic equation in n has one positive and one negative root, the positive root being θ, as given in (2.2.21). The largest value of n that contradicts (2.2.25) is $[\theta]$, and hence the theorem.

Following similar arguments as in Theorems 2.2.4 and 2.2.5, for orthogonal arrays of strength 3, Bose and Bush proved the following theorems:

Theorem 2.2.6. For an orthogonal array $(\lambda s^3, k, s, 3)$ we have the inequality

(2.2.26)
$$k \leq \left[\frac{\lambda s^2 - 1}{s - 1}\right] + 1.$$

Theorem 2.2.7. If $\lambda - 1$ is not divisible by $s - 1$, then for the orthogonal array $(\lambda s^3, k, s, 3)$ we have

(2.2.27)
$$k \leq \left[\frac{\lambda s^2 - 1}{s - 1}\right] - [\theta],$$

where θ is given by (2.2.21).

2.3. USE OF PROJECTIVE GEOMETRY IN THE CONSTRUCTION OF ORTHOGONAL ARRAYS

Rao (1946) considered constructions of hypercubes from PG(t, s). The orthogonal array $(s^t, (s^t - 1)/(s - 1), s, 2)$ of index s^{t-2}, which is same as the hypercube $[(s^t - 1)/(s - 1), s, t, 2]$ of strength 2, can be constructed from PG(t, s) in the following manner: in PG(t, s) the equation

$$(2.3.1) \qquad\qquad x_0 = 0$$

represents the $(t - 1)$-flat at infinity, and any other $(t - 1)$-flat can be represented by an equation of the form

$$(2.3.2) \qquad a_0 x_0 + a_1 x_1 + \cdots + a_t x_t = 0;$$

then a $(t - 2)$-flat at infinity can be represented by

$$(2.3.3) \quad x_0 = 0, \ a_1 x_1 + a_2 x_2 + \cdots + a_t x_t = 0 \qquad a_i \in \mathrm{GF}(s); \ i = 1, 2, \cdots, t.$$

We observe that there are $(s^t - 1)/(s - 1)$, $(t - 2)$-flats at infinity. From the $(t - 2)$-flat (2.3.3) there passes a pencil of s, $(t - 1)$-flats given by the following equation:

$$(2.3.4) \qquad \alpha_i x_0 + a_1 x_1 + a_2 x_2 + \cdots + a_t x_t = 0 \qquad \alpha_i \in \mathrm{GF}(s).$$

The $(t - 2)$-flat (2.3.3) is called the vertex of the pencil (2.3.4) and corresponds to a factor identified by $(a_1, a_2, \cdots, a_t)$. The $(t - 1)$-flat (2.3.4) with α_i as the coefficient of x_0 may be identified by α_i, which may correspond to the ith level $(i = 0, 1, \cdots, s - 1)$ of a factor, identified by $(a_1, a_2, \cdots, a_t)$, defining the vertex (2.3.3) of the pencil (2.3.4). Thus there are $m = (s^t - 1)/(s - 1)$ factors, each at s levels.

There are s^t finite points $(1, x_1, x_2, \cdots, x_t)$, and through each of the finite points there passes exactly one $(t - 1)$-flat from each of the $(s^t - 1)/(s - 1)$ pencils of $(t - 1)$-flats. We now identify the s^t finite points and the factors to the columns and rows of an array, and fill its (i, j)th position by k if the α_kth flat of the ith pencil passes through the jth finite point $(k = 0, 1, \cdots, s - 1;$ $i = 1, 2, \cdots, (s^t - 1)/(s - 1); j = 1, 2, \cdots, s^t)$. The array thus constructed can easily be verified to be an orthogonal array $(s^t, (s^t - 1)/(s - 1), s, 2)$ of index s^{t-2}.

Illustration 2.3.1. We illustrate this method of construction by constructing the orthogonal array $(8, 7, 2, 2)$ of index 2. This orthogonal array can be constructed from PG(3, 2). In PG(3, 2) each of the 15 points can be represented by (x_0, x_1, x_2, x_3), where $x_i = 0$ or 1 $(i = 0, 1, 2, 3)$ and not all the x_i coordinates are equal to zero. The vertices corresponding to the pencil (2.3.4) are $(1, 0, 0)$, $(0, 1, 0)$, $(0, 0, 1)$, $(1, 1, 0)$, $(1, 0, 1)$, $(0, 1, 1)$, and $(1, 1, 1)$, and we call them the factors A, B, C, D, E, F, and G, respectively. The eight finite points are $(1, 0, 0, 0)$, $(1, 1, 0, 0)$, $(1, 0, 1, 0)$, $(1, 0, 0, 1)$, $(1, 1, 1, 0)$, $(1, 1, 0, 1)$, $(1, 0, 1, 1)$, and $(1, 1, 1, 1)$, numbered in the serial order $1, 2, \cdots, 8$. Let us form an array of seven rows and eight columns, the rows corresponding to the seven factors, and the columns corresponding to the eight finite points, and fill the (i, j)th entry by the flat number of the ith pencil passing through the jth finite point $(i = A, B, \cdots, G; j = 1, 2, \cdots, 8)$. For example, the second point lies on the flat number 1 passing through the vertex D, and hence the entry in the second column under row D is 1. The completed arrangement constituting the orthogonal array $(8, 7, 2, 2)$ of index 2 will then be as follows:

	Finite Points							
Factor	1	2	3	4	5	6	7	8
A	0	1	0	0	1	1	0	1
B	0	0	1	0	1	0	1	1
C	0	0	0	1	0	1	1	1
D	0	1	1	0	0	1	1	0
E	0	1	0	1	1	0	1	0
F	0	0	1	1	1	1	0	0
G	0	1	1	1	0	0	0	1

To construct a hypercube of strength 3, or equivalently an orthogonal array of index s^{t-3} from $PG(t, s)$, we need to select the factors suitably. To get the orthogonal array of index s^{t-3}, it is necessary that three $(t - 1)$-flats belonging to three different pencils from the set of $(s^t - 1)/(s - 1)$ pencils intersect at s^{t-3} finite points. Any two vertices corresponding to two pencils intersect at a $(t - 3)$-flat at infinity. A third vertex has to be chosen so that it does not pass through the intersection of the previous two vertices; a fourth vertex has to be chosen so that it does not pass through the intersection of any two vertices already chosen; and so on. It is now easy to see that the conditions for an orthogonal array of index s^{t-3} are satisfied if the selected vertices are identified with factors.

Illustration 2.3.2. Let us construct an orthogonal array of index s^{t-3}, where $s = 2$ and $t = 3$. In $PG(3, 2)$ there are seven vertices at infinity, as shown in Illustration 2.3.1. Of these vertices, the vertices A, B, C, and G are such that any of them does not pass through the intersection of any pair of the other three vertices. We note this is the maximum possible number of factors. The orthogonal array $(8, 4, 2, 3)$ of index unity can be obtained from the orthogonal array $(8, 7, 2, 2)$ of Illustration 2.3.1 by writing down the rows corresponding to the factors A, B, C, and G.

This method can be generalized for constructing orthogonal arrays of index s^{t-d}, where $d > 3$, but we do not consider it here.

2.4. CONSTRUCTION OF ORTHOGONAL ARRAYS OF INDEX UNITY

In this section we show that orthogonal arrays $(s^t, s + 1, s, t)$ can be constructed when s is a prime or a prime power and $t < s$ [cf. Bush (1952b)]. When $t = 3$ and s is odd, the maximum number of constraints given in Lemma 2.2.1 is attained. However, if s is even, we show that another row orthogonal to the $s + 1$ rows can be constructed so that the maximum number of constraints is attained.

Let $\alpha_0 = 0, \alpha_1, \alpha_2, \cdots, \alpha_{s-1}$ be the elements of GF(s) and consider the s^t polynomials

$$(2.4.1) \qquad y_j(x) = a_{t-1} x^{t-1} + a_{t-2} x^{t-2} + \cdots + a_0 \qquad a_i \in \text{GF}(s);$$
$$i = 0, 1, \cdots, t - 1.$$

We fill a rectangle with s rows and s^t columns such that the (i, j)th cell is filled by the symbol u with the property

$$(2.4.2) \qquad y_j(\alpha_i) = \alpha_u.$$

To these s rows we add another row by putting the symbol u into the columns that are associated with the polynomials whose leading coefficient is α_u. The $s + 1$ rows so constructed can be seen to be an orthogonal array $(s^t, s + 1, s, t)$ that has the maximum number of constraints when $t = 3$. However, if s is even and $s = 2^n$, $t = 3$, yet another row can be added by putting the symbol v into the columns that have α_v as the coefficient of the term of degree 1, thereby constructing an orthogonal array $(2^{3n}, 2^n + 2, 2^n, 3)$ with the maximum number of constraints.

Illustration 2.4.1. Let us construct the orthogonal array $(27, 4, 3, 3)$ by using the above method. The elements of GF(3) are $0, 1, 2$, and the 27 polynomials are as follows:

$$(2.4.3) \quad \begin{aligned}
&y_1 = 0, \qquad y_2 = 1, \qquad y_3 = 2, \qquad y_4 = x, \qquad y_5 = x + 1, \\
&y_6 = x + 2, \qquad y_7 = 2x, \qquad y_8 = 2x + 1, \qquad y_9 = 2x + 2, \\
&y_{10} = x^2, \qquad y_{11} = x^2 + 1, \qquad y_{12} = x^2 + 2, \qquad y_{13} = x^2 + x, \\
&y_{14} = x^2 + x + 1, \qquad y_{15} = x^2 + x + 2, \qquad y_{16} = x^2 + 2x, \\
&y_{17} = x^2 + 2x + 1, \qquad y_{18} = x^2 + 2x + 2, \qquad y_{19} = 2x^2, \\
&y_{20} = 2x^2 + 1, \qquad y_{21} = 2x^2 + 2, \qquad y_{22} = 2x^2 + x, \\
&y_{23} = 2x^2 + x + 1, \qquad y_{24} = 2x^2 + x + 2, \qquad y_{25} = 2x^2 + 2x, \\
&y_{26} = 2x^2 + 2x + 1, \qquad y_{27} = 2x^2 + 2x + 2.
\end{aligned}$$

Then the four rows constituting the orthogonal array are as shown below:

Orthogonal array for Illustration 2.4.1

	1 2 3 4 5 6 7 8 9 10 11 12 13 14 15 16 17 18 19 20 21 22 23 24 25 26 27
0	0 1 2 0 1 2 0 1 2 0 1 2 0 1 2 0 1 2 0 1 2 0 1 2 0 1 2
1	0 1 2 1 2 0 2 0 1 1 2 0 2 0 1 0 1 2 2 0 1 0 1 2 1 2 0
2	0 1 2 2 0 1 1 2 0 1 2 0 0 1 2 2 0 1 2 0 1 1 2 0 0 1 2
A	0 0 0 0 0 0 0 0 0 1 1 1 1 1 1 1 1 1 2 2 2 2 2 2 2 2 2

We now construct an orthogonal array $(s^t, t + 1, s, t)$, where $s \leq t$, with the maximum number of constraints. Let us consider the s^t polynomials given by (2.4.1) and fill a rectangle with t rows and s^t columns such that the (i, j)th cell is filled by the subscript of the coefficient of x^{t-i} $(i = 1, 2, \cdots, t)$. To these t rows we add another row whose jth position is filled by the symbol u if $y_j(1) = \alpha_u$. The array so constructed can be seen to be an orthogonal array $(s^t, t + 1, s, t)$ with the maximum number of constraints.

Illustration 2.4.2. Let us construct the orthogonal array $(8, 4, 2, 3)$. The elements of GF(2) are 0, 1, and the eight polynomials are

$$(2.4.4) \quad y_1 = 0, \quad y_2 = 1, \quad y_3 = x, \quad y_4 = x + 1, \quad y_5 = x^2,$$
$$y_6 = x^2 + 1, \quad y_7 = x^2 + x, \quad y_8 = x^2 + x + 1.$$

Then the four rows constituting the orthogonal array are as follows:

	1	2	3	4	5	6	7	8
1	0	0	0	0	1	1	1	1
2	0	0	1	1	0	0	1	1
3	0	1	0	1	0	1	0	1
$y_j(1)$	0	1	1	0	1	0	0	1

2.5. METHOD OF DIFFERENCES IN THE CONSTRUCTION OF ORTHOGONAL ARRAYS

Let M be a module of m elements. On the lines of proof used for the module theorems for constructing the BIB designs of Section 5.7 the following theorem due to Bose and Bush (1952) can be established:

Theorem 2.5.1. If it is possible to choose r rows and s columns ($s = \lambda m$, λ an integer)

(2.5.1)
$$\begin{matrix} a_{11} & a_{12} & \cdots & a_{1s} \\ a_{21} & a_{22} & \cdots & a_{2s} \\ \cdot & \cdot & \cdots & \cdot \\ a_{r1} & a_{r2} & \cdots & a_{rs} \end{matrix}$$

with elements belonging to M such that among the differences of the corresponding elements of any two rows, each element of M occurs exactly λ times, then, by adding the elements of the module to the elements in (2.5.1) and reducing mod m, we can generate sm columns; this constitutes an orthogonal array (λm^2, r, m, 2) of index λ.

The s columns of (2.5.1) may be called the initial columns of the array.

The orthogonal arrays that can be constructed by the method of Theorem 2.5.1 are completely resolvable in the sense of the following definition:

Definition 2.5.1. An orthogonal array (λs^2, k, s, 2), where $\lambda = \alpha\beta$, is said to be β-resolvable if it is the juxtaposition of αs different arrays (βs, k, s, 1) of index β and strength 1. A 1-resolvable array is said to be completely resolvable.

By adding a new row with the element i written in λ component arrays ($i = 1, 2, \cdots, s$), we can increase the number of constraints at least by 1 if the array (λs^2, k, s, 2) is completely resolvable. Hence

Corollary 2.5.1.1. Subject to the conditions of Theorem 2.5.1, we can construct the orthogonal array (λm^2, $r + 1$, m, 2).

Applying the method of differences of Theorem 2.5.1, Bose and Bush proved the following:

Theorem 2.5.2. If λ and s are both powers of the same prime p, a completely resolvable orthogonal array (λs^2, λs, s, 2) can always be constructed.

Proof. Let $\lambda = p^u$ and $s = p^v$, where p is a prime and u, v are nonzero positive integers. Let the elements of $GF(p^{u+v})$ be expressed as polynomials of degree not greater than $u + v - 1$ and represent the elements in the lexicographic order; that is, if

$$(2.5.2) \quad \alpha_i = a_{n-1} x^{n-1} + \cdots + a_v x^v + a_{v-1} x^{v-1} + \cdots + a_1 x + a_0,$$

then $i = a_{n-1} \cdots a_1 a_0$ in the scale of numeration radix p, where $n = u + v$. Consider the module M consisting of p^v elements for which the coefficients of x^v and higher powers are zero, which implies the first p^v elements when expressed in the lexicographic order. We define a correspondence, mapping α_i of $GF(p^{u+v})$ as given in (2.5.2) to

$$(2.5.3) \quad \alpha_j = a_{v-1} x^{v-1} + \cdots + a_1 x + a_0$$

of M, the coefficients of x^{v-1} and lower powers of x for α_j being the same as the coefficients of the corresponding powers of x in α_i.

We now write down the multiplication table of the elements of $GF(p^{u+v})$ and replace each element by the corresponding element in M. We can verify that the $p^{u+v} = \lambda s$ columns so formed satisfy the necessary conditions of the initial columns. On developing these columns as indicated in Theorem 2.5.1, we get the orthogonal array $(\lambda s^2, \lambda s, s, 2)$.

Corollary 2.5.2.1. Subject to the conditions of Theorem 2.5.2, we can construct the array $(\lambda s^2, \lambda s + 1, s, 2)$.

Illustration 2.5.1. We construct the orthogonal array $(27, 9, 3, 2)$ by the method of Theorem 2.5.2. Here $s = 3$, $\lambda = 3$. The elements of $GF(3^2)$ expressed as powers of primitive root and as polynomials are

$$\begin{aligned}
& \alpha_0 = 0, \quad \alpha_1 = x^0 = 1, \quad \alpha_2 = x, \quad \alpha_3 = x^2 = 2x + 1, \\
(2.5.4) \quad & \alpha_4 = x^3 = 2x + 2, \quad \alpha_5 = x^4 = 2, \quad \alpha_6 = x^5 = 2x, \\
& \alpha_7 = x^6 = x + 2, \quad \alpha_8 = x^7 = x + 1.
\end{aligned}$$

The elements of M are α_0, α_1, and α_5, and the mapping of the elements of $GF(3^2)$ on the elements of M is

$$(2.5.5) \quad \begin{aligned}
\alpha_0, \quad \alpha_2, \quad \alpha_6 \quad &\rightarrow \quad \alpha_0 \\
\alpha_1, \quad \alpha_3, \quad \alpha_8 \quad &\rightarrow \quad \alpha_1 \\
\alpha_4, \quad \alpha_5, \quad \alpha_7 \quad &\rightarrow \quad \alpha_5
\end{aligned}$$

The multiplication table of the elements of $GF(3^2)$ is

	α_0	α_1	α_2	α_3	α_4	α_5	α_6	α_7	α_8
α_0	α_0	α_0	α_0	α_0	α_0	α_0	α_0	α_0	α_0
α_1	α_0	α_1	α_2	α_3	α_4	α_5	α_6	α_7	α_8
α_2	α_0	α_2	α_3	α_4	α_5	α_6	α_7	α_8	α_1
α_3	α_0	α_3	α_4	α_5	α_6	α_7	α_8	α_1	α_2
α_4	α_0	α_4	α_5	α_6	α_7	α_8	α_1	α_2	α_3
α_5	α_0	α_5	α_6	α_7	α_8	α_1	α_2	α_3	α_4
α_6	α_0	α_6	α_7	α_8	α_1	α_2	α_3	α_4	α_5
α_7	α_0	α_7	α_8	α_1	α_2	α_3	α_4	α_5	α_6
α_8	α_0	α_8	α_1	α_2	α_3	α_4	α_5	α_6	α_7

$(2.5.6)$

By mapping the elements on M as given by (2.5.5) and replacing α_0, α_1, and α_5 by 0, 1, and 2, respectively, we get the initial columns:

$$
\begin{array}{ccccccccc}
0 & 0 & 0 & 0 & 0 & 0 & 0 & 0 & 0 \\
0 & 1 & 0 & 1 & 2 & 2 & 0 & 2 & 1 \\
0 & 0 & 1 & 2 & 2 & 0 & 2 & 1 & 1 \\
0 & 1 & 2 & 2 & 0 & 2 & 1 & 1 & 0 \\
0 & 2 & 2 & 0 & 2 & 1 & 1 & 0 & 1 \\
0 & 2 & 0 & 2 & 1 & 1 & 0 & 1 & 2 \\
0 & 0 & 2 & 1 & 1 & 0 & 1 & 2 & 2 \\
0 & 2 & 1 & 1 & 0 & 1 & 2 & 2 & 0 \\
0 & 1 & 1 & 0 & 1 & 2 & 2 & 0 & 2 \\
\end{array}
$$

(2.5.7)

By developing these columns by adding the elements of GF(3) and reducing modulo 3, we construct an orthogonal array (27, 9, 3, 2) as below:

(2.5.8)

```
0 1 2 0 1 2 0 1 2 0 1 2 0 1 2 0 1 2 0 1 2 0 1 2 0 1 2
0 1 2 1 2 0 0 1 2 1 2 0 2 0 1 2 0 1 0 1 2 2 0 1 1 2 0
0 1 2 0 1 2 1 2 0 2 0 1 2 0 1 0 1 2 2 0 1 1 2 0 1 2 0
0 1 2 1 2 0 2 0 1 2 0 1 0 1 2 2 0 1 1 2 0 1 2 0 0 1 2
0 1 2 2 0 1 2 0 1 0 1 2 2 0 1 1 2 0 1 2 0 0 1 2 1 2 0
0 1 2 2 0 1 0 1 2 2 0 1 1 2 0 1 2 0 0 1 2 1 2 0 2 0 1
0 1 2 0 1 2 2 0 1 1 2 0 1 2 0 0 1 2 1 2 0 2 0 1 2 0 1
0 1 2 2 0 1 1 2 0 1 2 0 0 1 2 1 2 0 2 0 1 2 0 1 0 1 2
0 1 2 1 2 0 1 2 0 0 1 2 1 2 0 2 0 1 2 0 1 0 1 2 2 0 1
```

In this case the maximum number of constraints is 13. We now try to find four more rows that are orthogonal among themselves and orthogonal to the rows of (2.5.8). We construct the array (9, 4, 3, 2) of index unity from the MOLS of order 3. Clearly such an array is

$$
\begin{array}{ccccccccc}
0 & 0 & 0 & 1 & 1 & 1 & 2 & 2 & 2 \\
0 & 1 & 2 & 0 & 1 & 2 & 0 & 1 & 2 \\
0 & 1 & 2 & 1 & 2 & 0 & 2 & 0 & 1 \\
0 & 2 & 1 & 1 & 0 & 2 & 2 & 1 & 0 \\
\end{array}
$$

(2.5.9)

We now repeat each of the columns thrice to get the 27 columns.

(2.5.10)
$$
\begin{array}{ccccccccccccccccccccccccccc}
0&0&0&0&0&0&0&0&0&1&1&1&1&1&1&1&1&1&2&2&2&2&2&2&2&2&2\\
0&0&0&1&1&1&2&2&2&0&0&0&1&1&1&2&2&2&0&0&0&1&1&1&2&2&2\\
0&0&0&1&1&1&2&2&2&1&1&1&2&2&2&0&0&0&2&2&2&0&0&0&1&1&1\\
0&0&0&2&2&2&1&1&1&1&1&1&0&0&0&2&2&2&2&2&2&1&1&1&0&0&0
\end{array}
$$

The rows of (2.5.8) and (2.5.10) constitute the array $(27, 13, 3, 2)$ with the maximum number of constraints.

2.6. ORTHOGONAL ARRAYS $(2s^n, 2(s^n - 1)/(s - 1) - 1, s, 2)$

Addelman and Kempthorne (1961) gave an ingenious method of constructing orthogonal arrays $(2s^n, 2(s^n - 1)/(s - 1) - 1, s, 2)$. We present here this method for $n = 2$, and for its proof we refer to the original paper.

Let s be a prime or a prime power. We first construct an orthogonal array $(s^2, (s^2 - 1)/(s - 1), s, 2)$ of index unity with the factors represented by $x_1, x_2, x_1 + x_2, x_1 + 2x_2, \cdots, x_1 + (s - 1)x_2$ in the notation of Section 2.3. To these factors we add $(s^2 - 1)/(s - 1) - 1$ additional factors represented by $x_1^2 + x_2, x_1^2 + x_1 + x_2, x_1^2 + 2x_1 + x_2, \cdots, x_1^2 + (s - 1)x_1 + x_2$. These $2(s^2 - 1)/(s - 1) - 1$ factors form the first part, consisting of s^2 combinations. The factors in the second half are denoted by

$$
x_1, x_2, x_1 + x_2 + b_1, x_1 + 2x_2 + b_2, \cdots, x_1 + (s - 1)x_2 + b_{s-1},
$$
$$
kx_1^2 + x_2, kx_1^2 + k_1 x_1 + x_2 + c_1,
$$
$$
kx_1^2 + k_2 x_1 + x_2 + c_2, \cdots, kx_1^2 + k_{s-1} x_1 + x_2 + c_{s-1},
$$

where $b_1, b_2, \cdots, b_{s-1}, k, k_1, k_2, \cdots, k_{s-1}, c_1, c_2, \cdots, c_{s-1}$ are determined as follows:

If s is odd, we choose k to be a quadratic nonresidue of $GF(s)$. By letting $a_i = \alpha_1, \alpha_2, \cdots, \alpha_{s-1}$, the nonzero elements of $GF(s)$, we can determine $b_1, b_2, \cdots, b_{s-1}$ from the equation

(2.6.1)
$$
b_i = \frac{k - 1}{4ka_i}.
$$

Then, letting $d_i = \alpha_1, \alpha_2, \cdots, \alpha_{s-1}$, we determine $k_1, k_2, \cdots, k_{s-1}$ from the equation

(2.6.2)
$$
k_i = kd_i,
$$

and $c_1, c_2, \cdots, c_{s-1}$ from the equation

(2.6.3)
$$c_i = \frac{d_i^2(k - 1)}{4}.$$

However, when s is even, we let $k = 1$ and choose b_i to be any one of the 2^{m-1} elements of $GF(2^m)$ that is not given by $x_1^2 - (1/a_i)x_1$, where x_1 ranges over $GF(2^m)$. We also choose $k_i = d_i$ and c_i to be any one of the 2^{m-1} elements of $GF(2^m)$ that is not given by the values of $x_1^2 + d_i x_1$.

Illustration 2.6.1. Let us illustrate this method in the construction of the orthogonal array $(18, 7, 3, 2)$. Choosing the factors $x_1, x_2, x_1 + x_2, x_1 + 2x_2$, $x_1^2 + x_2, x_1^2 + x_1 + x_2, x_1^2 + 2x_1 + x_2$ and considering all the three levels for x_1 and x_2, we can write down the following first half of the array with nine assemblies:

x_1	0	0	0	1	1	1	2	2	2
x_2	0	1	2	0	1	2	0	1	2
$x_1 + x_2$	0	1	2	1	2	0	2	0	1
$x_1 + 2x_2$	0	2	1	1	0	2	2	1	0
$x_1^2 + x_2$	0	1	2	1	2	0	1	2	0
$x_1^2 + x_1 + x_2$	0	1	2	2	0	1	0	1	2
$x_1^2 + 2x_1 + x_2$	0	1	2	0	1	2	2	0	1

(2.6.4)

To the nine assemblies given by (2.6.4) we add nine more assemblies with factors

$$x_1, \quad x_2, \quad x_1 + x_2 + b_1, \quad x_1 + 2x_2 + b_2, \quad kx_1^2 + x_2,$$
$$kx_1^2 + k_1 x_1 + x_2 + c_1, \quad kx_1^2 + k_2 x_1 + x_2 + c_2,$$

where $b_1, b_2, k, k_1, k_2, c_1$, and c_2 are to be selected. Since 2 is a quadratic nonresidue of $GF(3)$, we select $k = 2$. For $a_i = 1, 2$ from (2.6.1) we get $b_1 = 2, b_2 = 1$. For $d_i = 1, 2$ from (2.6.2) we get $k_1 = 2, k_2 = 1$. Finally, from (2.6.3) $c_1 = 1, c_2 = 1$. Thus the seven factors for the second half are

$$x_1, \quad x_2, \quad x_1 + x_2 + 2, \quad x_1 + 2x_2 + 1, \quad 2x_1^2 + x_2,$$
$$2x_1^2 + 2x_1 + x_2 + 1, \quad 2x_1^2 + x_1 + x_2 + 1,$$

and the plan is

x_1	0	0	0	1	1	1	2	2	2
x_2	0	1	2	0	1	2	0	1	2
$x_1 + x_2 + 2$	2	0	1	0	1	2	1	2	0
$x_1 + 2x_2 + 1$	1	0	2	2	1	0	0	2	1
$2x_1{}^2 + x_2$	0	1	2	2	0	1	2	0	1
$2x_1{}^2 + 2x_1 + x_2 + 1$	1	2	0	2	0	1	1	2	0
$2x_1{}^2 + x_1 + x_2 + 1$	1	2	0	1	2	0	2	0	1

(2.6.5) appears to the left of the fourth row.

The 18 assemblies given by (2.6.4) and (2.6.5) constitute the required array.

2.7. PRODUCT OF ORTHOGONAL ARRAYS

Bush (1952a) proved the following theorem:

Theorem 2.7.1. The existence of orthogonal arrays (N_i, k_i, s_i, t) for $i = 1, 2, \cdots, m$ implies the existence of the orthogonal array (N, k, s, t), where $N = N_1 N_2 \cdots N_m$, $s = s_1 s_2 \cdots s_m$, and $k = \min(k_1, k_2, \cdots, k_m)$.

Proof. Let the orthogonal array (N_1, k_1, s, t) be denoted by the $k_1 \times N_1$ matrix $A = (a_{ij})$ and the array (N_2, k_2, s, t) be denoted by the $k_2 \times N_2$ matrix $B = (b_{ij})$. Let A_1 and B_1 denote the first k rows of A and B, respectively. Then form the $k \times N_1 N_2$ matrix:

$$
\begin{matrix}
(a_{11}, b_{11}) & \cdots & (a_{11}, b_{1N_2}) & \cdots & (a_{1N_1}, b_{11}) & \cdots & (a_{1N_1}, b_{1N_2}) \\
(a_{21}, b_{21}) & \cdots & (a_{21}, b_{2N_2}) & \cdots & (a_{2N_1}, b_{21}) & \cdots & (a_{2N_1}, b_{2N_2}) \\
\cdot & \cdots & \cdot & \cdots & \cdot & \cdots & \cdot \\
(a_{k1}, b_{k1}) & \cdots & (a_{k1}, b_{kN_2}) & \cdots & (a_{kN_1}, b_{k1}) & \cdots & (a_{kN_1}, b_{kN_2})
\end{matrix}
$$

which can be shown to be an orthogonal array $(N_1 N_2, k, s_1 s_2, t)$. From this array, by following the same procedure with (N_3, k_3, s_3, t), we get the array $(N_1 N_2 N_3, k, s_1 s_2 s_3, t)$. Continuing this procedure, we finally get the array (N, k, s, t).

2.8. EMBEDDING OF ORTHOGONAL ARRAYS

Seiden and Zemach (1966) showed that the orthogonal arrays $(\lambda 2^{t+1}, k + 1, 2, t + 1)$ can be constructed from the arrays $(\lambda 2^t, k, 2, t)$ with the maximum number of constraints if k is the maximum number of constraints of strength t. The particular case of this result for $t = 2$ was given by Seiden (1954).

For the orthogonal array $(s^2[(s-1)t+1], k, s, 2)$ the maximum number of constraints equals $s^2 t + s + 1 = k^*$, say. Then the orthogonal array $(s^2[(s-1)t+1], k^*, s, 2)$ will be called the maximal array, and the array $(s^2[(s-1)t+1], k, s, 2)$ is said to have a deficiency $d = k^* - k$. Shrikhande and Bhagwandas (1969) proved the following:

Theorem 2.3.1. An orthogonal array $(s^2[(s-1)t+1], k, s, 2)$ can be embedded into a maximal array if (a) the deficiency is 1 for any value of s or (b) the deficiency is 2 for $s = 2$ or 3.

For the proof we refer to the original paper of Shrikhande and Bhagwandas. The particular case when $s = 2$ reduces to the embedding problem of Hadamard matrices and was also tackled by the same authors (1970).

2.9. PARTIALLY BALANCED ARRAYS

The concept of orthogonal arrays was generalized by Chakravarti (1956) to partially balanced arrays. He made the following definition:

Definition 2.9.1. Let A be a $k \times N$ matrix with elements $0, 1, 2, \cdots,$ or $s - 1$. Consider the s^t ordered t-plet $(x_1, x_2, \cdots, x_t)$ that can be formed from a t-rowed submatrix of A and let there be associated a positive integer $\lambda(x_1, x_2, \cdots, x_t)$ that is invariant under permutations of $x_1, x_2, \cdots, x_t$. If for every t-rowed submatrix of A the s^t ordered t-plets $(x_1, x_2, \cdots, x_t)$ occur $\lambda(x_1, x_2, \cdots, x_t)$ times, the matrix A is called a partially balanced array of strength t in N assemblies, k constraints, and the specified $\lambda(x_1, x_2, \cdots, x_t)$ parameters.

We observe that a partially balanced array reduces to an orthogonal array if $\lambda(x_1, x_2, \cdots, x_t) = \lambda$ for all $x_1, x_2, \cdots, x_t$.

As an example we can see that, by omitting the first three assemblies in (2.5.8), we get an arrangement that can be seen to be a partially balanced array, with $s = 3$ symbols, $k = 9$ constraints, $N = 24$ assemblies, and the following λ-parameters:

(2.9.1)
$$\lambda(x_1, x_2) = 3 \quad \text{if } x_1 \text{ and } x_2 \text{ are distinct,}$$
$$= 2 \quad \text{if } x_1 \text{ and } x_2 \text{ are identical.}$$

Partially balanced arrays have the advantage that they can be constructed with fewer assemblies than the orthogonal arrays for given k, s, t parameters.

Let us consider partially balanced arrays of strength 2 with λ-parameters

(2.9.2)
$$\lambda(x_1, x_2) = 0 \quad \text{if } x_1 \text{ and } x_2 \text{ are identical,}$$
$$= \lambda \quad \text{if } x_1 \text{ and } x_2 \text{ are distinct.}$$

The columns of such arrays contain distinct elements, and every pair of distinct elements occurs in a constant number of columns. Such an arrangement is a balanced incomplete block design, which we study in detail in Chapter 5. Partially balanced arrays whose parameters are given by (2.9.2) can also be used as balanced designs for experiments involving sequences of treatments [cf. Patterson (1952)]. By relaxing the condition $\lambda(x_1, x_2) = \lambda$ whenever x_1 and x_2 are distinct and imposing certain weaker restrictions, we get the partially balanced incomplete block designs of Chapter 8. Considering the partially balanced arrays of strength t with λ-parameters

$$
(2.9.3) \quad
\begin{aligned}
\lambda(x_1, x_2, \cdots, x_t) &= 0 \quad &&\text{if at least two of the } x \text{ terms are identical,} \\
&= \lambda \quad &&\text{if all the } x \text{ terms are distinct,}
\end{aligned}
$$

we get the tactical configurations of Chapter 7.

For the construction of partially balanced arrays from tactical configurations and pairwise partially balanced designs we refer the reader to Chakravarti (1961). We also present in Section 7.5 a method of constructing partially balanced arrays from doubly balanced designs.

We conclude this chapter with the note that orthogonal arrays and partially balanced arrays play a vital role in the construction of symmetrical and asymmetrical confounded factorial experiments and fractionally replicated designs; we pursue this study in Chapters 13, 14, and 15. It may be added that multi-factorial designs derived from partially balanced arrays require fewer assemblies to accommodate a given number of factors. For an expository article in this connection we refer to Chakravarti (1963).

REFERENCES

1. Addelman, S., and Kempthorne, O. (1961). Some main effect plans and orthogonal arrays of strength two. *Ann. Math. Stat.*, **32**, 1167–1176.
2. Bose, R. C., and Bush, K. A. (1952). Orthogonal arrays of strength two and three. *Ann. Math. Stat.*, **23**, 508–524.
3. Bush, K. A. (1952a). A generalization of a theorem due to MacNeish. *Ann. Math. Stat.*, **23**, 293–295.
4. Bush, K. A. (1952b). Orthogonal arrays of index unity. *Ann. Math. Stat.*, **23**, 426–434.
5. Chakravarti, I. M. (1956). Fractional replication in asymmetrical factorial designs and partially balanced arrays. *Sankhyā*, **17**, 143–164.
6. Chakravarti, I. M. (1961). On some methods of constructions of partially balanced arrays. *Ann. Math. Stat.*, **32**, 1181–1185.
7. Chakravarti, I. M. (1963). Orthogonal and partially balanced arrays and their applications in design of experiments. *Metrika*, **7**, 231–243.
8. Kishen, K. (1942). On latin and hypergraecolatin cubes and hypercubes. *Current Science*, **11**, 98–99.

9. Kishen, K. (1949). On the construction of latin and hypergraecolatin cubes and hypercubes. *J. Indian Soc. Agric. Stat.*, **2**, 20–48.

10. Paley, R. E. A. C. (1933). On orthogonal matrices. *J. Math. Phys.*, **12**, 311–320.

11. Patterson, H. D. (1952). The construction of balanced designs for experiments involving sequences of treatments. *Biometrika*, **39**, 32–48.

12. Plackett, R. L., and Burman, J. P. (1943–1944). The design of optimum mulfactorial experiments. *Biometrika*, **33**, 305–325.

13. Rao, C. R. (1946). Hypercubes of strength " d " leading to confounded designs in factorial experiments. *Bull. Calcutta Math. Soc.*, **38**, 67–78.

14. Rao, C. R. (1947). Factorial experiments derivable from combinatorial arrangements of arrays. *J. Roy. Stat. Soc., Suppl.*, **9**, 128–139.

15. Saxena, P. N. (1960). On the latin cubes of the second order and the fourth replication of the three dimensional or cubic lattice designs. *J. Indian Soc. Agric. Stat.*, **12**, 100–140.

16. Seiden, E. (1954). On the problem of construction of orthogonal arrays. *Ann. Math. Stat.*, **25**, 151–156.

17. Seiden, E., and Zemach, R. (1966). On orthogonal arrays. *Ann. Math. Stat.*, **37**, 1355–1370.

18. Shrikhande, S. S. (1964) Generalized Hadamand matrices and orthogonal arrays of strength two. *Can. J. Math.*, **16**, 736–740.

19. Shrikhande, S. S., and Bhagwandas (1970). A note on embedding of Hadamard matrices. *Essays in Probability and Statistics*. University of North Carolina Press, pp. 673–688.

20. Shrikhande, S. S., and Bhagwandas (1969). A note on embedding of orthogonal arrays of strength two. *Combinatorial Mathematics and Its Applications*. University of North Carolina Press, pp. 256–273.

BIBLIOGRAPHY

Bose, R. C. On some connections between the design of experiments and information theory. *Bull. Intern. Stat. Inst.*, **38**, 257–271 (1961).

Bose, R. C., and Shrikhande, S. S. On the composition of balanced incomplete block designs. *Can. J. Math.*, **12**, 177–188 (1960).

Farrel, R. H., Kiefer, J., and Walburn, A. *Optimum Multivariate Designs*. Proceedings of the Fifth Berkeley Symposium on Mathematical Statistics and Probability, L. M. Lecam and J. Neyman, Eds., 1965.

Fisher, R. A. A system of confounding for factors with more than two alternatives giving completely orthogonal cubes and higher powers. *Ann. Eugenics*, **12**, 283–290 (1945).

Maurin, F. Sur une généralisation de la méthode des différences pour la construction de tableaux orthogonaux. *Compt. Rend.*, **259**, 4490–4491 (1964).

CHAPTER 3

Pairwise Balanced Designs and Mutually Orthogonal Latin Squares

3.1. INTRODUCTION

In Chapter 1 we have seen that the maximal number of MOLS of order s is $s - 1$ and a complete set of MOLS (consisting of $s - 1$ MOLS) can be constructed if s is a prime or a prime power. The problem of constructing MOLS when s is a composite number is considered in this chapter.

Let the prime decomposition of a composite number s be $p_1^{e_1} p_2^{e_2} \cdots p_m^{e_m}$. Define

(3.1.1) $$n(s) = \min (p_1^{e_1}, p_2^{e_2} \cdots, p_m^{e_m}) - 1.$$

MacNeish (1922) and Mann (1942) showed that a set of $n(s)$ MOLS of order s can always be constructed, and we give this method of construction in Section 3.2. If $N(s)$ denotes the maximum possible number of MOLS of order s, then clearly

(3.1.2) $$N(s) \geqq n(s),$$

which is the MacNeish–Mann theorem. MacNeish even conjectured that the upper bound of $N(s)$ is $n(s)$ and hence $N(s) = n(s)$. This conjecture implies that

(3.1.3) $$N(s) = 1 \qquad \text{if } s \equiv 2 \pmod 4,$$

which is known as Euler's conjecture and was made in 1782. When $s = 2$,

32

trivially (3.1.3) is true. Euler conjectured (3.1.3) when he failed to construct two MOLS of order 6 for providing an arrangement of 36 army officers belonging to six different ranks and six different regiments in a square array of order 6 so that each rank and regiment is represented exactly once in every row and column. By complete enumeration Tarry (1900–1901) verified the correctness of the conjecture for $s = 6$. Petersen (1901–1902), Wernicke (1910), and MacNeish (1922) made some erroneous attempts to prove Euler's conjecture. Levi (1942) spotted the error in the argument presented by Peterson and MacNeish, and Wernicke's bid was foiled by MacNeish.

Parker (1958) showed that if there exists a balanced incomplete block (BIB) design (see Section 5.1) with the parameters $v = s$, $\lambda = 1$, and k a prime or a prime power, then $N(s) > n(s)$, thereby disproving MacNeish's conjecture. Though this cast serious doubt on the validity of Euler's conjecture, it did not disprove it. Later Bose and Shrikhande (1959a,b,c) introduced a general class of designs, called pairwise balanced designs of index λ, which share with BIB designs the property that every pair of symbols occurs together in λ sets but differ from BIB designs in not having the constant set sizes. With the help of these designs Bose and Shrikhande proved the falsity of Euler's conjecture for an infinity of values of $s \geq 22$. By using the method of differences, Parker (1959) showed that $N(s) \geq 2$ for $s = (3q - 1)/2$, where q is a prime or a prime power $\equiv 3 \pmod 4$, which disproves Euler's conjecture for $s = 10$. Bose, Shrikhande, and Parker (1960) presented methods of constructing MOLS that disprove Euler's conjecture for every $s > 6$.

Chowla, Erdos, and Straus (1960), following the lines of argument presented by Bose, Shrikhande, and Parker, proved that

$$(3.1.4) \qquad\qquad N(s) > \tfrac{1}{3}s^{1/91}$$

for all sufficiently large s (the lower bound of s being unknown). The lower bound (3.1.4) on $N(s)$ was improved by Wang (1966) to

$$(3.1.5) \qquad\qquad N(s) > s^{1/24}$$

for all sufficiently large s.

From the embedding theorem of Bruck (1963), given in Section 9.5, we have that if $N(s) < s - 1$, then

$$(3.1.6) \qquad\qquad N(s) < (s - 1) - (2s)^{1/4}.$$

We discuss the pairwise balanced designs in Section 3.3 and devote the subsequent sections for the construction of MOLS, in particular showing the falsity of Euler's conjecture.

3.2. MacNEISH–MANN THEOREM

In this section we give a method of constructing $n(s)$ MOLS of order s, where $n(s)$ is given by (3.1.1). Let s be a composite number with prime decomposition $s = p_1^{e_1} p_2^{e_2} \cdots p_m^{e_m}$. Let us construct the system of s elements

$$(3.2.1) \qquad \gamma = (g_1, g_2, \cdots, g_m),$$

where $g_i \in \mathrm{GF}(p_i^{e_i})$, $i = 1, 2, \cdots, m$. Let us define addition and multiplication by the rules

$$(3.2.2) \qquad \begin{aligned} \gamma_1 + \gamma_2 &= (g_1, g_2, \cdots, g_m) + (h_1, h_2, \cdots, h_m) \\ &= (g_1 + h_1, g_2 + h_2, \cdots, g_m + h_m) \end{aligned}$$

and

$$(3.2.3) \qquad \gamma_1 \gamma_2 = (g_1 h_1, g_2 h_2, \cdots, g_m h_m),$$

where in the right-hand sides of (3.2.2) and (3.2.3) the operations in each component are as defined in the corresponding Galois field. The system so constructed is not a field, since, for instance, the element $(0, 1, \cdots, 1)$ has no multiplicative inverse. All the points that have no zero among their coordinates possess inverses.

Let $0, g_i^{(1)} = 1, g_i^{(2)}, \cdots, g_i^{(p_i^{e_i}-1)}$ be the elements of $\mathrm{GF}(p_i^{e_i})$ for every $i = 1, 2, \cdots, m$, and let $n(s)$ be as given by (3.1.1). Then

$$(3.2.4) \qquad \gamma_j = [g_1^{(j)}, g_2^{(j)}, \cdots, g_m^{(j)}] \qquad 0 < j \le n(s)$$

possesses inverses, as does $\gamma_j - \gamma_i$ if $i \ne j$. Now we number the points γ in such a way that $0 = \gamma_0 = (0, 0, \cdots, 0)$ and the next $n(s)$ elements are given by (3.2.4) and form the $n(s)$ arrays L_j, whose (r, t)th cell is filled by the element $\gamma_r + \gamma_j \gamma_t$ for $j = 1, 2, \cdots, n(s)$; $r, t = 0, 1, \cdots, s - 1$. We can easily verify that, for a given j, L_j is a latin square and $L_1, L_2, \cdots, L_{n(s)}$ are the required $n(s)$ MOLS of order s. We illustrate this method in the construction of a pair of MOLS of order 15. Clearly $n(15) = 2$. Let us take the γ elements

$$\begin{array}{llll} \gamma_0 = (0, 0), & \gamma_1 = (1, 1), & \gamma_2 = (2, 2), & \gamma_3 = (0, 1), \\ \gamma_4 = (0, 2), & \gamma_5 = (0, 3), & \gamma_6 = (0, 4), & \gamma_7 = (1, 0), \\ \gamma_8 = (1, 2), & \gamma_9 = (1, 3), & \gamma_{10} = (1, 4), & \gamma_{11} = (2, 0), \\ \gamma_{12} = (2, 1), & \gamma_{13} = (2, 3), & \gamma_{14} = (2, 14). \end{array}$$

The required two MOLS are then

$$L_1 = \begin{array}{ccccccccccccccc}
0 & 1 & 2 & 3 & 4 & 5 & 6 & 7 & 8 & 9 & 10 & 11 & 12 & 13 & 14 \\
1 & 2 & 5 & 8 & 9 & 10 & 7 & 12 & 13 & 14 & 11 & 3 & 4 & 6 & 0 \\
2 & 5 & 10 & 13 & 14 & 11 & 12 & 4 & 6 & 0 & 3 & 8 & 9 & 7 & 1 \\
3 & 8 & 13 & 4 & 5 & 6 & 0 & 1 & 9 & 10 & 7 & 12 & 2 & 14 & 11 \\
4 & 9 & 14 & 5 & 6 & 0 & 3 & 8 & 10 & 7 & 1 & 2 & 13 & 11 & 12 \\
5 & 10 & 11 & 6 & 0 & 3 & 4 & 9 & 7 & 1 & 8 & 13 & 14 & 12 & 2 \\
6 & 7 & 12 & 0 & 3 & 4 & 5 & 10 & 1 & 8 & 9 & 14 & 11 & 2 & 13 \\
7 & 12 & 4 & 1 & 8 & 9 & 10 & 11 & 2 & 13 & 14 & 0 & 1 & 5 & 6 \\
8 & 13 & 6 & 9 & 10 & 7 & 1 & 2 & 14 & 11 & 12 & 4 & 5 & 0 & 3 \\
9 & 14 & 0 & 10 & 7 & 1 & 8 & 13 & 11 & 12 & 2 & 5 & 6 & 3 & 4 \\
10 & 11 & 3 & 7 & 1 & 8 & 9 & 14 & 12 & 2 & 13 & 6 & 0 & 4 & 5 \\
11 & 3 & 8 & 12 & 2 & 13 & 14 & 0 & 4 & 5 & 6 & 7 & 1 & 9 & 10 \\
12 & 4 & 9 & 2 & 13 & 14 & 11 & 3 & 5 & 6 & 0 & 1 & 8 & 10 & 7 \\
13 & 6 & 7 & 14 & 11 & 12 & 2 & 5 & 0 & 3 & 4 & 9 & 10 & 1 & 8 \\
14 & 0 & 1 & 11 & 12 & 2 & 13 & 6 & 3 & 4 & 5 & 10 & 7 & 8 & 9 \\
\end{array}$$

$$L_2 = \begin{array}{ccccccccccccccc}
0 & 2 & 10 & 4 & 6 & 3 & 5 & 11 & 14 & 12 & 13 & 7 & 8 & 1 & 9 \\
1 & 5 & 11 & 9 & 7 & 8 & 10 & 3 & 0 & 4 & 6 & 12 & 13 & 2 & 14 \\
2 & 10 & 3 & 14 & 12 & 13 & 11 & 8 & 1 & 9 & 7 & 4 & 6 & 5 & 0 \\
3 & 13 & 7 & 5 & 0 & 4 & 6 & 12 & 11 & 2 & 14 & 1 & 9 & 8 & 10 \\
4 & 14 & 1 & 6 & 3 & 5 & 0 & 2 & 12 & 13 & 11 & 8 & 10 & 9 & 7 \\
5 & 11 & 8 & 0 & 4 & 6 & 3 & 13 & 2 & 14 & 12 & 9 & 7 & 10 & 1 \\
6 & 12 & 9 & 3 & 5 & 0 & 4 & 14 & 13 & 11 & 2 & 10 & 1 & 7 & 8 \\
7 & 4 & 14 & 8 & 10 & 1 & 9 & 0 & 6 & 3 & 5 & 11 & 2 & 12 & 13 \\
8 & 6 & 12 & 10 & 1 & 9 & 7 & 4 & 3 & 5 & 0 & 2 & 14 & 13 & 11 \\
9 & 0 & 2 & 7 & 8 & 10 & 1 & 5 & 4 & 6 & 3 & 13 & 11 & 14 & 12 \\
10 & 3 & 13 & 1 & 9 & 7 & 8 & 6 & 5 & 0 & 4 & 14 & 12 & 11 & 2 \\
11 & 8 & 6 & 2 & 14 & 12 & 13 & 7 & 10 & 1 & 9 & 0 & 4 & 3 & 5 \\
12 & 9 & 0 & 13 & 11 & 2 & 14 & 1 & 7 & 8 & 10 & 1 & 5 & 4 & 6 \\
13 & 7 & 4 & 11 & 2 & 14 & 12 & 9 & 8 & 10 & 11 & 5 & 0 & 6 & 3 \\
14 & 1 & 5 & 12 & 13 & 11 & 2 & 10 & 9 & 7 & 8 & 6 & 3 & 0 & 4 \\
\end{array}$$

where γ_i is simply written as i ($i = 0, 1, \cdots, 14$).

3.3 PAIRWISE BALANCED DESIGNS

Bose and Shrikhande (1960) made the following definition:

Definition 3.3.1. An arrangement of v symbols in b sets will be called a pairwise balanced design of index λ and type $(v; k_1, k_2, \cdots, k_m)$ if each set contains $k_1, k_2, \cdots,$ or k_m symbols that are all distinct ($k_i \leq v$, $k_i \neq k_j$) and every pair of distinct symbols occurs in exactly λ sets of the design.

If b_i is the number of sets of size k_i ($i = 1, 2, \cdots, m$), then

(3.3.1)
$$b = \sum_{i=1}^{m} b_i, \qquad \lambda v(v - 1) = \sum_{i=1}^{m} b_i k_i(k_i - 1).$$

Pairwise balanced designs of index unity play a vital role in disproving Euler's conjecture. The design formed by the b_i sets of size k_i is called the ith equiblock component D_i of the pairwise balanced design D. A class of subsets of D_i is said to be of type I if every symbol occurs in that class exactly k_i times and of type II if every symbol occurs exactly once. Clearly the number of sets in a class of type I is v, and of type II, v/k_i. The v sets of the type I class can be written out as columns, and, by using systems of distinct representatives (SDR) of Chapter 6, the symbols can be so arranged that every symbol occurs exactly once in each row. When so written out, the sets are said to be in standard form. The component D_i is called separable if the sets can be divided into classes of type I or type II, or both. The design D is called separable if each equiblock component D_i is separable. The sets $D_1, D_2, \cdots, D_l\,(l < m)$ are said to be clear if the $\sum_{i=1}^{l} b_i$ sets are disjoint.

In Chapter 11, we study another class of designs, known as symmetrical unequal block (SUB) arrangements, which form a subclass of pairwise balanced designs. All the SUB arrangements constructed there can be used as pairwise balanced designs.

We denote a BIB design with the parameters v, b, r, k, and $\lambda = 1$ by BIB$(v; k)$. Clearly BIB$(v; k)$ is a pairwise balanced design $(v; k)$ of index unity. We can further see that the symmetrical BIB designs and the resolvable BIB designs are separable. If a BIB$(v; k)$ exists, then, by omitting one symbol, we get a pairwise balanced design $(v - 1; k, k - 1)$ of index unity. Again from a BIB$(v; k)$, by omitting x symbols ($2 \leq x \leq k$) occurring in the same block, we get a pairwise balanced design $(v - x; k, k - 1, k - x)$ of index unity. We also note that, by omitting three symbols not occurring in the same set of a BIB$(v; k)$, we get a pairwise balanced design $(v - 3; k, k - 1, k - 2)$ of index unity. If BIB$(v; k)$ is a resolvable design, then, by adding a new symbol θ_i to each set of the ith replication ($i = 1, 2, \cdots, x; 1 < x \leq r$) and adding a new set of symbols $(\theta_1, \theta_2, \cdots, \theta_x)$, we get a unit-index pairwise balanced design $(v + x; k + 1, k, x)$ if $x < r$ and $(v + r; k + 1, r)$ if $x = r$.

If s is a prime or a prime power, by treating the points of EG(m, s) as symbols and lines of the geometry as sets, we get BIB(s^m, s), which is clearly resolvable. If we add a new symbol to each set of the first replication, we get a pairwise balanced design $(s^m + 1; s + 1, s)$ of index unity.

By using PG$(2, 2^3)$ and EG$(2, 3^2)$, Bose and Shrikhande (1959b) constructed pairwise balanced designs $(73 - x; 7, 8, 9)$ and $(81 - x; 7, 8, 9)$ for $x \leq 10$.

For the definition of group-divisible (GD) designs we refer to Definition 8.4.1. A group-divisible design with the parameters $v = mn, b, r, k, \lambda_1 = 0$, $\lambda_2 = 1$ is denoted by GD$(v; k, n; 0, 1)$. If GD$(v; k, n; 0, 1)$ exists, then, by adding m new sets corresponding to the groups of the association scheme, we get a pairwise balanced design $(v; k, n)$ of index unity if $k \neq n$. The sets of size n form a clear equiblock component.

From the GD$(v; k, n; 0, 1)$ by adding a new symbol θ to each of the m sets of size n corresponding to the groups of the association scheme, we get a pairwise balanced design $(v + 1; k, n + 1)$. Again, from the resolvable GD$(v; k, n; 0, 1)$ we can construct pairwise balanced designs $(v + 1; k + 1, k, n)$ if $x = 1$, $(v + x; k + 1, k, n, x)$ if $1 < x < r$, and $(v + x; k + 1, n, x)$ if $x = r$ by adding a new symbol θ_i to each set of the ith replicate $(x \leq r)$ and adjoining a set of the new symbols if $x > 1$.

3.4. APPLICATION OF PAIRWISE BALANCED DESIGNS IN THE CONSTRUCTION OF MOLS

All the results of this section are due to Bose and Shrikhande (1959a,b,c) and Bose, Shrikhande, and Parker (1960). The main result in this connection is the following theorem:

Theorem 3.4.1. Let D be a pairwise balanced design $(v; k_1, k_2, \cdots, k_m)$ of index unity in which the equiblock components $D_1, D_2, \cdots, D_l (l < m)$ are a clear set. Let there be $q_i - 1$ MOLS of order k_i and let

$$(3.4.1) \qquad q = \min (q_1 + 1, q_2 + 1, \cdots, q_l + 1, q_{l+1}, \cdots, q_m).$$

Then there exist at least $q - 2$ MOLS of order v.

Proof. Let δ_{ij} denote the jth set of the equiblock component D_i $(j = 1, 2, \cdots, b_i; i = 1, 2, \cdots, m)$. Let A_{ij} be an orthogonal array of strength 2 and index unity, with the symbols of δ_{ij}. Let $P(\delta_{ij})$ be a matrix of order $q_i \times k_i(k_i - 1)$ whose elements are the symbols of δ_{ij}, such that any ordered pair of symbols $\begin{pmatrix} \theta_u \\ \theta_w \end{pmatrix} \theta_u \neq \theta_w$ occurs as a column exactly once in any two-row submatrix of $P(\delta_{ij})$, and is constructed in the following manner:

Let the $q_i - 1$ MOLS of order k_i, with the symbols of δ_{ij}, be written in semistandard form. We form another square L with the symbol θ_u in each position of its uth column ($u = 1, 2, \cdots, k_i$). Then we form a $q_i \times k_i^2$ matrix whose rth row contains the symbols of the rth square listed from the first row to k_ith row. Now we delete the first k_i columns to get the required matrix $P(\delta_{ij})$.

Let Δ_1 be a matrix of order $q \times \sum_{i=1}^{l} b_i k_i^2$, obtained by considering the first q rows of A_{ij}, for $j = 1, 2, \cdots, b_i$; $i = 1, 2, \cdots, l$. Let Δ_2 be a matrix of order $q \times \sum_{i=l+1}^{m} b_i k_i(k_i - 1)$, obtained by considering the first q rows of $P(\delta_{ij})$ for $j = 1, 2, \cdots, b_i$; $i = l+1, l+2, \cdots, m$. Let Δ_3 be the $q \times (v - \sum_{i=1}^{l} b_i k_i)$ matrix whose tth column contains in every position the tth symbol of the symbols not occurring in $D_1, D_2, \cdots, D_l$. Then we can easily verify that

(3.4.2) $$\Delta = (\Delta_1, \Delta_2, \Delta_3)$$

is an orthogonal array $(v^2, q, v, 2)$ from which $q - 2$ MOLS of order v can be constructed.

Using the pairwise balanced designs discussed in the preceding section, we can state the following corollaries to the main theorem:

Corollary 3.4.1.1. The existence of a BIB$(v; k)$ implies that

(3.4.3) $N(v - 1) \geqq \min [N(k), 1 + N(k - 1)] - 1$;

(3.4.4) $N(v - x) \geqq \min [N(k), N(k - 1), 1 + N(k - x)] - 1$ \qquad if $2 \leqq x < k$;

(3.4.5) $N(v - 3) \geqq \min [N(k), N(k - 1), 1 + N(k - 2)] - 1$.

Corollary 3.4.1.2. The existence of a resolvable BIB$(v; k)$ implies that

(3.4.6) $N(v + x) \geqq \min [N(k), N(k + 1), 1 + N(x)] - 1$ \qquad if $1 < x \leqq r - 2$;

(3.4.7) $N(v + r - 1) \geqq \min [1 + N(k), N(k + 1), 1 + N(r - 1)] - 1$;

(3.4.8) $N(v + r) \geqq \min [N(k + 1), 1 + N(r)] - 1$;

(3.4.9) $N(v + 1) \geqq \min [N(k), N(k + 1)] - 1$.

Corollary 3.4.1.3. If s and $s + 1$ are both primes, then

(3.4.10) $$N(s^m + 1) \geqq s - 2.$$

Corollary 3.4.1.4.

(3.4.11) $$N(73 - x) \geqq 5 \qquad \text{if } x \leqq 10.$$

Corollary 3.4.1.5.

(3.4.12) $$N(81 - x) \geqq 5 \qquad \text{if } x \leqq 10.$$

Corollary 3.4.1.6. The existence of a $GD(v; k, n; 0, 1)$ implies that

(3.4.13) $N(v) \geqq \min [N(k), 1 + N(n)] - 1;$

(3.4.14) $N(v + 1) \geqq \min [N(k), N(n + 1)] - 1.$

Corollary 3.4.1.7. The existence of a $GD(v; k, n; 0, 1)$ implies that

(3.4.15) $N(v - 1) \geqq \min [N(k), N(k - 1), 1 + N(n), 1 + N(n - 1)] - 1.$

Corollary 3.4.1.8. The existence of a $BIB(v; k)$ with sets of type I implies that

(3.4.16) $N(v + r) \geqq \min [N(k + 1), 1 + N(r)] - 1.$

Corollary 3.4.1.9. The existence of a resolvable $GD(v; k, n; 0, 1)$ with r replications implies that

(3.4.17) $N(v + 1) \geqq \min [N(k), N(k + 1), 1 + N(n)] - 1;$

(3.4.18) $N(v + x) \geqq \min [N(k), N(k + 1), 1 + N(n), 1 + N(x)] - 1$

$$\text{if } 1 < x < r;$$

(3.4.19) $N(v + r) \geqq \max \{\min [N(k + 1), 1 + N(n), 1 + N(r)],$
$$\min [N(k + 1), N(n + 1), 1 + N(k), 1 + N(r)]\} - 1;$$

(3.4.20) $N(v + r + 1) \geqq \min [N(k + 1), N(n + 1), 1 + N(r + 1)] - 1.$

Theorem 3.4.2. Let there exist a separable pairwise balanced design $(v; k_1, k_2, \cdots, k_m)$ of index unity and suppose that there exist $q_i - 1$ MOLS of order k_i. If

(3.4.21) $q = \min (q_1, q_2, \cdots, q_m),$

then there exist at least $q - 1$ MOLS of order v.

Proof. With the notation used in the proof of Theorem 3.4.1, let Δ^* denote the first q rows of the matrices $P(\delta_{ij})$ for $j = 1, 2, \cdots, b_i; i = 1, 2, \cdots, m$. Then Δ^* is of order $q \times v(v - 1)$, and every possible distinct pair of symbols occurs together exactly once in any two-row submatrix of Δ^*. Let Δ_0 be a $q \times v$ matrix whose tth column contains the tth symbol ($t = 1, 2, \cdots, v$) in every position. Then (Δ_0, Δ^*) is an orthogonal array $(v^2, q, v, 2)$. When the pairwise balanced design is separable, we can show easily that the columns of Δ^* can be so arranged into $v - 1$ submatrices $\Delta^{(1)}, \Delta^{(2)}, \cdots, \Delta^{(v-1)}$ that each of the symbols occurs exactly once in every row. If α_i' denotes a $1 \times v$ row vector each element of which is the ith symbol, then

(3.4.22)
$$\begin{bmatrix} \alpha_v', & \alpha_1', & \cdots, & \alpha_{v-1}' \\ \Delta_0, & \Delta^{(1)}, & \cdots, & \Delta^{(v-1)} \end{bmatrix}$$

is an orthogonal array $(v^2, q + 1, v, 2)$; this is equivalent to the existence of $q - 1$ MOLS of order v. Hence the theorem.

As any easy consequence to the above theorem we have the following corollary:

Corollary 3.4.2.1. The existence of a symmetrical BIB($v; k$) implies that

$$(3.4.23) \qquad N(v) \geqq N(k).$$

A symmetrical BIB($s^2 + s + 1; s + 1$) exists (Chapter 5) when s is a prime or a prime power. Therefore

Corollary 3.4.2.2. If s is a prime or a prime power, then

$$(3.4.24) \qquad N(s^2 + s + 1) \geqq N(s + 1).$$

Since a resolvable BIB($s^3 + 1; s + 1$) exists when s is a prime (Bose, 1959), we have

Corollary 3.4.2.3. If s is a prime or a prime power, then

$$(3.4.25) \qquad N(s^3 + 1) \geqq N(s + 1).$$

Since a resolvable BIB($2^{m-1}(2^m - 1); 2^{m-1}$) exists (Bose and Shrikhande, 1959b), we have

Corollary 3.4.2.4.
$$(3.4.26) \qquad N\big(2^{m-1}(2^m - 1)\big) \geqq N(2^{m-1}).$$

Corollary 3.4.2.5. The existence of a symmetrical or resolvable

$$GD(v; k, n; 0, 1)$$

implies that

$$(3.4.27) \qquad N(v) \geqq \min \big[N(k), N(n)\big].$$

From the existence of symmetrical, regular GD($s^2 - 1; s, s - 1; 0, 1$), when s is a prime or a prime power, the pairwise balanced design ($s^2 - 1; s, s - 1$) can be constructed, and clearly it is separable. Hence

Corollary 3.4.2.6.
$$(3.4.28) \qquad N(s^2 - 1) \geqq N(s - 1)$$

when s is a prime or a prime power.

Using Theorem 2.7.1, the following theorem due to Bush (1952a) can be easily obtained:

Theorem 3.4.3. If $v = v_1 v_2 \cdots v_m$, then

$$(3.4.29) \qquad N(v) \geqq \min \big[N(v_1), N(v_2), \cdots, N(v_m)\big].$$

3.5. USE OF THE METHOD OF DIFFERENCES IN THE CONSTRUCTION OF MOLS

Let M be a module of u elements denoted by $0, 1, \cdots, u - 1$ and let X be a set of indeterminates denoted by $x_1, x_2, \cdots, x_w$. Any difference $i - j$, where $i, j = 0, 1, \cdots, u - 1$, is called a difference of type MM. Difference $i - x_k = x_k$ $(i = 0, 1, \cdots, u - 1)$ is called a difference of type MX, and difference $x_k - i = x_k$ $(i = 0, 1, \cdots, u - 1)$ is called a difference of type XM.

Theorem 3.5.1. Let $N(w) \geq 2$ and $t = u + 2w - 1$. Let there exist a $4 \times t$ matrix A with elements of M and X such that the t differences formed from the columns of any two-row submatrix satisfy the condition that all nonzero differences of type MM, all differences of type MX, and all differences of type XM be repeated exactly once. Let $A_i = A + iE_{4,t}$ for $i = 0, 1, \cdots, u - 1$, where $x_k + i = x_k$ $(k = 1, 2, \cdots, w; i = 0, 1, \cdots, u - 1)$. Put

$$A^* = (A_0, A_1, \cdots, A_{u-1}).$$

Let B be an orthogonal array of four constraints of strength 2 and index unity with the symbols of X and let

$$D = \begin{bmatrix} 0 & 1 \cdots u - 1 \\ 0 & 1 \cdots u - 1 \\ 0 & 1 \cdots u - 1 \\ 0 & 1 \cdots u - 1 \end{bmatrix}$$

Then the matrix (A^*, B, D) is an orthogonal array $((u + w)^2, 4, u + w, 2)$, of index unity, that gives at least two MOLS of order $u + w$.

The proof can be constructed along lines similar to those used in Theorem 5.7.1 and is left as an exercise.

From the above theorem we have the following:

Theorem 3.5.2. If m is odd, there exist at least two MOLS of order $3m + 1$. In particular, if $m = 4t + 3$, then $N(12t + 10) \geq 2$.

Proof. Let M be a module of $2m + 1$ elements $0, 1, 2, \cdots, 2m$ and let $x_1, x_2, \cdots, x_m$ be m indeterminates. Then the $4m$ columns

(	0,	i,	$2m - i + 1$,	x_i)',
(	i,	0,	x_i,	$2m - i + 1$)',
(	$2m - i + 1$,	x_i,	0,	i)',
(	x_i,	$2m - i + 1$,	i,	0)',

(3.5.1)

for $i = 1, 2, \cdots, m$, satisfy the requirements of the matrix A given in Theorem 3.5.1. All the $(3m + 1)^2$ assemblies of the orthogonal array can be constructed by the method given in that theorem.

Bose, Shrikhande, and Parker (1960) constructed two MOLS of order 10 by the method of Theorem 3.5.2. These are shown below.

A pair of 10 × 10 orthogonal squares

00	67	58	49	91	83	75	12	24	36
76	11	07	68	59	92	84	23	35	40
85	70	22	17	08	69	93	34	46	51
94	86	71	33	27	18	09	45	50	62
19	95	80	72	44	37	28	56	61	03
38	29	96	81	73	55	47	60	02	14
57	48	39	90	82	74	66	01	13	25
21	32	43	54	65	06	10	77	88	99
42	53	64	05	16	20	31	89	97	78
63	04	15	26	30	41	52	98	79	87

Theorem 3.5.3

$$N(14) \geqq 2.$$

Proof. Let M be a module of 11 elements $0, 1, \cdots, 10$ and let x_1, x_2, x_3 be three indeterminates. The cyclic permutations of the four columns

$$(3.5.2) \qquad \begin{array}{l} (0, \ 1, 4, 6)'; \quad (x_1, 0, 4, 1)'; \\ (x_2, 0, 6, 2)'; \quad (x_3, 0, 9, 8)' \end{array}$$

satisfy the requirements of the matrix A of Theorem 3.5.1, and at least two MOLS of order 14 can be constructed by the method indicated in that theorem.

Theorem 3.5.4

$$N(18) \geqq 2.$$

Proof. Let M be a module of 15 elements $0, 1, \cdots, 14$ and let x_1, x_2, x_3 be three indeterminates. The cyclic permutations of the columns

$$(3.5.3) \qquad \begin{array}{l} (0, 1, 4, 8)'; \quad (0, 2, 7, 6)'; \quad (x_1, 0, 6, 14)'; \\ (x_2, 0, 10, 6)'; \quad (x_3, 0, 12, 10)' \end{array}$$

satisfy the requirements of the matrix A of Theorem 3.5.1, and at least two MOLS of order 18 can be constructed by the method of that theorem.

Theorem 3.5.5

$$N(26) \geq 2.$$

Proof. Let M be a module of 23 elements $0, 1, \cdots, 22$ and let x_1, x_2, x_3 be three indeterminates. The cyclic permutations of the columns

(3.5.4)
$$\begin{array}{lll}
(0, \quad 3, \quad 8, 12)'; & (0, \quad 6, 20, 16)'; & (0, \quad 2, 12, 7)'; \\
(0, \quad 1, 16, \quad 2)'; & (x_1, 0, 20, 19)'; & (x_2, 0, 17, 6)'; \\
(x_3, 0, \quad 8, 21)'
\end{array}$$

satisfy the requirements of the matrix A of Theorem 3.5.1, and at least two MOLS of order 26 can be constructed by the method of that theorem.

The A-matrices of Theorems 3.5.3 and 3.5.5 were given by Parker (1959) and the A-matrix of Theorem 3.5.4 is supposed to be new. All these A-matrices were constructed by trial and error. A systematic study of the method of differences in constructing MOLS remains to be done. The above idea can be generalized to construct t MOLS of order $v(t \leq v - 1)$ and this method is yet to be developed completely.

3.6. FALSITY OF EULER'S CONJECTURE

We have seen the existence of at least two MOLS of order 10, 14, 18, 22, 26, etc., in Section 3.5. By using various pairwise balanced designs and methods of Section 3.4, Bose and Shrikhande (1959a,b,c) and Bose, Shrikhande, and Parker (1960) showed that there exists at least a pair of MOLS of order $4t + 2$, where t is an integer other than 0 or 1. For order 6 it was known that a pair of MOLS does not exist. If a positive integer $s > 2$ is called Eulerian if two orthogonal latin squares of order s do not exist, then 6 is the only Eulerian number.

3.7. GREATEST LOWER BOUND ON THE NUMBER OF MOLS OF ORDER $s \leq 100$

Table 3.7.1 lists $n(s)$ and the greatest lower bound of $N(s)$ so far known for all $s \leq 100$, where s is not a prime or a prime power. The method of construction is also indicated.

Table 3.7.1

s	n(s)	N(s) ≥	Method of Construction
6	1	1	Eulerian number
10	1	2	Theorem 3.5.2 ($m = 3$)
12	2	5	Johnson, Dulmage, and Mendelsohn (1961)
14	1	2	Theorem 3.5.3
15	2	2	MacNeish–Mann theorem
18	1	2	Theorem 3.5.4
20	3	3	MacNeish–Mann theorem
21	2	4	Corollary 3.4.2.2 ($s = 4$)
22	1	2	Theorem 3.5.2 ($m = 7$)
24	2	3	Corollary 3.4.2.6 ($s = 5$)
26	1	2	Theorem 3.5.5
28	3	3	MacNeish–Mann theorem
30	1	2	Corollary 3.4.1.2, (3.4.7) ($v = 21$, $k = 3$)
33	2	3	Corollary 3.4.1.8 ($v = 25$), $r = 8$)
34	1	2	Theorem 3.5.2 ($m = 11$)
35	4	4	MacNeish–Mann theorem
36	3	3	MacNeish–Mann theorem
38	1	2	Corollary 3.4.1.1, (3.4.5) ($v = 41$, $k = 5$)
39	2	3	Corollary 3.4.1.7 on GD(40; 5, 8; 0, 1)
40	4	4	MacNeish–Mann theorem
42	1	2	Corollary 3.4.1.1, (3.4.5) ($v = 45$, $k = 5$)
44	3	3	MacNeish–Mann theorem
45	4	4	MacNeish–Mann theorem
46	1	3	Shih (1965)
48	2	2	MacNeish–Mann theorem
50	1	5	Corollary 3.4.1.3 ($s = 7$, $m = 2$)
51	2	2	MacNeish–Mann theorem
52	3	3	MacNeish–Mann theorem
54	1	4	Corollary 3.4.1.2, (3.4.6) ($v = 49$, $k = 7$, $x = 5$)
55	4	4	MacNeish–Mann theorem
56	6	6	MacNeish–Mann theorem
57	2	7	Corollary 3.4.2.1 ($v = 57$, $k = 8$)
58	1	2	Theorem 3.5.2 ($m = 19$)
60	2	3	Example 9(iv) of Bose, Shrikhande, and Parker (1960)
62	1	2	Corollary 3.4.1.1, (3.4.5) ($v = 65$, $k = 5$)
63	6	6	MacNeish–Mann theorem
65	4	7	Corollary 3.4.1.8 ($v = 57$, $r = 8$)
66	1	5	Corollary 3.4.1.4 ($x = 7$)
68	3	5	Corollary 3.4.1.4 ($x = 5$)
69	2	5	Corollary 3.4.1.4 ($x = 4$)
70	1	6	Corollary 3.4.1.1, (3.4.5) ($v = 73$, $k = 9$)
72	7	7	MacNeish–Mann theorem
74	1	5	Corollary 3.4.1.5 ($x = 7$)
75	2	5	Corollary 3.4.1.6 ($x = 6$)
76	3	5	Corollary 3.4.1.6 ($x = 5$)

44

Table 3.7.1—(Continued)

s	$n(s)$	$N(s) \geq$	Method of Construction
77	6	6	MacNeish–Mann theorem
78	1	6	Corollary 3.4.1.1, (3.4.5) ($v = 81$, $k = 9$)
80	4	7	Corollary 3.4.2.6 ($s = 9$)
82	1	4	Example 9(i) of Bose, Shrikhande, and Parker (1960)
84	2	5	Corollary 3.4.1.9 on GD(77; 7, 11; 0, 1) ($x = 7$)
85	4	5	Corollary 3.4.1.9 on GD(77; 7, 11; 0, 1) ($x = 8$)
86	1	5	Corollary 3.4.1.9 on GD (77; 7, 11; 0, 1) ($x = 9$)
87	2	2	MacNeish–Mann theorem
88	7	7	MacNeish–Mann theorem
90	1	4	Murthy (1965)
92	3	5	Corollary 3.4.1.9 on GD(91; 7, 13; 0, 1) ($x = 1$)
93	2	3	Shih (1965)
94	1	5	Murthy (1965)
95	4	6	Example 9(ii) of Bose, Shrikhande, and Parker (1960)
96	2	6	Corollary 3.4.1.9 on GD(88; 8, 11; 0, 1) ($x = 8$)
98	1	5	Corollary 3.4.1.9 on GD(91; 7, 13; 0, 1) ($x = 7$)
99	8	8	MacNeish–Mann theorem
100	3	4	Corollary 3.4.1.9 on GD(91; 7, 13; 0, 1) ($x = 9$)

REFERENCES

1. Bose, R. C. (1959). On the application of finite projective geometry for deriving a certain series of balanced Kirkman arrangements. *Bull. Calcutta Math. Soc.*, Golden Jubilee number, 341–354.
2. Bose, R. C., and Shrikhande, S. S. (1959a). On the falsity of Euler's conjecture about the non-existence of two orthogonal latin squares of order $4t + 2$. *Proc. Natl. Acad. Sci. U.S.*, **45**, 734–737.
3. Bose, R. C., and Shrikhande, S. S. (1959b). *On the Construction of Sets of Pairwise Orthogonal Latin Squares and the Falsity of a Conjecture of Euler.* Mimeo Series No. 222, Institute of Statistics, University of North Carolina, Chapel Hill.
4. Bose, R. C., and Shrikhande, S. S. (1959c). *On the Construction of Pairwise Orthogonal Sets of Latin Squares and the Falsity of a Conjecture of Euler—II.* Mimeo Series No. 225, Institute of Statistics, University of North Carolina, Chapel Hill.
5. Bose, R. C., and Shrikhande, S. S. (1960). On the composition of balanced incomplete block designs. *Can. J. Math.*, **12**, 177–188.
6. Bose, R. C., Shrikhande, S. S., and Parker, E. T. (1960). Further results on the construction of mutually orthogonal latin squares and the falsity of Euler's conjecture. *Can. J. Math.*, **12**, 189–203.
7. Bruck, R. H. (1963). Finite nets, II: Uniqueness and imbedding. *Pacific J. Math.*, **13**, 421–457.

8. Bush, K. A. (1952a). A generalization of a theorem due to MacNeish. *Ann. Math. Stat.*, **23**, 293–295.

9. Bush, K. A. (1952b). Orthogonal arrays of index unity. *Ann. Math. Stat.*, **23**, 426–434.

10. Chowla, S., Erdos, P., and Straus, E. G. (1960). On the maximal number of pairwise orthogonal latin squares of a given order. *Can. J. Math.*, **12**, 204–208.

11. Euler, L. (1782). Recherches sur une nouvelle espece des quarres magiques. *Verh. Zeeuwsch Genoot. Wetenschappen*, **9**, 85–239.

12. Johnson, D. M., Dulmage, A. L., and Mendelsohn, N. S. (1961). Orthomorphisms of Abelian groups and orthogonal latin squares, I. *Can. J. Math.*, **13**, 356–372.

13. Levi, F. W. (1942). *Finite Geometrical Systems*. University of Calcutta.

14. MacNeish, H. F. (1922). Eulers squares. *Ann. Math.*, **23**, 221–227.

15. Mann, H. B. (1942). The construction of orthogonal latin squares. *Ann. Math. Stat.*, 13, 418–423.

16. Murthy, J. S. (1965). On the construction of mutually orthogonal latin squares of non-prime power orders. *J. Indian Soc. Agric. Stat.*, **17**, 224–229.

17. Parker, E. T. (1958). Construction of some sets of pairwise orthogonal latin squares. *Am. Math. Soc. Notices*, **5**, 815 (abstract).

18. Parker, E. T. (1959). Construction of some sets of mutually orthogonal latin squares. *Proc. Am. Math. Soc.*, **10**, 946–951.

19. Peterson, J. (1901–1902). Les 36 officieurs. *Ann. Math.*, 413–427.

20. Shih, C.-C. (1965). A method of constructing the orthogonal latin squares. *Shuxue Jinzhan*, **8**, 98–104.

21. Tarry, G. (1900). Le probleme de 36 officieurs. *Compt. Rend. Assoc. Franc. Av. Sci.*, **1**, 122–123.

22. Tarry, G. (1901). Le probleme de 36 officieurs. *Compt. Rend. Assoc. Franc. Av. Sci.*, **2**, 170–203.

23. Wang, Y. (1966). On the maximal number of pairwise orthogonal latin squares of order s; an application of the Sieve method. *Acta Math. Sinica*, **16**, 400–410.

24. Wernicke, P. (1910). Das Problem der 36 offiziere. *Ber. Deuts. Math. Ver.*, **19**, 264–267.

BIBLIOGRAPHY

Barra, J. R. Carres latins et Eulerians. *Rev. Int. Stat. Inst.*, **33**, 16–23 (1965).

Bose, R. C. On the construction of balanced incomplete block designs. *Ann. Eugenics*, **9**, 353–399 (1939).

Bose, R. C., Chakravarti, I. M., and Knuth, D. E. On methods of constructing sets of mutually orthogonal latin squares using a computer. *Technometrics*, **2**, 507–516 (1960).

Bose, R. C., Chakravarti, I. M., and Knuth, D. E. On methods of constructing sets of mutually orthogonal latin squares using a computer–II. *Technometrics*, **3**, 111–118 (1961).

Bose, R. C., and Connor, W. S. Combinatorial properties of group divisible incomplete block designs. *Ann. Math. Stat.*, **23**, 367–383 (1952).

Bose, R. C., Shrikhande, S. S., and Bhattacharya, K. N. On the construction of group divisible incomplete block designs. *Ann. Math. Stat.*, **24**, 167–195 (1953).

Hall, M., Jr. *Combinatorial Theory*. Blaisdell (1967).

Johnson, D. M., Dulmage, A. L., and Mendelsohn, N. S. Orthogonal latin squares. *Can. Math. Bull.*, **2**, 211–216 (1959).

MacNeish, H. F. Das Problem der 36 Offiziere. *Ber. Deuts. Mat. Ver.*, **30**, 151–153 (1921).

Parker, E. T. Nonextendibility conditions on mutually orthogonal latin squares. *Proc. Am. Math. Soc.*, **13**, 219–221 (1962).

Parker, E. T. On orthogonal latin squares. *Proc. Symp. Pure. Math.*, **6**, 43–46 (1962).

Parker, E. T. Computer investigations of orthogonal latin squares of order 10. *Proc. Symp. Appl. Math.*, **15**, 73–81 (1963).

Rogers, K. A note on orthogonal latin squares. *Pacific J. Math.*, **14**, 1395–1397 (1964).

Shrikhande, S. S. Combinatorial problems. Lecture notes of summer course held at Indian Statistical Institute, Calcutta, 1961.

Shrikhande, S. S. Some recent developments on mutually orthogonal latin squares. *Math. Student*, **31**, 167–177 (1963–1964).

Wang, Y. A note on the maximal number of pairwise orthogonal latin squares of a given order. *Sci. Sinica*, **13**, 841–843 (1964).

CHAPTER 4

General Properties of Incomplete Block Designs

4.1. INTRODUCTION

In experimenting for various reasons it will not be possible to use large-size blocks accommodating all the treatments in each block.† If all the treatments are used in each block, we say that it is a complete block design. The randomized block design is an example of this type. When we cannot use a complete block design, we use an incomplete block design; that is, a design in which the number of plots in a block is less than the number of treatments.

Let v treatments be applied to plots arranged in b blocks of sizes $k_1, k_2, \cdots, k_b$. Let the ith treatment occur in r_i blocks and the ith and i'th treatments occur together in $\lambda_{ii'}$ blocks $(i, i' = 1, 2, \cdots, v; i \neq i')$. Let T_i be the total yields for the ith treatment and B_j the total yield for the jth block $(i = 1, 2, \cdots, v; j = 1, 2, \cdots, b)$. Put $\mathbf{T}' = (T_1, T_2, \cdots, T_v)$ and $\mathbf{B}' = (B_1, B_2, \cdots, B_b)$. In the analysis and combinatorial problems of incomplete block designs the incidence matrix $N = (n_{ij})$ of order $v \times b$, where n_{ij} is the number of times the ith treatment occurs in the jth block, plays a key role. If $n_{ij} = 1$ or 0, the design is called a binary design; we study only binary designs in this chapter.

If $\hat{\mathbf{t}}$ is the estimate of the column vector $\mathbf{t}$ of treatment effects, it is well known that the normal equations for $\hat{\mathbf{t}}$ are given by

(4.1.1) $$\mathbf{Q} = C\hat{\mathbf{t}},$$

where

(4.1.2) $$\mathbf{Q} = \mathbf{T} - N \operatorname{diag} [k_1^{-1}, k_2^{-1}, \cdots, k_b^{-1}]\mathbf{B}$$

† In discussing the analysis of designs we use the terms "treatments" for symbols or elements and "blocks" for sets.

and

(4.1.3) $C = \operatorname{diag}\left[r_1, r_2, \cdots, r_v\right] - N \operatorname{diag}\left[k_1^{-1}, k_2^{-1}, \cdots, k_b^{-1}\right] N'.$

[See Kempthorne (1952) and Chakrabarti (1962). For a brief exposition please refer to Appendix B.]

The matrix C defined in (4.1.3) is well known as the C-matrix of incomplete block designs and is very useful in various studies of incomplete blocks, to which we devote this chapter.

4.2. CONNECTEDNESS

If l is a $v \times 1$ column vector of constants and the linear parametric function $l't$ is estimable [cf. Kempthorne (1952) and Chakrabarti (1962)], then the best linear unbiased estimator (BLUE) of $l't$ is given by $l'\hat{t}$, where $\hat{t}$ is any solution of (4.1.1), and the variance of the estimate is $l'Dl\sigma^2$, where σ^2 is the constant variance of the observations and D is such that

(4.2.1) $$\hat{t} = DQ.$$

In fact D is the generalized inverse [cf. C. R. Rao (1962)] of the C-matrix.

Definition 4.2.1. The linear parametric function $l't$ is said to be a contrast if $l'E_{v,1} = 0$. It is said to be an elementary contrast if l has only two nonzero elements 1 and -1.

Elementary contrasts of treatment effects show the comparison of treatments involved in them and it is very desirable to estimate all the elementary contrasts. A design satisfying this property is called a connected design.

Definition 4.2.2. A design is said to be connected if all elementary contrasts are estimable. Otherwise, it is called a disconnected design.

Another definition of connectedness is as follows:

Definition 4.2.3. A design is said to be connected if, given any two treatments θ and ϕ, it is possible to construct a chain of treatments $\theta = \theta_0$, $\theta_1, \theta_2, \cdots, \theta_n = \phi$ such that every consecutive pair of treatments in the chain occurs together in a block.

We can easily prove the following:

Theorem 4.2.1. Definitions 4.2.2 and 4.2.3 are equivalent.

The C-matrix of an incomplete block design satisfies

(4.2.2) $$CE_{v,1} = 0.$$

Hence C is singular and has $E_{v,1}$ as a characteristic vector corresponding to its zero root. Thus the rank of the C-matrix is at most $v - 1$. In fact the rank of the C-matrix is related to the connectedness property of the design, as given by the following:

Theorem 4.2.2. An incomplete block design is connected if and only if the rank of its C-matrix is exactly $v - 1$.

Proof. Let the design be connected and let the $v - 1$ independent contrasts $t_i - t_j$ $(j = 1, 2, \cdots, v; j \neq i)$ be estimated by $\mathbf{l}_1' \mathbf{Q}, \mathbf{l}_2' \mathbf{Q}, \cdots, \mathbf{l}_{v-1}' \mathbf{Q}$. Then we see that

$$(4.2.3) \qquad \mathbf{l}_i' C \neq \mathbf{0} \quad i = 1, 2, \cdots, v - 1,$$

which means that the $\mathbf{l}_i'$ are characteristic vectors corresponding to the nonzero characteristic root of C and hence are orthogonal to $E_{v,\,1}$, a characteristic vector corresponding to the zero root of C. But the $\mathbf{l}_i'$ vectors are all independent and are $v - 1$ in number; since C is of order v, the rank of C is $v - 1$.

Conversely, let us assume that the C-matrix is of rank $v - 1$. Let $\xi_1, \xi_2, \cdots, \xi_{v-1}$ be a set of orthonormal characteristic vectors corresponding to the nonzero roots $\theta_1, \theta_2, \cdots, \theta_{v-1}$ of C. Then

$$(4.2.4) \qquad \mathscr{E}(\xi_i' \mathbf{Q}) = \xi_i' C \mathbf{t} = \theta_i \xi_i' \mathbf{t},$$

where $\mathscr{E}$ stands for mathematical expectation and $\xi_i' \mathbf{t}$ $(i = 1, 2, \cdots, v - 1)$ is estimated by $\xi_i' \mathbf{Q}/\theta_i$. All the ξ_i are orthogonal to $E_{v,\,1}$. Since in a contrast $\mathbf{l}' \mathbf{t}, \mathbf{l}$ is orthogonal to $E_{v,\,1}, \mathbf{l}$ lies in the vector space generated by the ξ_i $(i = 1, 2, \cdots, v - 1)$. Let

$$(4.2.5) \qquad \mathbf{l} = \sum_{i=1}^{v-1} a_i \xi_i.$$

Then

$$(4.2.6) \qquad \mathscr{E} \sum_{i=1}^{v-1} \frac{a_i \xi_i' \mathbf{Q}}{\theta_i} = \mathbf{l}' \mathbf{t}.$$

Thus $\mathbf{l}' \mathbf{t}$ is estimable, and the proof is complete.

We now prove certain interesting results about the structure of the C-matrix of a connected design.

Theorem 4.2.3. In a connected design the diagonal elements of the C-matrix are all positive. Further principal minors of all orders $(1, 2, \cdots, v - 1)$ of C are all positive.

Proof. Since C is of rank $v - 1$ and $\sigma^2 C$ is the dispersion matrix of $\mathbf{Q}$ [cf. Kempthorne (1952) and Chakrabarti (1962)], C is positive semidefinite.

Hence none of the diagonal elements of C can be negative. Let, if possible, the ith diagonal element of C be zero. Consider the vector ξ whose ith element is its only nonzero element equal to 1. Then

(4.2.7) $\xi' C \xi = 0,$

implying that ξ is also a characteristic vector corresponding to the zero root of C. Since ξ and $E_{v,1}$ are independent and both are characteristic vectors corresponding to the zero root of C, the rank of C is at most $v - 2$ and the design is disconnected, contrary to hypothesis. Thus none of the diagonal elements is negative or zero.

The second part of the theorem can be similarly proved by suitably selecting ξ and is left as an exercise to the reader.

Theorem 4.2.4. In a connected design the cofactors of all elements of C have the same positive value.

Proof. Let $C = (c_{ij})$ and let C_{ij} be the cofactor of c_{ij} in C. Let $C^* = (C_{ji})$. It is well known that

(4.2.8) $CC^* = 0_{v,v}.$

Since the design is connected, a nonzero scalar multiple of $E_{v,1}$ is a characteristic vector corresponding to the zero root. Hence each column of C^* contains identical elements and because C^* is symmetrical, in view of Theorem 4.2.3, it is positive scalar multiple of $E_{v,v}$. Hence the result.

It may be added that the result of Theorem 4.2.4 holds even for disconnected designs with a slight modification. In that case all the cofactors of C will have a value of zero.

The results of Theorems 4.2.3 and 4.2.4 were proved by Chakrabarti (1963).

4.3. BALANCING IN CONNECTED AND DISCONNECTED DESIGNS

In experimenting it is desirable to estimate all the elementary contrasts with the same precision. In Section 4.5 we show that designs having this property are the most efficient among the class of designs of the same size. This leads us to the concept given by the following definition:

Definition 4.3.1. A design is said to be balanced if all the elementary contrasts are estimated with the same precision.

The following necessary and sufficient condition for balancing in connected designs is due to V. R. Rao (1958):

Theorem 4.3.1. A connected design is balanced if and only if all the nonzero characteristic roots of C are equal.

Proof. Let $\xi_1, \xi_2, \cdots, \xi_{v-1}$ be a set of orthonormal characteristic vectors corresponding to the nonzero roots $\theta_1, \theta_2, \cdots, \theta_{v-1}$ of the C-matrix. Then the spectral decomposition of C is

$$(4.3.1) \qquad\qquad C = \sum_{i=1}^{v-1} \theta_i \xi_i \xi_i'.$$

We can easily see that

$$(4.3.2) \qquad\qquad \hat{t} = \left\{ \sum_{i=1}^{v-1} \theta_i^{-1} \xi_i \xi_i' \right\} Q$$

is a solution of the normal equations (4.1.1) with the dispersion matrix

$$(4.3.3) \qquad\qquad \sigma^2 \left\{ \sum_{i=1}^{v-1} \theta_i^{-1} \xi_i \xi_i' \right\}.$$

The average variance of all elementary contrasts can be seen to be

$$(4.3.4) \qquad\qquad \frac{2\sigma^2}{v(v-1)} \left\{ v \operatorname{tr} \left(\sum_{i=1}^{v-1} \theta_i^{-1} \xi_i \xi_i' \right) - E_{1,v} \left(\sum_{i=1}^{v-1} \theta_i^{-1} \xi_i \xi_i' \right) E_{v,1} \right\}$$
$$= 2\sigma^2 \frac{\sum_{i=1}^{v-1} \theta_i^{-1}}{v-1}.$$

Let the connected design be balanced. Then all the elementary contrasts are estimated with variance $2\sigma^2 (\sum_{i=1}^{v-1} \theta_i^{-1})/(v-1)$. Furthermore,

$$(4.3.5) \quad \operatorname{var}(\hat{t}_i - \hat{t}_j) = \operatorname{var}(\hat{t}_i - \hat{t}_k) + \operatorname{var}(\hat{t}_j - \hat{t}_k) - 2 \operatorname{cov}(\hat{t}_i - \hat{t}_k, \hat{t}_j - \hat{t}_k)$$

and hence

$$(4.3.6) \qquad\qquad \operatorname{cov}(\hat{t}_i - \hat{t}_k, \hat{t}_j - \hat{t}_k) = \sigma^2 \frac{\sum_{i=1}^{v-1} \theta_i^{-1}}{v-1}.$$

Consider the linear parametric function $\xi_i' t$. The variance of its estimate on one hand is

$$(4.3.7) \qquad\qquad \sigma^2 \xi_i' \left(\sum_{i=1}^{v-1} \theta_i^{-1} \xi_i \xi_i' \right) \xi_i = \sigma^2 \theta_i^{-1}.$$

On the other hand,

$$(4.3.8) \qquad\qquad \xi_i' t = \sum_{j=2}^{v} a_j (t_j - t_1),$$

where $\sum_{j=2}^{v} a_j^2 + (\sum_{j=2}^{v} a_j)^2 = 1$. Now

$$\text{var}(\xi_i' \mathbf{t}) = \left(\sum_{j=2}^{v} a_j^2 \right) \left\{ 2\sigma^2 \frac{\sum_{j=1}^{v-1} \theta_j^{-1}}{v-1} \right\}$$

(4.3.9)

$$+ \left(\sum_{\substack{j,j'=2 \\ j \neq j'}}^{v} a_j a_{j'} \right) \left\{ \sigma^2 \frac{\sum_{j=1}^{v-1} \theta_j^{-1}}{v-1} \right\} = \sigma^2 \frac{\sum_{j=1}^{v-1} \theta_j^{-1}}{v-1}.$$

Equating (4.3.7) and (4.3.9), we get

(4.3.10) $$\theta_i^{-1} = \frac{\sum_{j=1}^{v-1} \theta_j^{-1}}{v-1}.$$

Since this result holds for every i, we have

(4.3.11) $$\theta_1 = \theta_2 = \cdots = \theta_{v-1},$$

and thus the condition is necessary.

When the nonzero characteristic roots of C are all equal (say θ), each of the elementary contrasts is estimated with a variance $2\sigma^2 \theta^{-1}$ and the design is balanced. The proof is thus complete.

For a balanced design the C-matrix is of the form

(4.3.12) $$C = \theta(I_v - v^{-1} E_{v,v}).$$

Theorem 4.3.2. An equireplicated, equiblock-sized balanced design has its incidence matrix N, satisfying

(4.3.13) $$NN' = (r - \lambda)I_v + \lambda E_{v,v},$$

where

(4.3.14) $$\lambda(v - 1) = r(k - 1).$$

Proof. By equating the trace of the C-matrix to $(v - 1)\theta$, where θ is the nonzero root of the multiplicity $(v - 1)$ of C, we get

(4.3.15) $$\theta = \frac{vr - b}{v - 1}.$$

Thus

(4.3.16) $$C = rI_v - k^{-1} NN' = \frac{vr - b}{v - 1} (I_v - v^{-1} E_{v,v}),$$

from which we get

(4.3.17) $$NN' = (r - \lambda)I_v + \lambda E_{v,v}.$$

The incomplete block design as defined in the above theorem is known as a balanced incomplete block (BIB) design. We formally define it as follows:

Definition 4.3.2. A BIB design is an arrangement of v symbols in b sets each containing $k(< v)$ symbols, meeting the following conditions:

1. Every symbol occurs at most once in a set.
2. Every symbol occurs in exactly r sets.
3. Every pair of symbols occurs together in λ sets.

A BIB design is said to be symmetrical if $v = b$ and consequently $r = k$.

The terms v, b, r, k, λ are known as the parameters of a BIB design. The incidence matrix N of a BIB design satisfies (4.3.13) and has all the nonzero roots of its C-matrix equal; thus it follows from Theorem 4.3.1 that Definition 4.3.1 is satisfied. We study BIB designs in detail in the next chapter.

It may be noted that the BIB designs of Definition 4.3.2 are not the only balanced designs defined in Definition 4.3.1. In fact a design consisting of blocks of two or more BIB designs with the same v and different k terms is balanced as defined in Definition 4.3.1 but does not form a BIB design. Theorem 4.3.2 only asserts that among the class of binary, equireplicated, equiblock-sized designs, a BIB design is the only balanced design. To distinguish the balanced designs of Definition 4.3.1 from BIB designs, we call the former variance-balanced designs. To make things worse, pairwise balanced designs (Chapter 3) and symmetrical unequal-block arrangements (Chapter 11) share property 3 of Definition 4.3.2 of BIB designs but are not balanced in the sense of Definition 4.3.1. They may, however, be called combinatorially-balanced designs, as every pair of symbols occurs together in the same number of sets in these designs.

There exist balanced designs with unequal block sizes and unequal numbers of replicates; John (1964) has given one such example.

We prove another property of balancing in relation to BIB designs; this is due to Bhaskararao (1966).

Theorem 4.3.3. An equireplicated, balanced design with $b = v$ is a symmetrical BIB design.

Proof. As in the proof of Theorem 4.3.2, we can find that the nonzero characteristic root θ of C of multiplicity $v - 1$ is given by (4.3.15). Thus

$$(4.3.18) \quad C = rI_v - N \operatorname{diag}[k_1^{-1}, k_2^{-1}, \cdots, k_b^{-1}]N' = \frac{vr - b}{v - 1}(I_v - v^{-1}E_{v,v}).$$

Hence

$$(4.3.19) \quad N \operatorname{diag}[k_1^{-1}, k_2^{-1}, \cdots, k_b^{-1}]N' = (v - 1)^{-1}[(v - r)I_v + (r - 1)E_{v,v}].$$

Since $v = b$, N is a square matrix; furthermore, from (4.3.19) we see that N diag $(k_1^{-1}, k_2^{-1}, \cdots, k_b^{-1}) N'$ is nonsingular. Hence inverting both sides of (4.3.19) and simplifying, we get

$$(4.3.20) \quad \text{diag } [k_1, k_2, \cdots, k_b] = \frac{v-1}{v-r} \left[N'N - \frac{r-1}{r(v-1)} N' E_{v,v} N \right].$$

From the definition of N we have

$$(4.3.21) \quad N'N = (\mu_{ij})$$

and

$$(4.3.22) \quad N' E_{v,v} N = (k_i k_j),$$

where μ_{ij} is the number of symbols in common between the ith and the jth sets. Hence from (4.3.20) we get

$$(4.3.23) \quad k_i = \frac{v-1}{v-r} k_i - \frac{r-1}{r(v-r)} k_i^2$$

and

$$(4.3.24) \quad 0 = \frac{v-1}{v-r} \mu_{ij} - \frac{r-1}{r(v-r)} k_i k_j$$

for $i, j = 1, 2, \cdots, v; i \neq j$. Thus $k_i = r$ for every i and $\mu_{ij} = r(r-1)/(v-1)$ for all i, j. Hence $N'N = (r - \lambda)I_v + \lambda E_{v,v}$, and the design with incidence matrix N' is a symmetrical BIB design, and from Theorem 5.2.1, to be proved in the next chapter, the design with incidence matrix N is a symmetrical BIB design. The proof is complete.

The definition of balancing as given in Definition 4.3.1 will not hold for disconnected designs because not all elementary contrasts are estimable. Vartak (1963a) has modified the definition of balancing to hold for disconnected as well as connected designs as follows:

Definition 4.3.3. A design is said to be balanced if every normalized estimable linear function of the treatment effects of the design can be estimated with the same variance.

It is easy to verify that Definitions 4.3.1 and 4.3.3 are identical for connected designs. Analogously to Theorem 4.3.1 we can prove the following theorem, whose proof is left as an exercise to the reader:

Theorem 4.3.4. A design is balanced if and only if all the nonzero characteristic roots of its C-matrix are equal.

4.4. KRONECKER-PRODUCT DESIGNS

The concept of the Kronecker product of matrices was used in experimental designs for the first time by Vartak (1955). A Kronecker product of designs was defined by Vartak as follows:

Definition 4.4.1. If N_1 is the incidence matrix of a design D_1 and N_2 is the incidence matrix of another design D_2, the design with incidence matrix $N_1 \times N_2$, where $\times$ is the Kronecker product of matrices, is called the Kronecker product of designs D_1 and D_2.

The meaning of this definition is as follows: Let D_i be the design in v_i symbols arranged in b_i sets of sizes k_i, each symbol being replicated r_i times ($i = 1, 2$). The $v_1 v_2$ symbols in the Kronecker product of designs are the ordered pair of symbols (α, β), α belonging to D_1 and β belonging to D_2. The $b_1 b_2$ sets are formed by taking any set of D_1 and forming ordered pairs of these symbols with the symbols occurring in any set of D_2. For example, if D_1 and D_2 are

$$
\begin{array}{cc}
D_1 & D_2 \\
\{1, 2\} & \{1, 2, 4, 5\} \\
\{1, 3\} & \{3, 4, 1, 2\} \\
\{2, 3\} & \{5, 2, 3, 1\}
\end{array}
$$

then the Kronecker product of designs D_1 and D_2 is

$$
\begin{array}{l}
\{(1, 1), \quad (1, 2), \quad (1, 4), \quad (1, 5), \quad (2, 1), \quad (2, 2), \quad (2, 4), \quad (2, 5)\} \\
\{(1, 3), \quad (1, 4), \quad (1, 1), \quad (1, 2), \quad (2, 3), \quad (2, 4), \quad (2, 1), \quad (2, 2)\} \\
\{(1, 5), \quad (1, 2), \quad (1, 3), \quad (1, 1), \quad (2, 5), \quad (2, 2), \quad (2, 3), \quad (2, 1)\} \\
\{(1, 1), \quad (1, 2), \quad (1, 4), \quad (1, 5), \quad (3, 1), \quad (3, 2), \quad (3, 4), \quad (3, 5)\} \\
(4.4.1) \quad \{(1, 3), \quad (1, 4), \quad (1, 1), \quad (1, 2), \quad (3, 3), \quad (3, 4), \quad (3, 1), \quad (3, 2)\} \\
\{(1, 5), \quad (1, 2), \quad (1, 3), \quad (1, 1), \quad (3, 5), \quad (3, 2), \quad (3, 3), \quad (3, 1)\} \\
\{(2, 1), \quad (2, 2), \quad (2, 4), \quad (2, 5), \quad (3, 1), \quad (3, 2), \quad (3, 4), \quad (3, 5)\} \\
\{(2, 3), \quad (2, 4), \quad (2, 1), \quad (2, 2), \quad (3, 3), \quad (3, 4), \quad (3, 1), \quad (3, 2)\} \\
\{(2, 5), \quad (2, 2), \quad (2, 3), \quad (2, 1), \quad (3, 5), \quad (3, 2), \quad (3, 3), \quad (3, 1)\}
\end{array}
$$

The following two theorems, due to Vartak (1960), show the structure of the sets (also known as block structure) of the Kronecker product of designs:

Theorem 4.4.1. If in design D_1 there exists a pair of sets with m_1 symbols in common and in design D_2 a pair of sets with m_2 symbols in common, then in their Kronecker product there exists a pair of sets with $m_1 m_2$ symbols in common.

Proof. Let N_1, N_2, and N be the incidence matrices of D_1, D_2, and their Kronecker product, respectively. Then

$$(4.4.2) \qquad\qquad N = N_1 \times N_2$$

and hence

(4.4.3) $$N'N = (N_1'N_1) \times (N_2'N_2).$$

From the statement of the theorem m_1 is an element of $N_1'N_1$ and m_2 is an element of $N_2'N_2$, which implies that there is a pair of sets with $m_1 m_2$ symbols, in common.

Theorem 4.4.2. Let the sets of D_i be of size k_i ($i = 1, 2$). Consider a set $B^{(i)}$ of design D_i and let $b_j^{(i)}$ be the number of sets of D_i with j symbols in common with the set $B^{(i)}$ ($j = 1, 2, \cdots, k_i; i = 1, 2$). Then there exists a set B in the Kronecker product D of D_1 and D_2 such that $b_\gamma^{(0)}$ sets of D have each γ symbols in common with B, where $b_\gamma^{(0)}$ is the coefficient of $\{\gamma\}$ in the expression

(4.4.4) $$\left[\sum_{j=0}^{k_1} b_j^{(1)} \{\alpha\} \right] \left[\sum_{j=0}^{k_2} b_j^{(2)} \{\beta\} \right],$$

where the symbols $\{\gamma\}$ obey the ordinary laws of algebra, namely,

(4.4.5) $$a\{\alpha\} + b\{\alpha\} = (a +) b\{\alpha\},$$
$$\{\alpha\}\{\beta\} = \{\beta\}\{\alpha\} = \{\alpha\beta\},$$
$$(a\{\alpha\})(b\{\beta\}) = ab\{\alpha\beta\}.$$

Proof. If N_i is the incidence matrix of D_i, then $N_i'N_i$ contains a row of the form

(4.4.6) $$\boldsymbol{\rho_i} = (\overbrace{0, 0, \cdots, 0,}^{b_0^{(i)}} \overbrace{1, 1, \cdots, 1,}^{b_1^{(i)}} \cdots, \overbrace{k_i, k_i, \cdots, k_i}^{b_{k_i}^{(i)}})$$

for $i = 1, 2$. Then $(N_1 \times N_2)'(N_1 \times N_2)$ contains a row $\boldsymbol{\rho} = \boldsymbol{\rho_1} \times \boldsymbol{\rho_2}$. We can now easily verify that in $\boldsymbol{\rho}$ the integer γ occurs $b_\gamma^{(0)}$ times, as given in (4.4.4).

Theorem 4.4.3. A necessary and sufficient condition for the Kronecker product of designs D_1 and D_2 to be connected is that D_1 and D_2 be both connected.

Proof. Let D_1 and D_2 be connected designs. Consider two symbols (α_1, β_1) and (α_2, β_2) in the Kronecker product, where α_1, α_2 are symbols in D_1 and β_1, β_2 are symbols in D_2. Since D_1 is connected, there exists a chain of symbols

(4.4.7) $$\alpha_1 = \alpha^0, \alpha^1, \alpha^2, \cdots, \alpha^m = \alpha_2$$

such that every pair of consecutive symbols occurs together in a set. Similarly there exists another chain of symbols

(4.4.8) $$\beta_1 = \beta^0, \beta^1, \beta^2, \cdots, \beta^n = \beta_2$$

such that every pair of consecutive symbols occurs together in a set of D_2. Then in the chain

(4.4.9)
$$(\alpha_1, \beta_1) = (\alpha^0, \beta^0), (\alpha^0, \beta^1), (\alpha^0, \beta^2), \cdots, (\alpha^0, \beta^n),$$
$$(\alpha^1, \beta^n), (\alpha^2, \beta^n), \cdots, (\alpha^m, \beta^n) = (\alpha_2, \beta_2)$$

every pair of consecutive symbols occurs together in a set of the Kronecker product of D_1 and D_2.

The converse also can be established on similar lines, and the proof is complete.

An alternative proof through the C-matrices of the designs D_1, D_2, and their Kronecker product is given by Vartak (1963b).

4.5. EFFICIENCY FACTOR

With the availability of a large number of designs of the same size it becomes essential to have a yardstick for comparing the various designs. The efficiency factor, as defined below, provides such a measure.

Definition 4.5.1. The efficiency factor of a design is defined as a ratio

(4.5.1)
$$E = \frac{\overline{V}_R}{\overline{V}},$$

where $\overline{V}$ is the average variance of the estimated intrablock treatment elementary contrasts for the design under consideration and $\overline{V}_R$ that for a randomized block design using the same number of experimental units, $\overline{V}_R$ and $\overline{V}$ being computed on the assumption that the intrablock error variance is same in both cases.

It can be easily seen that the average variance of the estimates of elementary contrasts in a connected design is

(4.5.2)
$$\overline{V} = 2\sigma^2 \frac{\sum_{i=1}^{v-1} \theta_i^{-1}}{v - 1},$$

where $\theta_1, \theta_2, \cdots, \theta_{v-1}$ are the nonzero characteristic roots of the C-matrix of the design. Furthermore,

(4.5.3)
$$\overline{V}_R = \frac{2\sigma^2}{\bar{r}},$$

where $\bar{r}$ is the average number of replications of the treatments in the design under consideration. Then

(4.5.4)
$$E = \frac{2\sigma^2/\bar{r}}{2\sigma^2 \sum_{i=1}^{v-1} \theta_i^{-1}/(v - 1)} = \frac{\delta}{\bar{r}}$$

where δ is the harmonic mean of the nonzero characteristic roots of C. Since the harmonic mean of a set of positive quantities cannot exceed the arithmetic mean, we have

(4.5.5)
$$\delta \leqq \frac{\sum_{i=1}^{v-1} \theta_i}{v-1} = \frac{\operatorname{tr}(C)}{v-1} = \frac{v\bar{r} - b}{v-1}.$$

Thus

(4.5.6)
$$E \leqq \frac{v\bar{r} - b}{\bar{r}(v-1)}.$$

We summarize this result in the following theorem:

Theorem 4.5.1. The efficiency factor E of any connected design satisfies inequality (4.5.6.)

The equality sign in (4.5.6) holds when and only when $\theta_1 = \theta_2 = \cdots = \theta_{v-1}$, in which case the design is a balanced design. Thus we have

Theorem 4.5.2. The most efficient connected design is the balanced design.

This property of a BIB design was shown by Kiefer (1958), Kshirsagar (1958), and Mote (1958).

4.6. α-RESOLVABILITY AND AFFINE α-RESOLVABILITY

In this section we restrict our study to equireplicated and equiblock-sized designs. Shrikhande and Raghavarao (1963) have generalized the concept of resolvability and affine resolvability of Bose (1942) to α-resolvability and affine α-resolvability as follows:

Definition 4.6.1. An incomplete block design with the parameters v, b, r, k is said to be α-resolvable if the sets can be grouped into t classes $S_1, S_2, \cdots, S_t$, each of β sets, such that in each class every symbol is replicated α times. We then have

(4.6.1)
$$v\alpha = k\beta, \qquad b = t\beta, \qquad r = t\alpha.$$

Definition 4.6.2. An α-resolvable incomplete block design is called affine α-resolvable if any pair of sets of the same class intersects in q_1 symbols and any pair of sets from different classes intersects in q_2 symbols.

A 1-resolvable and an affine 1-resolvable design may be simply called resolvable and affine resolvable design, respectively.

We now prove the following theorem:

Theorem 4.6.1. In an affine α-resolvable design k^2/v is an integer.

Proof. Consider the first set in class S_1 of an affine α-resolvable design.

By definition this has q_2 symbols in common with each set of the classes $S_2, S_3, \cdots, S_t$. Hence

(4.6.2) $$\beta(t-1)q_2 = k(r-\alpha),$$

from which we get $q_2 = k^2/v$. Since q_2 is an integer, k^2/v is an integer.

The following theorem is due to Shrikhande and Raghavarao (1963):

Theorem 4.6.2. If an α-resolvable incomplete block design has any two of the following properties, then it has the third:

1. Affine α-resolvability.
2. Balance.
3. $b - v = t - 1$.

We leave the proof of this theorem to the reader as an exercise on matrix theory.

4.7. AN INEQUALITY FOR BALANCED DESIGNS

We now prove an inequality for balanced designs, first given by Atiqullah (1961); a simplified proof for this was given by Raghavarao (1962).

Theorem 4.7.1. For balanced, equireplicated incomplete block designs

(4.7.1) $$b \geq v.$$

Proof. We can easily see that the C-matrix of a balanced design is

(4.7.2) $$\frac{vr - b}{v - 1}(I_v - v^{-1}E_{v,v}).$$

Hence

$$N \operatorname{diag}(k_1^{-1}, k_2^{-1}, \cdots, k_b^{-1})N' = rI_v - \frac{vr - b}{v - 1}(I_v - v^{-1}E_{v,v})$$

(4.7.3)
$$= \frac{b - r}{v - 1}I_v + \frac{vr - b}{v(v - 1)}E_{v,v}.$$

Since the matrix on the right-hand side of (4.7.3) is nonsingular, it follows that

$$v = \operatorname{rank}\{N \operatorname{diag}(k_1^{-1}, k_2^{-1}, \cdots, k_b^{-1})N'\}$$

(4.7.4)
$$= \operatorname{rank}\{N \operatorname{diag}(k_1^{-1/2}, k_2^{-1/2}, \cdots, k_b^{-1/2})\}$$

$$= \operatorname{rank}(N) \leq b.$$

Inequality (4.7.1) was obtained for BIB designs by Fisher (1940).

For α-resolvable incomplete block designs

$$(4.7.5) \qquad \text{rank}(N) \leq b - t + 1,$$

and hence (4.7.1) can be improved for such designs to

$$(4.7.6) \qquad b \geq v + t - 1.$$

Theorem 4.7.2. For balanced, equireplicated, α-resolvable incomplete block designs inequality (4.7.6) holds.

Inequality (4.7.6) was obtained for resolvable BIB designs by Bose (1942).

REFERENCES

1. Atiqullah, M. (1961). On a property of balanced designs. *Biometrika*, **48**, 215–218.
2. Bose, R. C. (1942). A note on the resolvability of balanced incomplete block designs. *Sankhyā*, **6**, 105–110.
3. Bhaskararao, M. (1966). A note on equireplicate balanced designs with $b = v$. *Calcutta Stat. Assn. Bull.*, **15**, 43–44.
4. Chakrabarti, M. C. (1962). *Mathematics of Design and Analysis of Experiments.* Asia Publishing House, Bombay.
5. Chakrabarti, M. C. (1963). On the C-matrix in design of experiments. *J. Indian Stat. Assn.*, **1**, 8–23.
6. Fisher, R. A. (1940). An examination of the different possible solutions of a problem in incomplete blocks. *Ann. Eugenics*, **10**, 52–75.
7. John, P. W. M. (1964). Balanced designs with unequal number of replications. *Ann. Math. Stat.*, **35**, 897–899.
8. Kempthorne, O. (1952). *The Design and Analysis of Experiments.* Wiley, New York.
9. Kiefer, J. (1958). On the nonrandomized optimality and randomized non-optimality of symmetrical designs. *Ann. Math. Stat.*, **29**, 675–699.
10. Kshirsagar, A. M. (1958). A note on incomplete block designs. *Ann. Math. Stat.*, **29**, 907–910.
11. Mote, V. L. (1958). On a minimax property of balanced incomplete block designs. *Ann. Math. Stat.*, **29**, 910–914.
12. Raghavarao, D. (1962). On balanced unequal block designs. *Biometrika*, **49**, 561–562.
13. Rao, C. R. (1962). A note on generalized inverse of a matrix with applications to problems in mathematical statistics. *J. Roy. Stat. Soc.*, **B24**, 152–158.
14. Rao, V. R. (1958). A note on balanced designs. *Ann. Math. Stat.*, **29**, 290–294.
15. Shrikhande, S. S., and Raghavarao, D. (1963). Affine α-resolvable incomplete block designs. *Contributions to Statistics.* Presented to Prof. P. C. Mahalanobis on the occasion of his 70th birthday. Pergamon Press, pp. 471–480.
16. Vartak, M. N. (1955). On an application of Kronecker product of matrices to statistical designs. *Ann. Math. Stat.*, **26**, 420–438.

17. Vartak, M. N. (1960). Relations among the blocks of the Kronecker product of designs. *Ann. Math. Stat.*, **31**, 772–778.
18. Vartak, M. N. (1963a). Disconnected balanced designs. *J. Indian Stat. Assn.*, **1**, 104–107.
19. Vartak, M. N. (1963b). Connectedness of Kronecker product designs. *J. Indian Stat. Assn.*, **1**, 215–218.

CHAPTER 5

Balanced Incomplete Block Designs

5.1. INTRODUCTION

In the preceding chapter we defined a balanced incomplete block (BIB) design as an arrangement of v symbols into b sets each of k ($< v$) symbols, satisfying the following conditions:

1. Every symbol occurs at most once in each set.
2. Every symbol occurs in exactly r sets.
3. Every pair of symbols occurs together in exactly λ sets.

The parameters of the BIB design are v, b, r, k, and λ, and they satisfy

$$(5.1.1) \qquad vr = bk$$

and

$$(5.1.2) \qquad \lambda(v - 1) = r(k - 1).$$

In fact (5.1.1) counts the number of places used in two ways; there being b sets, each containing k symbols, and v symbols, each occurring r times. To prove (5.1.2) consider the pairs containing a particular symbol θ. Here θ occurs in r sets and in each of these is paired with $k - 1$ other symbols. But θ must be paired with each of the remaining $v - 1$ symbols exactly λ times. Comparing these counts, we get (5.1.2).

We can easily verify that the following arrangement

$$(5.1.3)$$

$$
\begin{array}{llll}
(\ 0, & 1, & 3, & 9); \quad (\ 1, \quad 2, \quad 4, 10); \quad (\ 2, \quad 3, \quad 5, 11); \\
(\ 3, & 4, & 6, 12); \quad (\ 4, \quad 5, \quad 7, \ 0); \quad (\ 5, \quad 6, \quad 8, \ 1); \\
(\ 6, & 7, & 9, \ 2); \quad (\ 7, \quad 8, 10, \ 3); \quad (\ 8, \quad 9, 11, \ 4); \\
(\ 9, & 10, & 12, \ 5); \quad (10, 11, \ 0, \ 6); \quad (11, 12, \ 1, \ 7); \\
(12, & 0, & 2, \ 8)
\end{array}
$$

in 13 symbols 0, 1, $\cdots$, 12 in 13 sets is a BIB design with the parameters

(5.1.4) $v = 13 = b,$ $r = 4 = k,$ $\lambda = 1.$

A BIB design is said to be symmetrical if $v = b$ and consequently $r = k$.

Although particular series of BIB designs were known as early as 1847 by Kirkman, a systematic treatment was missing till Yates (1936) introduced them as statistical designs. From then rich contributions were made to these configurations mainly by Fisher (1940, 1942), Fisher and Yates (1963), and Bose (1939–1959) and now form a standard part of the theory.

Let N be the incidence matrix of a BIB design with parameters v, b, r, k, and λ. Then

(5.1.5) $NE_{b,1} = rE_{v,1};$

(5.1.6) $E_{1,v}N = kE_{1,b};$

and

(5.1.7) $NN' = (r - \lambda)I_v + \lambda E_{v,v}.$

Conversely, if N is a $v \times b$ matrix with elements 0 and 1 satisfying (5.1.6) and (5.1.7), it can easily be verified that it is the incidence matrix of a BIB design with parameters v, b, r, k, and λ, as noted by Majumdar (1953).

From (5.1.7) it follows that, since $r > \lambda$,

(5.1.8) $v = \text{rank } (NN') = \text{rank } (N) \leq b.$

Inequality (5.1.8.) is due to Fisher (1940). Though some other inequalities have been proposed for nonresolvable designs, Kishen and Rao (1952) showed that other known inequalities are implied by (5.1.8).

If the design is resolvable—that is, if the b sets are grouped into r classes of n sets each, such that each class forms a complete replication of all the v symbols—then (5.1.8) can be improved to

(5.1.9) $v \leq b - r + 1,$

as proved by Bose (1942b). Kishen and Rao (1952) showed that other known inequalities for resolvable BIB designs are implied by (5.1.9).

We devote this chapter for the study of combinatorial problems and constructions related to BIB designs.

5.2. BIB DESIGNS RELATED TO A GIVEN BIB DESIGN

Let $S_1, S_2, \cdots, S_b$ be the b sets forming a BIB design with the parameters v, b, r, k, and λ. Then, if S_i' is the complement of the set S_i, that is,

$S_i' = \{1, 2, \cdots, v\} - S_i$, then we can easily verify that the sets $S_1', S_2', \cdots, S_b'$ form another BIB design, known as the complement of the original design with the parameters

$$(5.2.1) \qquad \begin{aligned} v_1 &= v, & b_1 &= b, & r_1 &= b - r, \\ k_1 &= v - k, & \lambda_1 &= b - 2r + \lambda. \end{aligned}$$

We prove a very useful property of a symmetrical BIB design in the following theorem:

Theorem 5.2.1. In a symmetrical BIB design any two sets have exactly λ symbols in common.

Proof. Let N be the incidence matrix of a symmetrical BIB design. Then

$$(5.2.2) \qquad NE_{v,v} = E_{v,v}N = rE_{v,v}$$

and NN' is given by (5.1.7). Now

$$(5.2.3) \qquad \begin{aligned} N'NN' &= (r - \lambda)N' + \lambda N'E_{v,v} \\ &= \{(r - \lambda)I_v + \lambda E_{v,v}\}N', \end{aligned}$$

and since N' is nonsingular,

$$(5.2.4) \qquad N'N = (r - \lambda)I_v + \lambda E_{v,v},$$

which implies that any two sets of the design intersect in λ symbols.

Let D be a symmetrical BIB design consisting of the v sets $S_1, S_2, \cdots, S_v$ and with the parameters $v = b$, $r = k$, λ. Let us select one of these sets, say S_1, and delete from each of the $S_2, S_3, \cdots, S_v$ sets the λ symbols it has in common with S_1. The deleted sets $S_2^*, S_3^*, \cdots, S_v^*$ form what is called the residual design D^*, in which the parameters v_2, b_2, r_2, k_2, and λ_2 are given by

$$(5.2.5) \quad v_2 = v - k, \qquad b_2 = v - 1, \qquad r_2 = k, \qquad k_2 = k - \lambda, \qquad \lambda_2 = \lambda.$$

This is easily seen, since exactly λ symbols are deleted from each of the $S_2, S_3, \cdots, S_v$ sets, and since the $v - k$ symbols of D not in S_1 remain unaltered in their occurrences in $S_2^*, S_3^*, \cdots, S_v^*$. We may also form the derived design D' by taking, as our sets $S_2', S_3', \cdots, S_v'$, the λ symbols that $S_2, S_3, \cdots, S_v$, respectively, have in common with S_1. This again can easily be shown to be a BIB design with the parameters

$$(5.2.6) \quad v_3 = k, \qquad b_3 = v - 1, \qquad r_3 = k - 1, \qquad k_3 = \lambda, \qquad \lambda_3 = \lambda - 1.$$

For the BIB design given in (5.1.3) the complementary design is

(2, 4, 5, 6, 7, 8, 10, 11, 12);	(0, 3, 5, 6, 7, 8, 9, 11, 12);
(0, 1, 4, 6, 7, 8, 9, 10, 12);	(0, 1, 2, 5, 7, 8, 9, 10, 11);
(1, 2, 3, 6, 8, 9, 10, 11, 12);	(0, 2, 3, 4, 7, 9, 10, 11, 12);
(5.2.7) (0, 1, 3, 4, 5, 8, 10, 11, 12);	(0, 1, 2, 4, 5, 6, 9, 11, 12);
(0, 1, 2, 3, 5, 6, 7, 10, 12);	(0, 1, 2, 3, 4, 6, 7, 8, 11);
(1, 2, 3, 4, 5, 7, 8, 9, 12);	(0, 2, 3, 4, 5, 6, 8, 9, 10);
(1, 3, 4, 5, 6, 7, 9, 10, 11),	

which has the parameters

$$(5.2.8) \qquad v_1 = 13 = b_1, \qquad r_1 = 9 = k_1, \qquad \lambda_1 = 6.$$

On omitting the first set, the derived and residual designs can be seen to be

	Residual Design	Derived Design
	0, 3, 9	5, 6, 7, 8, 11, 12
	0, 1, 9	4, 6, 7, 8, 10, 12
	0, 1, 9	2, 5, 7, 8, 10, 11
	1, 3, 9	2, 6, 8, 10, 11, 12
	0, 3, 9	2, 4, 7, 10, 11, 12
	0, 1, 3	4, 5, 8, 10, 11, 12
(5.2.9)	0, 1, 9	2, 4, 5, 6, 11, 12
	0, 1, 3	2, 5, 6, 7, 10, 12
	0, 1, 3	2, 4, 6, 7, 8, 11
	1, 3, 9	2, 4, 5, 7, 8, 12
	0, 3, 9	2, 4, 5, 6, 8, 10
	1, 3, 9	4, 5, 6, 7, 10, 12

The residual design and the derived design can be respectively verified to have the parameters

$$(5.2.10) \quad v_2 = 4, \qquad b_2 = 12, \qquad r_2 = 9, \qquad k_2 = 3, \qquad \lambda_2 = 6$$

and

$$(5.2.11) \quad v_3 = 9, \qquad b_3 = 12, \qquad r_3 = 8, \qquad k_3 = 6, \qquad \lambda_3 = 5.$$

One may be interested to know whether the above method of obtaining the residual and derived designs from a symmetrical BIB design has a converse. The answer is in the negative, as was shown by Bhattacharya (1945). Consider a symmetrical BIB design with the parameters

$$(5.2.12) \qquad v = b = 25, \qquad r = k = 9, \qquad \lambda = 3,$$

where every pair of sets has exactly three symbols in common. The residual design of (5.2.12) has the parameters

$$(5.2.13) \qquad v_2 = 16, \qquad b_2 = 24, \qquad r_2 = 9, \qquad k_2 = 6, \qquad \lambda_2 = 3.$$

If there is a method of constructing the symmetrical BIB design from the residual design, then any pair of sets of (5.2.13) must not have more than three symbols in common. However, the solution given by Bhattacharya [also reproduced in Hall (1964, 1967)] has a pair of sets having four symbols in common. Thus the residual and derived designs cannot always be embedded into the corresponding symmetrical BIB design. But Hall and Connor (1953) and Shrikhande (1960) showed that the embedding can always be done if $\lambda = 1$ or 2 [see also Hall (1967)].

We now prove the following result due to Bhagwandas (1965):

Theorem 5.2.2. Let N be the incidence matrix of a symmetrical BIB design with the parameters $v = b$, $r = k$, λ, and let the incidence matrix be written without loss of generality as

$$(5.2.14) \qquad N = \begin{bmatrix} 1 & E_{1, k-1} & 0_{1, v-k} \\ E_{k-1, 1} & N_1 & N_2 \\ 0_{v-k, 1} & N_3 & N_4 \end{bmatrix}.$$

Then

$$(5.2.15) \qquad N^* = \begin{bmatrix} N_1 & E_{k-1, v-k} - N_2 \\ E_{v-k, k-1} - N_3 & N_4 \end{bmatrix}$$

is the incidence matrix of another symmetrical BIB design if and only if N is the incidence matrix of a BIB design with the parameters

$$(5.2.16) \qquad v = b = 4m^2, \qquad r = k = 2m^2 \pm m, \qquad \lambda = m^2 \pm m.$$

The parameters of the design whose incidence matrix is N^* are given by

$$(5.2.17) \quad v_4 = 4m^2 - 1 = b_4, \qquad r_4 = 2m^2 - 1 = k_4, \qquad \lambda_4 = m^2 - 1.$$

Proof. Using the property that N is the incidence matrix of a BIB design with the parameters $v = b$, $r = k$, λ, we can verify the following relations very easily:

$$(5.2.18) \quad N_1 E_{k-1, 1} = (\lambda - 1)E_{k-1, 1}; \qquad E_{1, k-1} N_1 = (\lambda - 1)E_{1, k-1};$$

$$(5.2.19) \quad N_2 E_{v-k, 1} = (r - \lambda)E_{k-1, 1}; \qquad E_{1, k-1} N_2 = \lambda E_{1, v-k};$$

$$(5.2.20) \quad N_3 E_{k-1, 1} = \lambda E_{v-k, 1}; \qquad E_{1, v-k} N_3 = (r - \lambda)E_{1, k-1};$$

$$(5.2.21) \quad N_4 E_{v-k, 1} = (r - \lambda)E_{v-k, 1}; \qquad E_{1, v-k} N_4 = (r - \lambda)E_{1, v-k};$$

(5.2.22) $\quad N_1 N_1' + N_2 N_2' = (r - \lambda)I_{k-1} + (\lambda - 1)E_{k-1, k-1};$

(5.2.23) $\quad N_1 N_3' + N_2 N_4' = \lambda E_{k-1, v-k};$

(5.2.24) $\quad N_3 N_3' + N_4 N_4' = (r - \lambda)I_{v-k} + \lambda E_{v-k, v-k}.$

We can now verify that

(5.2.25)
$$N^* E_{v-1, 1} = (2r - 2\lambda - 1)E_{v-1, 1};$$
$$E_{1, v-1} N^* = (2r - 2\lambda - 1)E_{1, v-1};$$

(5.2.26) $\qquad N^* N^{*\prime} = (r - \lambda)I_{v-1} + (r - \lambda - 1)E_{v-1, v-1}$

if and only if

(5.2.27) $$v = 4(r - \lambda).$$

Thus if and only if the parameters of the design whose incidence matrix is N satisfy (5.2.27), the matrix N^* will be the incidence matrix of a BIB design with the parameters

$$v_4 = v - 1 = b_4, \qquad r_4 = 2r - 2\lambda - 1 = k_4, \qquad \lambda_4 = r - \lambda - 1.$$

In Chapter 12 we show that a necessary condition for the existence of a symmetrical BIB design is that $r - \lambda$ be a perfect square, when v is even. Thus for the existence of N, which makes N^* the incidence matrix of a symmetrical BIB design, $r - \lambda$ must be a perfect square, say m^2. Then we can see that v, r, and λ are given by (5.2.16). The parameters of the new design are

$$v_4 = b_4 = v - 1 = 4m^2 - 1;$$
(5.2.28) $\quad r_4 = k_4 = 2r - 2\lambda - 1 = 2(2m^2 \pm m) - 2(m^2 \pm m) - 1 = 2m^2 - 1;$
$$\lambda_4 = r - \lambda - 1 = 2m^2 \pm m - (m^2 \pm m) - 1 = m^2 - 1.$$

The theorem is thus established.

The following results due to Bhat and Shrikhande (1970) can be proved along similar lines:

Theorem 5.2.3. If N is the incidence matrix of a BIB design with the parameters $v = 4t + 3 = b, r = 2t + 1 = k, \lambda = t$ and if $\overline{N} = E_{v, v} - N$ is the incidence matrix of its complement, then

(5.2.29)
$$M = \begin{bmatrix} N & \overline{N} \\ E_{1, v} & 0_{1, v} \end{bmatrix}$$

is the incidence matrix of a BIB design with the parameters

(5.2.30)
$$v_5 = 4t + 4, \qquad b_5 = 8t + 6, \qquad r_5 = 4t + 3,$$
$$k_5 = 2t + 2, \qquad \lambda_5 = 2t + 1.$$

In fact in Section 7.4 we show that the design with incidence matrix M is a doubly balanced design.

Theorem 5.2.4. If N and $\overline{N}$ are as defined in Theorem 5.2.3, then

$$(5.2.31) \qquad M^* = \begin{bmatrix} N & \overline{N} & 0_{v,1} \\ E_{1,v} & 0_{1,v} & 0_{1,1} \\ N & N & E_{v,1} \end{bmatrix}$$

is the incidence matrix of a symmetrical BIB design with the parameters

$$(5.2.32) \qquad v_6 = 8t + 7 = b_6, \qquad r_6 = 4t + 3 = k_6, \qquad \lambda_6 = 2t + 1.$$

The designs discussed in Theorems 5.2.2, 5.2.3, and 5.2.4 have close connections with the orthogonal arrays of strength 2 in two symbols of Chapter 2 and also the Hadamard matrices of Section 17.4.

5.3. FAMILY (A) BIB DESIGNS

Definition 5.3.1. A BIB design with the parameters v, b, r, k, and λ is said to belong to the family (A) if

$$(5.3.1) \qquad b = 4(r - \lambda).$$

Shrikhande (1962), generalizing the result of Menon (1962), showed that the BIB designs belonging to the family (A) reproduce themselves under the composition given by the following theorem:

Theorem 5.3.1. If D_i is a BIB design with the parameters v_i, b_i, r_i, k_i, and λ_i belongs to the family (A) and N_i and $\overline{N}_i$ are the incidence matrices of D_i and D_i^*, where D_i^* is the complement of D_i $(i = 1, 2)$, then

$$(5.3.2) \qquad N = N_1 \times N_2 + \overline{N}_1 \times \overline{N}_2,$$

where $\times$ is the Kronecker product of matrices, is the incidence matrix of a BIB design D with the parameters

$$v = v_1 v_2, \qquad b = b_1 b_2, \qquad r = r_1 r_2 + (b_1 - r_1)(b_2 - r_2),$$

$$(5.3.3) \qquad k = k_1 k_2 + (v_1 - k_1)(v_2 - k_2), \qquad \lambda = r - \frac{b}{4},$$

and the design D also belongs to the same family (A).

Proof. From the structure of N in (5.3.2) we can easily see that it is a 0, 1 matrix of order $v_1 v_2 \times b_1 b_2$. Noting the properties of N_i ($i = 1, 2$), we can easily verify that

(5.3.4) $NE_{b_1 b_2, 1} = \{r_1 r_2 + (b_1 - r_1)(b_2 - r_2)\}E_{v_1 v_2, 1};$

(5.3.5) $E_{1, v_1 v_2} N = \{k_1 k_2 + (v_1 - k_1)(v_2 - k_2)\}E_{1, b_1 b_2};$

and

$$NN' = 4(r_1 - \lambda_1)(r_2 - \lambda_2)I_{v_1 v_2} + \left(r - \frac{b}{4}\right)E_{v_1 v_2, v_1 v_2}$$

(5.3.6)
$$+ (r_1 - \lambda_1)(b_2 - 4r_2 + 4\lambda_2)I_{v_1} \times E_{v_2, v_2}$$
$$+ (r_2 - \lambda_2)(b_1 - 4r_1 + 4\lambda_1)E_{v_1, v_1} \times I_{v_2}$$

$$= \frac{b}{4} I_{v_1 v_2} + \left(r - \frac{b}{4}\right)E_{v_1 v_2, v_1 v_2}.$$

Thus N is the incidence matrix of a BIB design with the parameters given in (5.3.3). The parameters given in (5.3.3) satisfy $b = 4(r - \lambda)$, and hence the design D also belongs to the family (A).

The parameters of a BIB design belonging to the family (A) satisfy three relations in all, and hence there are only two independent parameters. Shrikhande (1962) also showed that there are two subfamilies of BIB designs of family (A). These are denoted by $A_1(s, t)$, with the parameters

$$v = s^2, \qquad b = 2st, \qquad r = (s - 1)t, \qquad k = \frac{s(s - 1)}{2},$$

(5.3.7)
$$\lambda = \frac{(s - 2)t}{2},$$

where s and t are positive integers, s even, $2t \geqq s$, and $A_2(s, t)$, with the parameters

(5.3.8) $v = s^2, \qquad b = 4st, \qquad r = 2(s - 1)t, \qquad k = \frac{s(s - 1)}{2},$

$$\lambda = (s - 2)t,$$

where s and t are positive integers, s odd, $4t \geqq s$.

The composition Theorem 5.3.1 shows that the existence of $A_1(s_1, t_1)$ and $A_1(s_2, t_2)$ implies the existence of $A_1(s_1 s_2, 2t_1 t_2)$; the existence of $A_2(s_1, t_1)$ and $A_2(s_2, t_2)$ implies the existence of $A_2(s_1 s_2, 4t_1 t_2)$; and the existence of $A_1(s_1, t_1)$ and $A_2(s_2, t_2)$ implies the existence of $A_1(s_1 s_2, 4t_1 t_2)$.

5.4. AFFINE RESOLVABLE BIB DESIGNS

In the preceding chapter we saw that a resolvable BIB design is called affine resolvable if any two sets belonging to different classes have exactly k^2/v symbols in common. We prove the following:

Theorem 5.4.1. A resolvable BIB design is affine resolvable if and only if

$$(5.4.1) \qquad\qquad b = v + r - 1.$$

Proof. Let S_{ij} denote the jth set of the ith class $(i = 1, 2, \cdots, r; j = 1, 2, \cdots, n$, where $n = b/r)$. Let the set S_{11} have l_{ij} symbols in common with the set S_{ij} $(i = 2, 3, \cdots, r; j = 1, 2, \cdots, n)$. Then

$$(5.4.2) \qquad\qquad \sum_{i=2}^{r} \sum_{j=1}^{n} l_{ij} = k(r - 1)$$

and

$$(5.4.3) \qquad\qquad \sum_{i=2}^{r} \sum_{j=1}^{n} l_{ij}(l_{ij} - 1) = k(k - 1)(\lambda - 1).$$

Define

$$(5.4.4) \qquad\qquad l = \sum_{i=2}^{r} \sum_{j=1}^{n} \frac{l_{ij}}{n(r - 1)} = \frac{k^2}{v}.$$

Then we can show on simplification that

$$(5.4.5) \qquad\qquad \sum_{i=2}^{r} \sum_{j=1}^{n} (l_{ij} - l)^2 = \frac{k(v - k)(b - v - r + 1)}{n(v - 1)},$$

which will be zero if and only if $b = v + r - 1$, in which case $l_{ij} = l$, and the set S_{11} has k^2/v symbols in common with each of the sets S_{ij} $(i = 2, 3, \cdots, r; j = 1, 2, \cdots, n)$. This result holds irrespective of S_{11}, and hence the theorem is established.

The above theorem is a particular case, given by Bose (1942b), of Theorem 4.6.2, and the proof supplied here is an alternative one to that based on matrix theory.

As a consequence to Theorem 5.4.1 we have

Corollary 5.4.1.1. If the parameters of a BIB design satisfy (5.4.1) and k^2/v is not an integer, then the design is not resolvable.

As a consequence of Corollary 5.4.1.1, the following four designs with the parameters are not resolvable:

$$(5.4.6) \qquad v = 6, \qquad b = 10, \qquad r = 5, \qquad k = 3, \qquad \lambda = 2;$$

(5.4.7) $v = 10,$ $b = 18,$ $r = \; 9,$ $k = 5,$ $\lambda = 4;$

(5.4.8) $v = 28,$ $b = 36,$ $r = \; 9,$ $k = 7,$ $\lambda = 2;$

(5.4.9) $v = 14,$ $b = 26,$ $r = 13,$ $k = 7,$ $\lambda = 6.$

We now derive the parameters of an affine resolvable BIB design in terms of only two integral parameters n and t as follows:

Since $b = v + r - 1$, it follows that $b = r\{[(k-1)/\lambda] + 1\}$, and hence, using $vr = bk$ and $v = nk$, we get $n = [(k-1)/\lambda] + 1$. Thus $\lambda = (k-1)/(n-1)$ is an integer. Hence $k = \lambda(n-1) + 1$. Again, since k^2/v is an integer, k/n is also an integer. Thus

(5.4.10) $\lambda(n-1) + 1 \equiv 0(\mathrm{mod}\; n),$

or, equivalently,

(5.4.11) $\lambda \equiv 1(\mathrm{mod}\; n).$

Let $\lambda = nt + 1$. Then $k = n[(n-1)t + 1]$ and $v = nk = n^2[(n-1)t + 1]$. Again $r = (v-1)/(n-1) = n^2 t + n + 1$ and $b = nr = n(n^2 t + n + 1)$. Thus the parameters of an affine resolvable BIB design are

(5.4.12)
$$v = n^2[(n-1)t + 1], \qquad b = n(n^2 t + n + 1),$$
$$r = n^2 t + n + 1, \qquad k = n[(n-1)t + 1], \qquad \lambda = nt + 1.$$

5.5. SET STRUCTURE OF BIB DESIGNS

In Section 5.2 we saw that any two sets of a symmetrical BIB design have exactly λ symbols in common. We shall study the structure of the sets in any asymmetrical BIB design.

Let us choose any $t \leqq b$ sets of the given BIB design. Let the submatrix corresponding to the t chosen sets of the incidence matrix N of the design be denoted by N_0. We make the following definition:

Definition 5.5.1. The matrix $S_t = (S_{ju}) = N_0' N_0$ is called the structural matrix of the t chosen sets.

Obviously S_{ju} denotes the number of symbols in common between the jth and the uth sets of the t chosen sets. Let the columns of N be permuted so that the first t columns correspond to the t chosen sets. Then form

(5.5.1) $N_1 = \begin{bmatrix} N \\ I_t \quad 0_{t,\, b-t} \end{bmatrix},$

from which it follows that

(5.5.2) $$N_1 N_1' = \begin{bmatrix} NN' & N_0 \\ N_0' & I_t \end{bmatrix}.$$

Applying Theorem A.4.5, we get

(5.5.3) $$|N_1 N_1'| = kr^{-t+1}(r - \lambda)^{v-t-1}|C_t|,$$

where C_t has the elements

(5.5.4) $c_{jj} = (r - k)(r - \lambda);$ $c_{ju} = \lambda k - rS_{ju}$ $j \neq u; j, u = 1, 2, \cdots, t.$

We can see that

(5.5.5) $$C_t = \lambda k E_{t,t} + r(r - \lambda)I_t - rS_t.$$

Definition 5.5.2. The matrix C_t, as defined in (5.5.5), is called the characteristic matrix of the t chosen sets.

When P is a matrix with real elements of order $m \times n$ ($n \geq m$), it is well known that $|PP'| \geq 0$. Hence if $b \geq v + t$, then $|N_1 N_1'| \geq 0$. Furthermore, since the elements of N_1 are integers, if $b = v + t$, then $|N_1 N_1'|$ is a perfect integral square. Finally, if $b < v + t$, then $|N_1 N_1'| = 0$. Hence we get the following very useful theorem:

Theorem 5.5.1. If C_t is the characteristic matrix of any t chosen sets of a BIB design with the parameters v, b, r, k, and λ, then

1. $|C_t| \geq 0$ if $t < b - v$.
2. $|C_t| = 0$ if $t > b - v$.
3. $kr^{-t+1}(r - \lambda)^{v-t-1}|C_t|$ is a perfect square, if $t = b - v$.

To illustrate the kind of information that is contained in this theorem, consider the BIB design with the parameters

(5.5.6) $v = 10,$ $b = 15,$ $r = 9,$ $k = 6,$ $\lambda = 5.$

Let the symbols be denoted by letters and consider whether it is possible to fill up four sets in such a way that each set will have four symbols in common with each of the other three sets. One way in which this can be done is as follows:

(5.5.7)   (A   B   C   D   E   F);   (A   B   C   D   G   H);
          (A   B   C   E   G   I);   (A   B   E   F   G   H).

We can now ask whether these four sets can form part of the completed design with the parameters shown in (5.5.6). The characteristic matrix C_4 for the four sets in (5.5.7) can easily be seen to satisfy

$$(5.5.8) \qquad |C_4| = |18I_4 - 6E_{4,4}| = -6 \times 18^3 < 0,$$

contradicting statement 1 of Theorem 5.5.1. Hence the four sets given in (5.5.7) cannot form part of the plan of the BIB design with the parameters shown in (5.5.6).

As a consequence of Theorem 5.5.1, letting $t = 2$,

$$(5.5.9) \qquad |C_2| = (r - \lambda)^2 (r - k)^2 - (\lambda k - S_{12} r)^2 \geqq 0.$$

Thus we have the following theorem:

Theorem 5.5.2. For a BIB design the common symbols between the jth and the uth sets satisfy the inequality

$$(5.5.10) \qquad -(r - \lambda - k) \leqq S_{ju} \leqq \frac{2\lambda k + r(r - \lambda - k)}{r}.$$

For a symmetrical BIB design the above theorem gives $S_{ju} = \lambda$ and every pair of sets have λ common symbols. For a design with the parameters given in (5.2.13) Theorem 5.5.2 gives $0 \leqq S_{ju} \leqq 4$ and Bhattacharya's solution attains both the bounds.

The notions of structural and characteristic matrices and the results of Theorems 5.5.1 and 5.5.2 are due to Connor (1952). Theorem 5.5.1 is very powerful in checking whether any given t sets can form a part of the completed solution of a BIB design.

A number of other bounds for the number of symbols common to any two sets of a BIB design have been proposed by various investigators [see Parker (1963), Raghavarao (1963), Seiden (1963), Shrikhande and Singh (1963), and Trehan (1963)]. Chakrabarti (1963) showed that the bounds proposed by Connor in Theorem 5.5.2 are the best ones and can, of course, be slightly improved on in very particular cases.

We now derive bounds for the number of symbols common to any given m sets of a BIB design. It is clear that the number of symbols common to any m sets is not greater than unity if $m > \lambda$, and we obtain an upper bound for the number of symbols common to any m ($\leqq \lambda$) sets.

Let x be the number of symbols common to any m sets and let the ith set of the remaining $b - m$ sets have y_i symbols in common with these x symbols. Then

$$(5.5.11) \qquad \sum_{i=1}^{b-m} y_i = (r - m)x$$

and

(5.5.12)
$$\sum_{i=1}^{b-m} y_i(y_i - 1) = x(x - 1)(\lambda - m).$$

Hence

(5.5.13)
$$\sum_{i=1}^{b-m} y_i^2 = x[(r - \lambda) + x(\lambda - m)].$$

Now, applying Schwarz's inequality that $(\sum_{i=1}^{b-m} y_i)^2 \leq (b - m) \sum_{i=1}^{b-m} y_i^2$, we get

(5.5.14)
$$x \leq \frac{(b - m)(r - \lambda)}{(r - m)^2 - (\lambda - m)(b - m)}$$
$$= \frac{k(b - m)}{(r - m) + m(v - k)} = \gamma, \text{ say.}$$

Since x is an integer, $x \leq [\gamma]$, where $[\gamma]$ stands for the largest integer contained in γ. Hence the following:

Theorem 5.5.3. In a BIB design the number of symbols common to any m ($\leq \lambda$) sets is not greater than $[\gamma]$, where γ is given by (5.5.14).

When $m = 2$, we get the particular case given by

Corollary 5.5.3.1. The number of symbols common to any two sets of an asymmetrical BIB design with $\lambda \geq 2$ is not greater than $[\gamma_1]$, where

(5.5.15)
$$\gamma_1 = \frac{k(b - 2)}{(r - 2) + 2(v - k)}.$$

Though the bound of the above Corollary is not an improvement over the bound of Theorem 5.5.2, the technique leading to Theorem 5.5.3 throws an interesting sidelight. If γ of (5.5.14) is an integer, since the equality sign in Schwarz's inequality holds when and only when all the y_i terms are equal, it follows that if any m sets have exactly γ common symbols, say $v_1, v_2, \cdots, v_\gamma$, then each of the remaining sets contains exactly

(5.5.16)
$$\frac{k(r - m)}{(r - m) + m(v - k)} = \alpha, \text{ say,}$$

symbols out of the γ common symbols $v_1, v_2, \cdots, v_\gamma$. Hence

Theorem 5.5.4. If γ is an integer and any m sets of a BIB design have exactly γ symbols in common, $v_1, v_2, \cdots, v_\gamma$, then, by omitting the m sets that have $v_1, v_2, \cdots, v_\gamma$ as the common symbols and retaining these common symbols in the remaining $b - m$ sets, we get another BIB design with the parameters

(5.5.17) $v_1 = \gamma,$ $b_1 = b - m,$ $r_1 = r - m,$ $k_1 = \alpha,$ $\lambda_1 = \lambda - m.$

The results of Theorems 5.5.3 and 5.5.4 were obtained by Raghavarao (1963).

We had so far given bounds for the number of symbols common to given sets. We now derive a bound for the number of sets disjoint with a given set in a BIB design. This result, due to Majindar (1962), is given by the following theorem:

Theorem 5.5.5. A given set in a BIB design with the parameters v, b, r, k, and λ can never have more than $b - 1 - \left[(r - 1)^2 k / (r - \lambda - k + k\lambda) \right]$ disjoint sets with it. If some set has that many, then $(r - \lambda - k + k\lambda)/(r - 1)$ is a positive integer and each of the nondisjoint sets has $(r - \lambda - k + k\lambda)/(r - 1)$ symbols in common with it.

Proof. Let the set S_1 of the BIB design have m disjoint sets. Let the ith set of the remaining $b - m - 1$ sets have x_i symbols in common with S_1. Then

(5.5.18)
$$\sum_{i=1}^{b-m-1} x_i = (r - 1)k$$

and

(5.5.19)
$$\sum_{i=1}^{b-m-1} x_i(x_i - 1) = (\lambda - 1)k(k - 1).$$

Applying Schwarz's inequality, we get

(5.5.20)
$$m \leqq b - 1 - \frac{(r - 1)^2 k}{(r - \lambda - k + k\lambda)}.$$

The equality sign holds in (5.5.20) when all the x_i terms are equal, in which case

(5.5.21)
$$x_i = \frac{(r - \lambda - k + k\lambda)}{r - 1} \qquad i = 1, 2, \cdots, b - m - 1,$$

completing the proof of the theorem.

For another relation of set structure see Shah (1963).

5.6. CONSTRUCTION OF BIB DESIGNS THROUGH FINITE GEOMETRIES

So far we have studied the various combinatorial problems related to BIB designs. In this and the next sections we study the various methods of constructing them.

From the properties of finite projective geometry, as given in Appendix A, it is clear that if the points of the geometry $PG(n, s)$ are identified with the v symbols and the m-flats are identified with the b sets, we get a BIB design with the parameters

(5.6.1.)
$$v = \frac{s^{n+1} - 1}{s - 1}, \qquad b = \phi(n, m, s), \qquad r = \phi(n - 1, m - 1, s),$$
$$k = \frac{s^{m+1} - 1}{s - 1}, \qquad \lambda = \phi(n - 2, m - 2, s),$$

where

(5.6.2)
$$\phi(n, m, s) = \frac{(s^{n+1} - 1)(s^n - 1) \cdots (s^{n-m+1} - 1)}{(s^{m+1} - 1)(s^m - 1) \cdots (s - 1)}.$$

In particular, when s is a prime or a prime power, the following series of designs can be constructed:

(5.6.3) $v = s^2 + s + 1 = b, \qquad r = s + 1 = k, \qquad \lambda = 1;$

(5.6.4) $\begin{aligned} & v = (s + 1)(s^2 + 1), \qquad b = (s^2 + 1)(s^2 + s + 1), \\ & r = s^2 + s + 1, \qquad k = s + 1, \qquad \lambda = 1; \end{aligned}$

(5.6.5) $v = (s + 1)(s^2 + 1) = b, \qquad r = s^2 + s + 1 = k, \qquad \lambda = s + 1.$

Design (5.6.3) corresponds to $n = 2, m = 1$, design (5.6.4) corresponds to $n = 3, m = 1$, and design (5.6.5) corresponds to $n = 3, m = 2$.

As a numerical illustration let us construct the BIB design with the parameters given in (5.6.5) when $s = 2$. In $PG(3, 2)$ each point is represented by four coordinates (x_0, x_1, x_2, x_3), where each coordinate is either zero or unity and not all coordinates are zero. These points correspond to our symbols. The 15 planes (2-flats) have equations

(5.6.6) $\begin{aligned} & x_i & = 0 & \qquad i = 0, 1, 2, 3; \\ & x_i + x_j & = 0 & \qquad i, j = 0, 1, 2, 3; \quad i \neq j; \\ & x_i + x_j + x_k & = 0 & \qquad i, j, k = 0, 1, 2, 3; \quad i \neq j \neq k; \\ & x_0 + x_1 + x_2 + x_3 & = 0. \end{aligned}$

Thus writing down the points lying on these planes, we get the following solution:

$$
\begin{array}{l}
(0001, 0010, 0011, 0100, 0101, 0110, 0111) \\
(0001, 0010, 0011, 1000, 1001, 1010, 1011) \\
(0001, 0100, 0101, 1000, 1001, 1100, 1101) \\
(0010, 0100, 0110, 1000, 1010, 1100, 1110) \\
(0001, 0010, 0011, 1100, 1101, 1110, 1111) \\
(0001, 0100, 0101, 1010, 1011, 1110, 1111) \\
(0010, 0100, 0110, 1001, 1011, 1101, 1111) \\
(0001, 1000, 1001, 0110, 0111, 1110, 1111) \\
(0010, 1000, 1010, 0101, 0111, 1101, 1111) \\
(0100, 1000, 1100, 0011, 0111, 1011, 1111) \\
(0011, 0101, 0110, 1000, 1011, 1101, 1110) \\
(0011, 1001, 1010, 0100, 0111, 1101, 1110) \\
(0101, 1001, 1100, 0010, 0111, 1011, 1110) \\
(0110, 1010, 1100, 0001, 0111, 1011, 1101) \\
(0011, 0101, 0110, 1001, 1010, 1100, 1111)
\end{array}
$$

(5.6.7)

Again from Euclidean geometry, by identifying the points of EG(n, s) with the v symbols and identifying the m-flats of the geometry with the b sets we get BIB designs with the parameters

(5.6.8)
$$
\begin{aligned}
& v = s^n, \quad b = \phi(n, m, s) - \phi(n - 1, m, s) = s^{n-m} \phi(n - 1, m - 1, s), \\
& r = \phi(n - 1, m - 1, s), \quad k = s^m, \quad \lambda = \phi(n - 2, m - 2, s).
\end{aligned}
$$

In particular when s is a prime or a prime power, the following series of designs can be constructed:

(5.6.9) $v = s^2, \quad b = s(s + 1), \quad r = s + 1, \quad k = s, \quad \lambda = 1;$

(5.6.10)
$$
\begin{aligned}
& v = s^3, \quad b = s^2(s^2 + s + 1), \quad r = s^2 + s + 1, \quad k = s, \\
& \lambda = 1;
\end{aligned}
$$

(5.6.11)
$$
\begin{aligned}
& v = s^3, \quad b = s(s^2 + s + 1), \quad r = s^2 + s + 1, \quad k = s^2, \\
& \lambda = s + 1.
\end{aligned}
$$

Designs (5.6.9), (5.6.10), and (5.6.11) correspond respectively to $n = 2, m = 1$; $n = 3, m = 1$; and $n = 3, m = 2$.

As a numerical illustration let us construct the BIB design with the parameters given in (5.6.11) when $s = 2$. In EG(3, 2) each point is represented by three coordinates (x_1, x_2, x_3), where each coordinate is either zero or unity.

These points correspond to our eight symbols. The 14 planes (2-flats) have the equations

$$
\begin{aligned}
x_i &= 0 \text{ or } 1 & i &= 1, 2, 3; \\
x_i + x_j &= 0 \text{ or } 1 & i, j &= 1, 2, 3; \quad i \neq j; \\
x_1 + x_2 + x_3 &= 0 \text{ or } 1.
\end{aligned}
$$

(5.6.12)

Thus writing down the points lying on these planes, we get the solution

(5.6.13)
$$
\begin{array}{ll}
(000, 001, 010, 011); & (000, 001, 100, 101); \\
(000, 010, 100, 110); & (100, 101, 110, 111); \\
(010, 011, 110, 111); & (001, 011, 101, 111); \\
(000, 001, 110, 111); & (000, 010, 101, 111); \\
(000, 100, 011, 111); & (010, 011, 100, 101); \\
(001, 011, 100, 110); & (001, 101, 010, 110); \\
(000, 011, 101, 110); & (001, 010, 100, 111).
\end{array}
$$

For further reading, please refer to Bose (1939, 1959), Bose and Chakravarti (1966), and Yamamoto, Fukuda, and Hamaka (1966), where other series of BIB designs constructible by geometric methods are given.

It may be noted that certain designs, whose parameters are identical with the geometrical designs considered in this section, can also be constructed by other methods.

5.7. CONSTRUCTION OF BIB DESIGNS THROUGH THE METHOD OF SYMMETRICALLY REPEATED DIFFERENCES

Consider a module M, containing exactly m elements. To each element of the module let there correspond exactly n symbols, the symbols corresponding to the element u being denoted by $u_1, u_2, \cdots, u_n$. Thus there are exactly mn symbols. Symbols with the same subscript j may be said to belong to the jth class.

Let a set S contain $k = \sum_{i=1}^{n} p_i$ symbols, the number of symbols of the ith class contained in the set being p_i. Let the symbols contained in this set of the ith class be $a_i^{(1)}, a_i^{(2)}, \cdots, a_i^{(p_i)}$ and the symbols contained in this set of the jth class be $b_j^{(1)}, b_j^{(2)}, \cdots, b_j^{(p_j)}$.

The $p_i(p_i - 1)$ differences $a^{(u)} - a^{(w)}$ ($u \neq w = 1, 2, \cdots, p_i$) may be called pure differences of the type (i, i) arising from the set S. Again the $p_i p_j$ differences $a^{(u)} - b^{(w)}$ ($u = 1, 2, \cdots, p_i; w = 1, 2, \cdots, p_j$) may be called mixed differences of the type (i, j) arising from the set S. Clearly there are n types of different pure differences and $n(n - 1)$ types of mixed differences.

Now let us consider a class of t sets satisfying the following conditions:

1. If p_{il} denotes the number of symbols of the ith class in the lth set, then among the $\sum_{l=1}^{t} p_{il}(p_{il} - 1)$ pure differences of the type (i, i) arising from the t sets every nonzero element of M is repeated exactly λ times (independently of i).

2. Among the $\sum_{l=1}^{t} p_{il} p_{jl}$ mixed differences of the type (i, j) arising from the t sets every element of M is repeated exactly λ times (independently of i and j). When these conditions are satisfied, we say that in the t sets the differences are symmetrically repeated, each occurring λ times.

For example, consider M to be a module of residue classes (mod 5) and to every element u of the module let there correspond just three symbols denoted by u_1, u_2, u_3. It is then easy to verify that the differences arising from the seven sets $(0_1, 1_1, 0_2)$; $(0_2, 1_2, 2_3)$; $(0_3, 1_3, 2_1)$; $(0_1, 2_1, 3_2)$; $(0_2, 2_2, 0_3)$; $(0_3, 2_3, 0_1)$; $(0_1, 2_2, 1_3)$; are symmetrically repeated once, for among the pure differences of the type $(1, 1)$, $(2, 2)$ and $(3, 3)$ each of the elements $1, 2, 3, 4$ occurs once, and among the mixed differences of the type $(1, 2)$, $(2, 1)$, $(1, 3)$, $(3, 1)$, $(2, 3)$, and $(3, 2)$ each of the elements $0, 1, 2, 3, 4$ occurs once.

The following two theorems due to Bose (1939) provide a powerful method of constructing a large number of BIB designs:

Theorem 5.7.1.† Let M be a module containing the m elements; with each element we associate n symbols, the symbols associated with the element u being $u_1, u_2, \cdots, u_n$. Let it be possible to find a class of t sets $S_1, S_2, \cdots, S_t$ satisfying the following conditions:

1. Every set contains k symbols (the symbols contained in the same set being different from one another).

2. Among the kt symbols occurring in the t sets exactly r symbols should belong to each of the n classes. (Clearly $kt = nr$.)

3. The differences from the t sets are symmetrically repeated, each occurring λ times.

If θ is any element of M, from each set S_i ($i = 1, 2, \cdots, t$) we can form another set $S_{i, \theta}$ by adding θ to the symbols of S_i, keeping the class number unchanged. Then the mt sets

$$(5.7.1) \qquad\qquad S_{i, \theta} \qquad (i = 1, 2, \cdots, t; \; \theta \in M)$$

provide us with a BIB design with the parameters

$$(5.7.2) \qquad\qquad v = mn; \qquad b = nt; \qquad r, k, \lambda.$$

Proof. Corresponding to a symbol $x_i^{(u)}$ of the ith class of S_l we have a symbol $x_i^{(w)}$ in the set $S_{l, \theta}$, where $x^{(w)} = x^{(u)} + \theta$. As θ takes all values of

† The first fundamental theorem of the method of differences.

M, $x^{(w)}$ also takes these values in some order or other. Hence, among the symbols corresponding to $x_i^{(u)}$ in S_l, every symbol of the ith class is replicated exactly once. From condition 2, it follows that every symbol is replicated exactly r times in our design.

In order for a pure pair of symbols $x_i^{(u)}$ and $x_i^{(w)}$ to occur together in some set of our design it is necessary and sufficient that we be able to find the symbols $x_i^{(u')}$ and $x_i^{(w')}$ occurring together in one of the sets S_l together with an element $\theta \in M$, such that

(5.7.3) $$x^{(u)} = x^{(u')} + \theta, \qquad x^{(w)} = x^{(w')} + \theta.$$

Then

(5.7.4) $$x^{(u')} - x^{(w')} = x^{(u)} - x^{(w)} = \text{a fixed element of } M.$$

It follows from condition 3 that it is possible to find symbols $x_i^{(u')}$ and $x_i^{(w')}$ belonging to some set S_l and satisfying (5.7.4) in exactly λ ways. Then θ is determined uniquely by any one of the equations (5.7.3). Hence, given a pure pair of symbols, we find them occurring together in exactly λ sets of our design. The same can be similarly proved for mixed pairs of symbols. The parameters v, b, k need no explanation. Hence the theorem is established.

The sets $S_1, S_2, \cdots, S_t$, from which the complete solution for the design can be obtained, are called the initial sets.

As a numerical illustration of the method of construction indicated in Theorem 5.7.1, we construct a BIB design with the parameters

(5.7.5) $$v = 15, \qquad b = 35, \qquad r = 7, \qquad k = 3, \qquad \lambda = 1.$$

Let us consider the module M of the residue classes (mod 5) and assign three symbols to each of the elements of M, the symbols assigned to the element u being u_1, u_2, u_3. Then the seven sets

(5.7.6) $(0_1, 1_1, 0_2); \quad (0_2, 1_2, 2_3); \quad (0_3, 1_3, 2_1); \quad (0_1, 2_1, 3_2);$
$(0_2, 2_2, 0_3); \quad (0_3, 2_3, 0_1); \quad (0_1, 2_2, 1_3),$

satisfy all the conditions of Theorem 5.7.1 and hence the solution for (5.7.5) is

(5.7.7)
$(0_1, 1_1, 0_2); \quad (1_1, 2_1, 1_2); \quad (2_1, 3_1, 2_2); \quad (3_1, 4_1, 3_2);$
$(4_1, 0_1, 4_2); \quad (0_2, 1_2, 2_3); \quad (1_2, 2_2, 3_3); \quad (2_2, 3_2, 4_3);$
$(3_2, 4_2, 0_3); \quad (4_2, 0_2, 1_3); \quad (0_3, 1_3, 2_1); \quad (1_3, 2_3, 3_1);$
$(2_3, 3_3, 4_1); \quad (3_3, 4_3, 0_1); \quad (4_3, 0_3, 1_1); \quad (0_1, 2_1, 3_2);$
$(1_1, 3_1, 4_2); \quad (2_1, 4_1, 0_2); \quad (3_1, 0_1, 1_2); \quad (4_1, 1_1, 2_2);$
$(0_2, 2_2, 0_3); \quad (1_2, 3_2, 1_3); \quad (2_2, 4_2, 2_3); \quad (3_2, 0_2, 3_3);$
$(4_2, 1_2, 4_3); \quad (0_3, 2_3, 0_1); \quad (1_3, 3_3, 1_1); \quad (2_3, 4_3, 2_1);$
$(3_3, 0_3, 3_1); \quad (4_3, 1_3, 4_1); \quad (0_1, 2_2, 1_3); \quad (1_1, 3_2, 2_3);$
$(2_1, 4_2, 3_3); \quad (3_1, 0_2, 4_3); \quad (4_1, 1_2, 0_3).$

Along similar lines the following theorem can also be established:

Theorem 5.7.2.† Let the mn symbols be defined as in Theorem 5.7.1; these are called finite symbols, and to them we add a new infinite symbol ∞. Let it be possible to find a class of $t + s$ sets $S_1, S_2, \cdots, S_t; S_1', S_2', \cdots, S_s'$ satisfying the following conditions:

1. Each of the sets $S_1, S_2, \cdots, S_t$ contains exactly k of the finite symbols, and each of the sets $S_1', S_2', \cdots, S_s'$ contains ∞ and exactly $k - 1$ finite symbols (the symbols occurring in a set being different from one another).

2. Of the finite symbols of the ith class, $(ms - \lambda)$ occur in sets $S_1, S_2, \cdots, S_t$ and λ occur in sets $S_1', S_2', \cdots, S_s'$ for each $i = 1, 2, \cdots, n$. [Clearly $kt = n(ms - \lambda)$ and $(k - 1)s = n\lambda$.]

3. The differences arising from the finite symbols in the $(s + t)$ sets are symmetrically repeated, each occurring λ times.

If θ is any element of M, then from each set $S_i(S_j')$ we can form another set $S_{i,\theta}(S_{j,\theta}')$ by adding θ to the symbols of S_i, keeping the class number unchanged and subject to the further condition that $\theta + \infty = \infty$. Then the $m(s + t)$ sets

(5.7.8) $\quad S_{i,\theta} \quad (i = 1, 2, \cdots, t; \theta \in M)$ and $S_{j,\theta}' \quad (j = 1, 2, \cdots, s; \theta \in M)$

provide us with a BIB design with the parameters

(5.7.9) $\qquad v = mn + 1, \qquad b = m(s + t), \qquad r = ms, \qquad k, \lambda.$

As a numerical illustration of the method of construction indicated in Theorem 5.7.2, we construct a BIB design with the parameters

(5.7.10) $\qquad v = 8, \qquad b = 14, \qquad r = 7, \qquad k = 4, \qquad \lambda = 3.$

Let us consider a module M of seven elements $0, 1, \cdots, 6$, which we identify as the seven symbols of our design, to which we add the symbol ∞. We can easily verify that the two initial sets $(0, 1, 3, 6)$ and $(\infty, 0, 1, 3)$ satisfy all the conditions of Theorem 5.7.2; hence the solution of the BIB design with the parameters given in (5.7.10) is

$$
\begin{array}{ll}
(0, 1, 3, 6); & (\infty, 0, 1, 3); \\
(1, 2, 4, 0); & (\infty, 1, 2, 4); \\
(2, 3, 5, 1); & (\infty, 2, 3, 5); \\
(3, 4, 6, 2); & (\infty, 3, 4, 6); \\
(4, 5, 0, 3); & (\infty, 4, 5, 0); \\
(5, 6, 1, 4); & (\infty, 5, 6, 1); \\
(6, 0, 2, 5); & (\infty, 6, 0, 2).
\end{array}
$$

(5.7.11)

---

† The second fundamental theorem of the method of difference.

One may be interested to know whether the name "symmetrical BIB design" has originated from the symmetrical incidence matrix. We prove the following theorem:

Theorem 5.7.3. If a difference-set solution on a module of v elements exists for a symmetrical BIB design with the parameters $v = b, r = k, \lambda$, then its incidence matrix is symmetrical.

Proof. Let $(a_1, a_2, \cdots, a_k)$ be an initial set, which when developed in the usual way gives the design under consideration. Then denote this initial set by v and develop the sets, numbering them $v - 1, v - 2, \cdots, 1$. With this numbering of the sets we can easily verify that if the ith symbol occurs in the jth set, the jth symbol occurs in the ith set, which implies that the incidence matrix of the design is symmetrical.

It is to be remarked here that it is not known whether the above result holds true even if a difference-set solution does not exist. In this connection also see Bhat and Shrikhande (1970).

The methods of construction indicated in Theorems 5.7.1 and 5.7.2 are very useful in constructing many series of BIB designs; in the next two theorems we study two series of particular interest.

Theorem 5.7.4. The series of symmetrical BIB designs with the parameters

$$(5.7.12) \qquad v = b = 4t + 3, \qquad r = k = 2t + 1, \qquad \lambda = t$$

can always be constructed when $4t + 3$ is a prime or a prime power.

Proof. Let $4t + 3 = s$ and x be a primitive root of the field GF(s). Then

$$(5.7.13) \qquad x^{4t+2} = 1$$

and hence

$$(5.7.14) \qquad x^{2t+1} = -1.$$

Consider the set

$$(5.7.15) \qquad (x^0, x^2, x^4, \cdots, x^{4t}).$$

All the differences that arise from the set are

$$(5.7.16) \qquad \begin{aligned} &\pm(x^2 - x^0),\ \pm x^2(x^2 - x^0),\ \pm x^4(x^2 - x^0),\ \cdots,\ \pm x^{4t-2}(x^2 - x^0), \\ &\pm(x^4 - x^0),\ \pm x^2(x^4 - x^0),\ \pm x^4(x^4 - x^0),\ \cdots,\ \pm x^{4t-4}(x^4 - x^0), \\ &\pm(x^6 - x^0),\ \pm x^2(x^6 - x^0),\ \pm x^4(x^6 - x^0),\ \cdots,\ \pm x^{4t-6}(x^6 - x^0), \\ &\cdots,\ \pm(x^{4t} - x^0). \end{aligned}$$

Using the properties of x, we can easily show that

$$(5.7.17) \qquad x^{2t+i} - x^0 = x^{i-1}(x^{2t+2-i} - x^0),$$

and using (5.7.14) and (5.7.17) we can verify that each nonzero element of GF(s) occurs exactly t times in (5.7.16); hence in the set (5.7.15) the differences are symmetrically repeated λ times. Thus this initial set satisfies all the requirements of Theorem 5.7.1, and we get the required solution on developing this initial set.

Compare the above method of constructing a BIB design with that of constructing the Hadamard matrices given in Section 17.4.

We illustrate this theorem in the construction of a BIB design with the parameters

$$(5.7.18) \qquad v = b = 11, \qquad r = k = 5, \qquad \lambda = 2.$$

Since 2 is a primitive element of GF(11), the initial set is $(2^0, 2^2, 2^4, 2^6, 2^8)$, that is (1, 3, 4, 5, 9).

The complete solution can be obtained on expanding this initial set and is

$$(5.7.19)$$

$$
\begin{array}{lll}
(\,1, 3, \ 4, 5, 9); & (2, \ 4, 5, 6, 10); & (3, 5, 6, \ 7, 0); \\
(\,4, 6, \ 7, 8, 1); & (5, \ 7, 8, 9, \ 2); & (6, 8, 9, 10, 3); \\
(\,7, 9, 10, 0, 4); & (8, 10, 0, 1, \ 5); & (9, 0, 1, \ 2, 6); \\
(10, 1, \ 2, 3, 7); & (0, \ 2, 3, 4, \ 8). &
\end{array}
$$

Theorem 5.7.5. The series of BIB designs with the parameters

$$(5.7.20) \qquad v = 4t + 1, \qquad b = 8t + 2, \qquad r = 4t, \qquad k = 2t, \qquad \lambda = 2t - 1$$

can always be constructed when $4t + 1$ is a prime or a prime power.

Proof. Let x be a primitive root of GF(s), where $s = 4t + 1$ is a prime or a prime power. Then the two sets

$$(5.7.21) \qquad \begin{aligned} &(x^0, x^2, x^4, \cdots, x^{4t-2}), \\ &(x, x^3, x^5, \cdots, x^{4t-1}) \end{aligned}$$

can easily be shown to satisfy the conditions of Theorem 5.7.1; hence on development these two initial sets yield the required design.

Let us use the above method in the construction of a BIB design with the parameters

$$(5.7.22) \qquad v = 13, \qquad b = 26, \qquad r = 12, \qquad k = 6, \qquad \lambda = 5.$$

Since 7 is a primitive element of GF(13), the two initial sets are $(7^0, 7^2, 7^4, 7^6, 7^8, 7^{10})$ and $(7, 7^3, 7^5, 7^7, 7^9, 7^{11})$, that is (1, 3, 4, 9, 10, 12) and (2, 5, 6, 7, 8, 11).

Thus the complete solution is

(1, 3, 4, 9, 10, 12);	(2, 5, 6, 7, 8, 11);
(2, 4, 5, 10, 11, 0);	(3, 6, 7, 8, 9, 12);
(3, 5, 6, 11, 12, 1);	(4, 7, 8, 9, 10, 0);
(4, 6, 7, 12, 0, 2);	(5, 8, 9, 10, 11, 1);
(5, 7, 8, 0, 1, 3);	(6, 9, 10, 11, 12, 2);
(6, 8, 9, 1, 2, 4);	(7, 10, 11, 12, 0, 3);
(7, 9, 10, 2, 3, 5);	(8, 11, 12, 0, 1, 4);
(8, 10, 11, 3, 4, 6);	(9, 12, 0, 1, 2, 5);
(9, 11, 12, 4, 5. 7);	(10, 0, 1, 2, 3, 6);
(10, 12, 0, 5, 6, 8);	(11, 1, 2, 3, 4, 7);
(11, 0, 1, 6, 7, 9);	(12, 2, 3, 4, 5, 8);
(12, 1, 2, 7, 8, 10);	(0, 3, 4, 5, 6, 9);
(0, 2, 3, 8, 9, 11);	(1, 4, 5, 6, 7, 10).

(5.7.23)

Bruck (1955) generalized the method of differences for the construction of BIB designs. He showed that if a set of k different elements D $(a_1, a_2, \cdots, a_k)$ from a group G of order v satisfies either of the following conditions—

1. For every $\alpha \neq e$, $\alpha \in G$, there are exactly λ ordered pairs (a_i, a_j), $a_i, a_j \in D$, such that $a_i * a_j^{-1} = \alpha$;

2. For every $\alpha \neq e$, $\alpha \in G$, there are exactly λ ordered pairs (a_i, a_j), $a_i, a_j \in D$, such that $a_i^{-1} * a_j = \alpha$—

where e is the group identity and $*$ is the group operation, then the v sets D_α $(a_1 * \alpha, a_2 * \alpha, \cdots, a_k * \alpha)$ as α runs over the v elements of G give the solution of a symmetrical BIB design with the parameters $v = b$, $r = k$, λ.

As an example, let G be the Abelian group of 16 elements generated by a, b, c, d, where $a^2 = b^2 = c^2 = d^2 = e$. Then $D = (a, b, c, d, ab, cd)$ satisfies conditions 1 and 2 of the preceding paragraph with $\lambda = 2$. Hence the solution of the BIB design with the parameters

(5.7.24) $v = 16 = b$, $r = 6 = k$, $\lambda = 2$

is

$(a,$	$b,$	$c,$	$d,$	$ab,$	cd);	$(e,$	$ab,$	$ac,$	$ad,$	$b,$	acd);
$(ab,$	$e,$	$bc,$	$bd,$	$a,$	bcd);	$(b,$	$a,$	$abc,$	$abd,$	$e,$	$abcd$);
$(ac,$	$bc,$	$e,$	$cd,$	$abc,$	d);	$(c,$	$abc,$	$a,$	$acd,$	$bc,$	ad);
$(abc,$	$c,$	$b,$	$bcd,$	$ac,$	bd);	$(bc,$	$ac,$	$ab,$	$abcd,$	$c,$	abd);
$(ad,$	$bd,$	$cd,$	$e,$	$abd,$	c);	$(d,$	$abd,$	$acd,$	$a,$	$bd,$	ac);
$(abd,$	$d,$	$bcd,$	$b,$	$ad,$	bc);	$(bd,$	$ad,$	$abcd,$	$ab,$	$d,$	abc);
$(acd,$	$bcd,$	$d,$	$c,$	$abcd,$	e);	$(cd,$	$abcd,$	$ad,$	$ac,$	$bcd,$	a);
$(abcd,$	$cd,$	$bd,$	$bc,$	$acd,$	b);	$(bcd,$	$acd,$	$abd,$	$abc,$	$cd,$	ab).

We give several difference sets in miscellaneous exercises 9 to 24 at the end of this book. For a detailed account of difference sets the reader is referred to Hall, Jr. (1967). Many references at the end of this chapter contain applications of difference sets to the construction of BIB designs. For a selected reading see Bose (1939, 1942a), Bruck (1955), Elliott and Butson (1966), Mann (1967), Rao (1946), Sprott (1964), and Stanton and Sprott (1958).

5.8. STEINER'S TRIPLE SYSTEMS

The class of BIB designs for $k = 3$ and $\lambda = 1$ is usually known as Steiner's triple systems. Using the necessary conditions (5.1.1) and (5.1.2), it is easy to verify that there are only two series of Steiner's triple systems with the respective parameters

$$(5.8.1) \quad \begin{matrix} v = 6t + 3, & b = (3t + 1)(2t + 1), & r = 3t + 1, & k = 3, \\ \lambda = 1 \end{matrix}$$

and

$$(5.8.2) \quad v = 6t + 1, \quad b = t(6t + 1), \quad r = 3t, \quad k = 3, \quad \lambda = 1.$$

Bose (1939) called these the (T_1) and (T_2) series, respectively.

Steiner (1853) posed the problem whether the two series of BIB designs with the parameters given in (5.8.1) and (5.8.2) exist for every t, and Reiss (1859) affirmatively solved it. Later on Moore (1893) and Hanani (1961) gave recursive methods of constructing Steiner's triple systems for all t. In fact before Steiner this problem was posed and solved by Kirkman (1847). In his paper Kirkman posed another problem of constructing resolvable triple systems. Though resolvable triple systems can be constructed for particular parameter values, the general solution was unknown till recently, when Raychaudhuri and Wilson (1968) found it.

For a detailed account of showing the existence of Steiner's triple systems we refer to Hall, Jr. (1967). We shall content ourselves by giving the methods of constructing (T_1) and subseries of (T_2) by difference sets [cf. Bose (1939)].

Theorem 5.8.1. The series (T_1) exists for all t.

Proof. Consider a module M of $2t + 1$ elements denoted by $0, 1, 2, \cdots, 2t$. To each element there correspond three symbols u_1, u_2, u_3. We can easily verify that the following $3t + 1$ sets satisfy all the conditions of Theorem 5.7.1; hence the solution can be obtained by developing these initial sets in the usual way:

$$(5.8.3) \quad \begin{matrix} [1_1, 2t_1, 0_2]; [2_1, (2t-1)_1, 0_2]; \cdots, [t_1, (t+1)_1, 0_2]; \\ [1_2, 2t_2, 0_3]; [2_2, (2t-1)_2, 0_3]; \cdots, [t_2, (t+1)_2, 0_3]; \\ [1_3, 2t_3, 0_1]; [2_3, (2t-1)_3, 0_1]; \cdots, [t_3, (t+1)_3, 0_1]; \\ [0_1, 0_2, 0_3]. \end{matrix}$$

It may be noted that design 5.7.7 was obtained by using the above theorem.

Theorem 5.8.2. The (T_2) series can be constructed if $6t + 1$ is a prime or a prime power.

Proof. Let $s = 6t + 1$ be a prime or a prime power and x be a primitive element of GF(s). Then

$$(5.8.4) \qquad\qquad x^{6t} \quad = x^0 = 1;$$

$$(5.8.5) \qquad\qquad x^{3t} \quad = -1,$$

$$(5.8.6) \qquad\qquad x^{2t} + 1 = x^t.$$

Then in the t sets

$$(5.8.7) \qquad (x^0, x^{2t}, x^{4t}); (x, x^{2t+1}, x^{4t+1}); \cdots; (x^{t-1}, x^{3t-1}, x^{5t-1})$$

all the nonzero elements of GF(s) are repeated exactly once, and hence, from Theorem 5.7.1, we can observe that the design with the parameters given in (5.8.2) can be constructed in this case by developing the initial sets (5.8.7) in the usual way.

5.9. MISCELLANEOUS METHODS OF CONSTRUCTING BIB DESIGNS

The BIB design series with the parameters given in (5.6.9) and (5.6.3) were called orthogonal series BIB designs by Yates (1936). In particular, the series with the parameters given in (5.6.9) is known as the 0S1 series, and the series with the parameters given in (5.6.3) is known as the 0S2 series. The 0S1 series is the residual of the 0S2 series for a given s. Conversely, if a resolvable solution exists for the 0S1 series, the 0S2 series can be constructed from it.†
The BIB designs with the parameters given in (5.6.9) can be constructed with the help of complete sets of MOLS, as below.

Let the s^2 symbols be arranged in an $s \times s$ square L. Let us form the sets $S_1, S_2, \cdots, S_{s(s+1)}$ constituting the BIB design with the parameters given in (5.6.9) by writing the symbols as follows: The ith set contains the symbols occurring in the ith row of L ($i = 1, 2, \cdots, s$) and the $(s + j)$th set contains the symbols occurring in the jth column of L ($j = 1, 2, \cdots, s$). Let $L_1, L_2, \cdots, L_{s-1}$ be a complete set of MOLS of order s. Superimpose L_α on L, and the symbols corresponding to the kth letter of L_α constitute the set $S_{\{(\alpha+1)s+k\}}$ ($k = 1, 2, \cdots, s; \alpha = 1, 2, \cdots, s - 1$). The design so constructed can be shown, by using the properties of MOLS, to be an affine resolvable solution of the BIB design with the parameters given in (5.6.9).

In the above solution of (5.6.9) we add a new symbol θ_i to each set of the

---

† One can show that the solution of 0S1 is always resolvable.

ith replication $(i = 1, 2, \cdots, s + 1)$ and take a new set $(\theta_1, \theta_2, \cdots, \theta_{s+1})$. This will be seen to be the solution of (5.6.3). Thus we have the following theorem:

Theorem 5.9.1. The orthogonal series of BIB designs can be constructed with the help of a complete set of MOLS.

The following result is due to Shrikhande and Singh (1962):

Theorem 5.9.2. If there exists a BIB design with $\lambda = 1$ and $r = 2k + 1$, then we can construct a symmetrical BIB design with the parameters

$$(5.9.1) \qquad v^* = b^* = 4k^2 - 1, \qquad r^* = k^* = 2k^2, \qquad \lambda^* = k^2.$$

Proof. The parameters of a BIB design with $\lambda = 1$ and $r = 2k + 1$ are

$$(5.9.2) \quad v = k(2k - 1), \qquad b = 4k^2 - 1, \qquad r = 2k + 1, \qquad k = k, \qquad \lambda = 1.$$

Let N be its incidence matrix. Then

$$(5.9.3) \qquad\qquad NN' = 2kI_v + E_{v,v}.$$

Since $\lambda = 1$, we easily see that

$$(5.9.4) \qquad\qquad N'N = kI_b + M,$$

where M is a $(0, 1)$ symmetrical matrix of order $4k^2 - 1$. Now

$$(5.9.5) \qquad ME_{b,1} = 2k^2 E_{b,1}, \qquad E_{1,b} M = 2k^2 E_{1,b}$$

and

$$(5.9.6) \qquad\qquad MM' = k^2 I_b + k^2 E_{b,b}.$$

Hence M is the incidence matrix of a symmetrical BIB design with the parameters given in (5.9.1).

We now give an interesting recursive method of constructing BIB designs; it is due to Shrikhande and Raghavarao (1963). Let D_1 be a BIB design with the parameters $v_1, b_1, r_1, k_1, \lambda_1$ and let D_2 be a resolvable BIB design with the parameters $v_2 = k_2 v_1, b_2 = r_2 v_1, r_2, k_2, \lambda_2$. Let T_{ij} denote the jth set of the ith replication of D_2 $(i = 1, 2, \cdots, r_2; j = 1, 2, \cdots, v_1)$. Let M be the incidence matrix of D_1. Let M_i be the incidence matrix obtained from M by replacing each symbol j of D_1 by the set of symbols occurring in T_{ij} $(j = 1, 2, \cdots, v_1; i = 1, 2, \cdots, r_2)$. Then we can easily verify that

$$(5.9.7) \qquad\qquad N = (M_1 \quad M_2 \cdots M_{r_2})$$

is the incidence matrix of an α-resolvable BIB design D with the parameters

$$(5.9.8) \quad \begin{matrix} v = v_2, & b = b_1 r_2, & r = r_1 r_2, & k = k_1 k_2, & \alpha = r_1, \\ \beta = b_1, & t = r_2, & \lambda = r_1 \lambda_2 + \lambda_1(r_2 - \lambda_2). & & \end{matrix}$$

Hence we have the following theorem:

Theorem 5.9.3. The existence of a BIB design D_1 with the parameters $v_1, b_1, r_1, k_1, \lambda_1$ and of a resolvable BIB design D_2 with the parameters $v_2 = k_2 v_1, b_2 = r_2 v_1, r_2, k_2, \lambda_2$ implies the existence of an α-resolvable BIB design D with the parameters given in (5.9.8).

From the above theorem, noting the set structure of a symmetrical BIB design and an affine α-resolvable BIB design, we can easily establish:

Theorem 5.9.4. The existence of a symmetrical BIB design with the parameters

$$(5.9.9) \qquad v_1 = b_1 = n, \qquad r_1 = k_1, \lambda_1$$

and an affine resolvable BIB design with the parameters

$$(5.9.10) \quad \begin{aligned} v_2 = nk_2 = n^2[(n-1)m+1], \qquad b_2 = nr_2 = n(n^2 m + n + 1), \\ \lambda_2 = nm + 1 \end{aligned}$$

implies the existence of an affine α-resolvable BIB design with the parameters

$$(5.9.11) \quad \begin{aligned} v = v_2, \qquad b = b_2, \qquad r = r_1 r_2, \qquad k = k_1 k_2, \\ \lambda = r_1 \lambda_2 + \lambda_1 (r_2 - \lambda_2), \qquad \alpha = r_1, \qquad \beta = n, \qquad t = r_2. \end{aligned}$$

When $4m + 3$ is a prime or a prime power, there exist a symmetrical BIB design D_1 and an affine resolvable BIB design D_2, with the parameters

$$(5.9.12) \quad \begin{aligned} v_1 = b_1 = 4m + 3, \qquad r_1 = k_1 = 2m + 1, \qquad \lambda_1 = m \qquad \text{for } D_1; \\ v_2 = v_1{}^2, \qquad b_2 = v_1(v_1 + 1), \qquad r_2 = v_1 + 1, \qquad k_2 = v_1, \qquad \lambda_2 = 1 \\ \text{for } D_2. \end{aligned}$$

Hence we have the following corollary:

Corollary 5.9.4.1. The BIB design with the parameters

$$(5.9.13) \quad \begin{aligned} v = v_1{}^2, \qquad b = v_1(v_1 + 1), \qquad r = \frac{(v_1{}^2 - 1)}{2}, \\ k = \frac{v_1(v_1 - 1)}{2}, \qquad \lambda = \frac{(v_1 + 1)(v_1 - 2)}{4} \end{aligned}$$

exists when $v_1 = 4m + 3$ is a prime or a prime power.

When $4m + 1$ is a prime or a prime power, the following designs exist:

$$(5.9.14) \quad \begin{aligned} v_1 = 4m + 1, \qquad b_1 = 2v_1, \qquad r_1 = 4m, \qquad k_1 = 2m, \\ \lambda_1 = 2m - 1 \qquad\qquad\qquad\qquad\qquad \text{for } D_1; \\ v_2 = v_1{}^2, \qquad b_2 = v_1(v_1 + 1), \qquad r_2 = v_1 + 1, \qquad k_2 = v_1, \\ \lambda_2 = 1 \qquad\qquad\qquad\qquad\qquad\qquad \text{for } D_2. \end{aligned}$$

Hence

Corollary 5.9.4.2. The BIB design with the parameters

$$(5.9.15) \quad v = v_1{}^2, \qquad b = 2v_1(v_1 + 1), \qquad r = v_1{}^2 - 1, \qquad k = \frac{v_1(v_1 - 1)}{2},$$

$$\lambda = \frac{(v_1 + 1)(v_1 - 2)}{2}$$

exists when $v_1 = 4m + 1$ is a prime or a prime power.

It can easily be seen that the BIB designs with the parameters given in (5.9.13) and (5.9.15) belong to the family (A) discussed in Section 5.3.

In the end we remark that of the v symbols, all possible combinations of k symbols constituting the sets trivially form a BIB design known as an irreducible design, with the parameters

$$(5.9.16) \quad v = v, \qquad b = \binom{v}{k}, \qquad r = \binom{v-1}{k-1}, \qquad k = k, \qquad \lambda = \binom{v-2}{k-2}.$$

5.10. CONCLUDING REMARKS

In this chapter we have mainly considered direct methods of constructing BIB designs. For recursive methods of constructing BIB designs the reader is referred to Hanani (1960a,b; 1961; 1964) and Yalavigi (1968). While the conditions (5.1.1), (5.1.2), and (5.1.8) on the parameters v, b, r, k, and λ of BIB designs are not necessarily sufficient for the existence of the corresponding designs, Hanani (1960, 1961, 1964) showed their sufficiency for the existence of the designs when $k \leq 4$ or $k = 5$ and $\lambda = 1$, 4, or 20.

Ryser (1952) and Majindar (1966a,b,c) studied the conditions under which an integral matrix will be the incidence matrix of a BIB design.

Two BIB designs D_1 and D_2 with the same parameters are said to be isomorphic if there exists a bijection of the set of symbols of D_1 into that of D_2 such that under this bijection the class of sets of D_1 is mapped into the class of sets of D_2. Otherwise they are said to be nonisomorphic. Nonisomorphic solutions of BIB designs were studied by Atiqullah (1958), Bhat (1970), Bhat and Shrikhande (1970), and Nandi (1946a,b).

Fisher and Yates originally listed all parameter combinations and their solutions for BIB designs in the range v, $b \leq 100$ and r, $k \leq 10$ in their tables. This list was extended by Rao (1961) to include parameter combinations with r, $k \leq 15$ and further extended by Sprott (1962) to include the cases r, $k \leq 20$. We list all parameter combinations of existing designs in the range v, $b \leq 100$, r, $k \leq 15$ and indicate the method of their construction in Table 5.10.1,

Table 5.10.1 *Parameters of BIB designs with* $v, b \leqq 100, r, k \leqq 15$
and their solutions

Series	v	b	r	k	λ	Solution
1	4	6	3	2	1	Irreducible
2	4	4	3	3	2	Irreducible
3	5	10	4	2	1	Irreducible
4	5	5	4	4	3	Irreducible
5	5	10	6	3	3	Irreducible
6	6	15	5	2	1	Irreducible
7	6	10	5	3	2	Residual of Series 29
8	6	6	5	5	4	Irreducible
9	6	15	10	4	6	Irreducible
10	7	7	3	3	1	PG(2, 2): 1-flats as sets
11	7	7	4	4	2	Complement of Series 10
12	7	21	6	2	1	Irreducible
13	7	7	6	6	5	Irreducible
14	8	28	7	2	1	Irreducible
15	8	14	7	4	3	Difference set: $(\infty, x^0, x^2, x^4); (0, x^1, x^3, x^5),$ $x \in GF(7)$
16	8	8	7	7	6	Irreducible
17	9	12	4	3	1	EG(2, 3): 1-flats as sets
18	9	36	8	2	1	Irreducible
19	9	18	8	4	3	Difference set: $(x^0, x^2, x^4, x^6); (x, x^3, x^5, x^7);$ $x \in GF(3^2)$
20	9	12	8	6	5	Complement of Series 17
21	9	9	8	8	7	Irreducible
22	9	18	10	5	5	Complement of Series 19
23	10	15	6	4	2	Residual of Series 47
24	10	45	9	2	1	Irreducible
25	10	30	9	3	2	Difference set: $(0_1, 3_1, 1_2); (1_1, 2_1, 1_2);$ $(1_2, 4_2, 4_1); (0_2, 4_2, 1_2); (0_1, 3_1, 4_2);$ $(1_1, 2_1, 4_2)$ mod 5
26	10	18	9	5	4	Residual of Series 55
27	10	15	9	6	5	Complement of Series 23
28	10	10	9	9	8	Irreducible
29	11	11	5	5	2	Difference set: $(x^0, x^2, x^4, x^6, x^8), x \in GF(11)$
30	11	11	6	6	3	Complement of Series 29
31	11	55	10	2	1	Irreducible
32	11	11	10	10	9	Irreducible
33	11	55	15	3	3	Difference set: $(0, x^0, x^5); (0, x^1, x^6);$ $(0, x^2, x^7); (0, x^3, x^8); (0, x^4, x^9), x \in GF(11)$
34	12	44	11	3	2	Difference set: $(0, 1, 3); (0, 1, 5); (0, 4, 6);$ $(\infty, 0, 3)$ mod 11
35	12	33	11	4	3	Difference set: $(0, 1, 3, 7); (0, 2, 7, 8);$ $(\infty, 0, 1, 3)$, mod 11

Table 5.10.1—(Continued)

Series	v	b	r	k	λ	Solution
36	12	22	11	6	5	Difference set: $(0, 1, 3, 7, 8, 10)$; $(\infty, 0, 5, 6, 8, 10)$ mod 11
37	13	13	4	4	1	PG(2, 3): 1-flats as sets
38	13	26	6	3	1	Difference set: (x^0, x^4, x^8); (x, x^5, x^9), $x \in GF(13)$
39	13	13	9	9	6	Complement of Series 37
40	13	26	12	6	5	Difference set: $(x^0, x^2, x^4, x^6, x^8, x^{10})$; $(x^1, x^3, x^5, x^7, x^9, x^{11})$, $x \in GF(13)$
41	13	39	15	5	5	Difference set: $(0, x^0, x^3, x^6, x^9)$; $(0, x^1, x^4, x^7, x^{10})$; $(0, x^2, x^5, x^8, x^{11})$, $x \in GF(13)$
42	15	35	7	3	1	Difference set: $(1_1, 4_1, 0_2)$; $(2_1, 3_1, 0_2)$; $(1_2, 4_2, 0_3)$; $(2_2, 3_2, 0_3)$; $(1_3, 4_3, 0_1)$; $(2_3, 3_3, 0_1)$; $(0_1, 0_2, 0_3)$ mod 5
43	15	15	7	7	3	Complement of Series 44
44	15	15	8	8	4	Using Theorem 5.9.2 on Series 6
45	15	35	14	6	5	Difference set: $(\infty, 0_1, 0_2, 1_2, 2_2, 4_2)$; $(\infty, 0_1, 3_1, 5_1, 6_1, 0_2)$; $(0_1, 1_1, 3_1, 0_2, 2_2, 6_2)$; $(0_1, 1_1, 3_1, 1_2, 5_2, 6_2)$; $(0_1, 4_1, 5_1, 0_2, 1_2, 3_2)$, mod 7
46	16	20	5	4	1	EG(2, 4): 1-flats as sets
47	16	16	6	6	2	Complement of Series 49
48	16	24	9	6	3	Residual of Series 67
49	16	16	10	10	6	Using Theorem 5.3.1 on Series 2 with itself
50	16	80	15	3	2	Difference set: (x^0, x^5, x^{10}); (x^1, x^6, x^{11}); (x^2, x^7, x^{12}); (x^3, x^8, x^{13}); (x^4, x^9, x^{14}), $x \in GF(4^2)$
51	16	48	15	5	4	Difference set: $(x^0, x^3, x^6, x^9, x^{12})$; $(x^1, x^4, x^7, x^{10}, x^{13})$; $(x^2, x^5, x^8, x^{11}, x^{14})$, $x \in GF(4^2)$
52	16	40	15	6	5	Sets of Series 47 and 48 together
53	16	30	15	8	7	EG(4, 2); 3-flats as sets
54	19	57	9	3	1	Difference set: (x^0, x^6, x^{12}); (x^1, x^7, x^{13}); (x^2, x^8, x^{14}), $x \in GF(19)$
55	19	19	9	9	4	Difference set: $(x^0, x^2, x^4, x^6, x^8, x^{10}, x^{12}, x^{14}, x^{16})$, $x \in GF(19)$
56	19	19	10	10	5	Complement of Series 55
57	19	57	12	4	2	Difference set: $(0, x^0, x^6, x^{12})$; $(0, x^1, x^7, x^{13})$; $(0, x^2, x^8, x^{14})$, $x \in GF(19)$
58	21	21	5	5	1	PG(2, 4): 1-flats as sets
59	21	70	10	3	1	Difference set: $(1_1, 6_1, 0_2)$; $(2_1, 5_1, 0_2)$; $(3_1, 4_1, 0_2)$; $(1_2, 6_2, 0_3)$; $(2_2, 5_2, 0_2)$; $(3_2, 4_2, 0_2)$; $(1_3, 6_3, 0_1)$; $(2_3, 5_3, 0_1)$; $(3_3, 4_3, 0_1)$; $(0_1, 0_2, 0_3)$, mod 7

Table 5.10.1—(Continued)

Series	v	b	r	k	λ	Solution
60	21	30	10	7	3	Residual of Series 76
61	21	42	12	6	3	Difference set: $(0_1, 5_1, 1_2, 4_2, 2_3, 3_3)$; $(0_1, 1_1, 3_1, 0_2, 1_2, 3_2)$; $(0_2, 5_2, 1_3, 4_3, 2_1, 3_1)$; $(0_2, 1_2, 3_2, 0_3, 1_3, 3_3)$; $(0_3, 5_3, 1_1, 4_1, 2_2, 3_2)$; $(0_3, 1_3, 3_3, 0_1, 1_1, 3_1)$, mod 7
62	21	35	15	9	6	Residual of Series 80
63	22	77	14	4	2	Difference set: $(x_1^0, x_1^3, x_2^\alpha, x_2^{\alpha+3})$; $(x_1^1, x_1^4, x_2^{\alpha+1}, x_2^{\alpha+4})$; $(x_1^2, x_1^5, x_2^{\alpha+2}, x_2^{\alpha+5})$; $(x_2^0, x_2^3, x_3^\alpha, x_3^{\alpha+3})$; $(x_2^1, x_2^4, x_3^{\alpha+1}, x_3^{\alpha+4})$; $(x_2^2, x_2^5, x_3^{\alpha+2}, x_3^{\alpha+5})$; $(x_3^0, x_3^3, x_1^\alpha, x_1^{\alpha+3})$; $(x_3^1, x_3^4, x_1^{\alpha+1}, x_1^{\alpha+4})$; $(x_3^2, x_3^5, x_1^{\alpha+2}, x_1^{\alpha+5})$; $(\infty, 0_1, 0_2, 0_3)$; $(\infty, 0_1, 0_2, 0_3)$, $x \in GF(7)$
64	23	23	11	11	5	Difference set: $(x^0, x^2, x^4, x^6, x^8, x^{10}, x^{12}, x^{14}, x^{16}, x^{18}, x^{20})$, $x \in GF(23)$
65	25	30	6	5	1	EG(2, 5): 1-flats as sets
66	25	50	8	4	1	Difference set: $(0, x^0, x^8, x^{16})$; $(0, x^2, x^{10}, x^{18})$, $x \in GF(5^2)$
67	25	25	9	9	3	Trial-and-error solution; refer to Fisher and Yates (1963) statistical tables
68	25	100	12	3	1	Difference set: (x^0, x^8, x^{16}); (x^1, x^9, x^{17}): (x^2, x^{10}, x^{18}); (x^3, x^{11}, x^{19}), $x \in GF(5^2)$
69	26	65	15	6	3	Difference set: $(0_1, 3_1, 1_2, 2_2, 0_3, 3_3)$; $(0_1, 3_1, 1_3, 2_3, 1_4, 2_4)$; $(0_2, 3_2, 1_3, 2_3, 0_4, 3_4)$; $(0_2, 3_2, 1_4, 2_4, 1_5, 2_5)$; $(0_3, 3_3, 1_4, 2_4, 0_5, 3_5)$; $(0_3, 3_3, 1_5, 2_5, 1_1, 2_1)$; $(0_4, 3_4, 1_5, 2_5, 0_1, 3_1)$; $(0_4, 3_4, 1_1, 2_1, 1_2, 2_2)$; $(0_5, 3_5, 1_1, 2_1, 0_2, 3_2)$; $(0_5, 3_5, 1_2, 2_2, 1_3, 2_3)$; $(\infty, 0_1, 3_2, 1_3, 4_4, 2_5)$; $(\infty, 0_1, 2_2, 4_3, 1_4, 3_5)$; $(\infty, 0_1, 0_2, 0_3, 0_4, 0_5)$, mod 5
70	27	39	13	9	4	EG(3, 3): 2-flats as sets
71	27	27	13	13	6	Difference set: $(x^0, x^2, x^4, x^6, x^8, x^{10}, x^{12}, x^{14}$ $x^{16}, x^{18}, x^{20}, x^{22}, x^{24})$, $x \in GF(3^3)$
72	28	63	9	4	1	Difference set: $(x_1^0, x_1^4, x_2^\alpha, x_2^{\alpha+4})$; $(x_1^2, x_1^6, x_2^{\alpha+2}, x_2^{\alpha+6})$; $(x_2^0, x_2^4, x_3^\alpha, x_3^{\alpha+4})$; $(x_2^2, x_2^6, x_3^{\alpha+2}, x_3^{\alpha+6})$; $(x_3^0, x_3^4, x_1^\alpha, x_1^{\alpha+4})$; $(x_3^2, x_3^6, x_1^{\alpha+2}, x_1^{\alpha+6})$; $(\infty, 0_1, 0_2, 0_3)$, $x \in GF(3^2)$
73	28	36	9	7	2	Residual of Series 81
74	29	58	14	7	3	Difference set: $(x^0, x^4, x^8, x^{12}, x^{16}, x^{20}, x^{24})$; $(x^1, x^5, x^9, x^{13}, x^{17}, x^{21}, x^{25})$, $x \in GF(29)$
75	31	31	6	6	1	PG(2, 5): 1-flats as sets

Table 5.10.1—(Continued)

Series	v	b	r	k	λ	Solution
76[a]	31	31	10	10	3	Difference set: $(A, 0_1, 5_1, 1_2, 4_2, 2_3, 3_3, 2_4, 4_4, 5_4)$; $(B, 0_1, 3_1, 1_2, 2_2, 4_3, 6_3, 1_4, 3_4, 4_4)$; $(C, 0_1, 1_1, 3_2, 5_2, 2_3, 6_3, 0_4, 2_4, 3_4)$; $(0_1, 1_1, 3_1, 0_2, 1_2, 3_2, 0_3, 1_3, 3_3, 6_4)$, mod 7 $(A, B, C, 0_1, 1_1, 2_1, 3_1, 4_1, 5_1, 6_1)$; $(A, B, C, 0_2, 1_2, 2_2, 3_2, 4_2, 5_2, 6_2)$; $(A, B, C, 0_3, 1_3, 2_3, 3_3, 4_3, 5_3, 6_3)$
77	31	93	15	5	2	Difference set: $(x^0, x^6, x^{12}, x^{18}, x^{24})$; $(x^1, x^7, x^{13}, x^{19}, x^{25})$; $(x^2, x^8, x^{14}, x^{20}, x^{26})$, $x \in GF(31)$
78	31	31	15	15	7	PG(4, 2): 3-flats as sets
79	33	44	12	9	3	Residual of Series 85
80	36	36	15	15	6	Shrikhande and Singh (1962), corollary 2 on page 27
81	37	37	9	9	2	Difference set: $(x^0, x^4, x^8, x^{12}, x^{16}, x^{20}, x^{24}, x^{28}, x^{32})$, $x \in GF(37)$
82	40	40	13	13	4	PG(3, 3): 3-flats as sets
83	41	82	10	5	1	Difference set: $(x^0, x^8, x^{16}, x^{24}, x^{32})$; $(x^2, x^{10}, x^{18}, x^{26}, x^{34})$, $x \in GF(41)$
84	45	99	11	5	1	Difference set: $(x_1^0, x_1^4, x_3^\alpha, x_3^{\alpha+4}, 0_2)$; $(x_1^2, x_1^6, x_3^{\alpha+2}, x_3^{\alpha+6}, 0_2)$; $(x_2^0, x_2^4, x_4^\alpha, x_4^{\alpha+4}, 0_3)$; $(x_2^2, x_2^6, x_4^{\alpha+2}, x_4^{\alpha+6}, 0_3)$; $(x_3^0, x_3^4, x_5^\alpha, x_5^{\alpha+4}, 0_4)$; $(x_3^2, x_3^6, x_5^{\alpha+2}, x_5^{\alpha+6}, 0_4)$; $(x_4^0, x_4^4, x_1^\alpha, x_1^{\alpha+4}, 0_5)$; $(x_4^2, x_4^6, x_1^{\alpha+2}, x_1^{\alpha+6}, 0_5)$; $(x_5^0, x_5^4, x_2^\alpha, x_2^{\alpha+4}, 0_1)$; $(x_5^2, x_5^6, x_2^{\alpha+2}, x_2^{\alpha+6}, 0_1)$; $(0_1, 0_2, 0_3, 0_4, 0_5)$, $x \in GF(3^2)$
85	45	45	12	12	3	Shrikhande and Singh (1962), example 4 on page 28
86	49	56	8	7	1	EG(2, 7): 1-flats as sets
87	57	57	8	8	1	PG(2, 7): 1-flats as sets
88	64	72	9	8	1	EG(2, 8): 1-flats as sets
89	73	73	9	9	1	PG(2, 8): 1-flats as sets
90	81	90	10	9	1	EG(2, 9): 1-flats as sets
91	91	91	10	10	1	PG(2, 9): 1-flats as sets

$x \in GF$ (s) should be interpreted as x is a primitive root of GF (s) and $\alpha \neq 0$.

[a] The first four sets should be developed mod 7, while keeping A, B, C unaltered in the development. Thus we get 28 sets, and the last three sets complete the whole design.

which lists the parameters in increasing order of v, r, and k, respectively. In this range the solutions of BIB designs with the parameters

$$v = 46, \qquad b = 69, \qquad r = 9, \qquad k = 6, \qquad \lambda = 1$$

and

$$v = 51, \qquad b = 85, \qquad r = 10, \qquad k = 6, \qquad \lambda = 1$$

are still open problems.

REFERENCES

1. Atiqullah, M. (1958). On configurations and nonisomorphism of some incomplete block designs. *Sanhhyā*, **20**, 227–248.
2. Bhagwandas (1965). A note on balanced incomplete block design. *J. Indian Stat. Assn.*, **3**, 41–45.
3. Bhat, V. N. (1970). Nonisomorphic solutions of some balanced incomplete block designs II. *J. Comb. Theory*, **5**.
4. Bhat, V. N., and Shrikhande, S. S. (1970). Nonisomorphic solutions of some balanced incomplete block designs I. *J. Comb. Theory*, **5**, 174–191.
5. Bhattacharya, K. N. (1945). On a new symmetrical balanced incomplete block design. *Bull. Calcutta Math. Soc.*, **36**, 91–96.
6. Bose, R. C. (1939). On the construction of balanced incomplete block dseigns. *Ann. Eugen.*, **9**, 353–399.
7. Bose, R. C. (1942a). On some new series of balanced incomplete block designs. *Bull. Calcutta Math. Soc.*, **34**, 17–31.
8. Bose, R. C. (1942b). A note on the resolvability of incomplete block designs. *Sankhyā*, **6**, 105–110.
9. Bose, R. C. (1942c). A note on two series of balanced incomplete block designs. *Bull. Calcutta Math. Soc.*, **34**, 129–130.
10. Bose, R. C. (1947). On a resolvable series of balanced incomplete block designs. *Sankhyā*, **8**, 249–256.
11. Bose, R. C. (1949). A note on Fisher's inequality for balanced incomplete block designs. *Ann. Math. Stat.*, **20**, 619–620.
12. Bose, R. C. (1959). On the application of finite projective geometry for deriving a certain series of balanced Kirkman arrangements. *Calcutta Math. Soc. Golden Jubilee Vol.*, 341–354.
13. Bose, R. C., and Chakravarti, I. M. (1966). Hermitian varieties in a finite projective space PG(N, q^2). *Can. J. Math.*, **18**, 1161–1182.
14. Bruck, R. H. (1955). Difference sets in a finite group. *Trans. Am. Math. Soc.*, **78**, 464–481.
15. Chakrabarti, M. C. (1963). Answer to the query regarding the bounds for common number of treatments between blocks of balanced incomplete block designs. *J. Indian Stat. Soc.*, **1**, 230–234.
16. Connor, W. S., Jr. (1952). On the structure of balanced incomplete block designs. *Ann. Math. Stat.*, **23**, 57–71.

17. Elliott, J. E. H., and Butson, A. T. (1966). Relative difference sets. *Illinois J. Math.*, **10**, 517–531.
18. Fisher, R. A. (1940). An examination of the different possible solutions of a problem in incomplete blocks. *Ann. Eugen.*, **10**, 52–75.
19. Fisher, R. A. (1942). New cyclic solutions of problems in incomplete blocks. *Ann. Eugen.*, **11**, 290–299.
20. Fisher, R. A., and Yates, F. (1963). *Statistical Tables for Biological, Agricultural and Medical Research*. Oliver and Boyd, London.
21. Hall, M., Jr. (1964). Block designs. *Applied Combinatorial Mathematics*. J. Wiley, New York, pp. 369–405.
22. Hall, M., Jr. (1967). *Combinatorial Theory*. Blaisdell Publishing Company.
23. Hall, M., Jr., and Connor, W. S. (1953). An embedding theorem for balanced incomplete block designs. *Can. J. Math.*, **6**, 35–41.
24. Hanani, H. (1960a). On quadruple systems. *Can. J. Math.*, **12**, 145–157.
25. Hanani, H. (1960b). A note on Steiner triple systems. *Math. Scand.*, **8**, 154–156.
26. Hanani, H. (1961). The existence and construction of balanced incomplete block designs. *Ann. Math. Stat.*, **32**, 361–386.
27. Hanani, H. (1964). On covering of balanced incomplete block designs. *Can. J. Math.*, **16**, 615–625.
28. Kirkman, T. P. (1847). On a problem in combinations. *Cambridge and Dublin Math. J.*, **2**, 191–204.
29. Kishen, K., and Rao, C. R. (1952). An examination of various inequality relations among parameters of the balanced incomplete block designs. *J. Indian Soc. Agric. Stat.*, **2**, 137–144.
30. Majindar, K. N. (1962). On the parameters and intersection of blocks of balanced incomplete block designs. *Ann. Math. Stat.*, **33**, 1200–1205.
31. Majindar, K. N. (1966a). On integer matrices and incidence matrices of certain combinatorial configurations. I. Square matrices. *Can. J. Math.*, **18**, 1–5.
32. Majindar, K. N. (1966b). On integer matrices and incidence matrices of certain combinatorial configurations. II. Rectangular matrices. *Can. J. Math.*, **18**, 6–8.
33. Majindar, K. N. (1966c). On integer matrices and incidence matrices of certain combinatorial configurations. III. Rectangular matrices. *Can. J. Math.*, **18**, 9–17.
34. Majumdar, K. N. (1953). On some theorems in combinatorics relating to incomplete block designs. *Ann. Math. Stat.*, **24**, 377–389.
35. Mann, H. B. (1967). Recent advances in difference sets. *Am. Math. Monthly*, **74**, 229–235.
36. Menon, P. K. (1962). On difference sets whose parameters satisfy a certain relation. *Proc. Am. Math. Soc.*, **13**, 739–745.
37. Moore, E. H. (1893). Concerning triple systems. *Math. Ann.*, **43**, 271–285.
38. Nandi, H. K. (1946a). Enumeration of nonisomorphic solutions of balanced incomplete block designs. *Sankhyā*, **7**, 305–312.
39. Nandi, H. K. (1946b). A further note on nonisomorphic solutions of incomplete block designs. *Sankhyā*, **7**, 313–316.
40. Parker, E. T. (1963). Remarks on balanced incomplete block designs. *Proc. Am. Math. Soc.*, **14**, 729–730.

41. Raghavarao, D. (1963). A note on the block structure of BIB designs. *Calcutta Stat. Assn. Bull.*, **12**, 60–62.
42. Rao, C. R. (1946). Difference sets and combinatorial arrangements derivable from finite geometries. *Proc. Natl. Inst. Sci.*, **12**, 123–135.
43. Rao, C. R. (1961). A study of BIB designs with replications 11 to 15. *Sankhyā*, **A23**, 117–127.
44. Raychaudhuri, D. K., and Wilson, R. M. (1968). Solution of Kirkman's schoolgirl problem. Proc. Am. Math. Soc. Symposium on Combinatorics.
45. Reiss, M. (1859). Über eine Steinersche kombinatorische Aufgabe, welche im 45sten Bände dieses Journals, Seite 181, gestellt worden ist. *J. Reine Angew. Math.*, **56**, 326–344.
46. Ryser, H. J. (1952). Matrices with integer elements in combinatorial investigations. *Am. J. Math.*, **74**, 769–773.
47. Seiden, E. (1963). A supplement to Parker's remarks on balanced incomplete block designs. *Proc. Am. Math. Soc.*, **14**, 731–732.
48. Shah, S. M. (1963). On the upper bound for the number of blocks in balanced incomplete block designs having a given number of treatments common with a given block. *J. Indian Stat. Assn.*, **1**, 219–220.
49. Shrikhande, S. S. (1960). Relations between certain incomplete block designs. *Contributions to Probability and Statistics.* Stanford University Press, Stanford, pp. 388–395.
50. Shrikhande, S. S. (1962). On a two parameter family of balanced incomplete block designs. *Sankhyā*, **A24**, 33–40.
51. Shrikhande, S. S., and Raghavarao, D. (1963). A method of construction of incomplete block designs. *Sankhyā*, **A25**, 399–402.
52. Shrikhande, S. S., and Singh, N. K. (1962). On a method of constructing symmetrical balanced incomplete block designs. *Sankhyā*, **A24**, 25–32.
53. Shrikhande, S. S., and Singh, N. K. (1963). A note on balanced incomplete block designs. *J. Indian Stat. Assn.*, **1**, 97–101.
54. Sprott, D. A. (1962). Listing of BIB designs from $r = 16$ to 20. *Sankhyā* **A24**, 203–204.
55. Sprott, D. A. (1964). Generalizations arising from a family of difference sets. *J. Indian Stat. Assn.*, **2**, 197–209.
56. Stanton, R. G., and Sprott, D. A. (1958). A family of difference sets. *Can. J. Math.*, **10**, 73–77.
57. Steiner, J. (1853). Kombinatorische Aufgabe. *J. Reine Angew. Math.*, **45**, 181–182.
58. Trehan, A. M. (1963). On the bounds of the common treatments between blocks of balanced incomplete block designs. *J. Indian Stat. Assn.*, **1**, 102–103.
59. Yalavigi, C. C. (1968). A series of balanced incomplete block designs. *Ann. Math. Stat.*, **39**, 681–683.
60. Yamamoto, S., Fukuda, T., and Hamaka, N. (1966). On finite geometries and cyclically generated incomplete block designs. *J. Sci. Hiroshima Univ.*, Ser. A-I Math., **30**, 137–149.
61. Yates, F. (1936). Incomplete randomized blocks. *Ann. Eugen.*, **7**, 121–140.

BIBLIOGRAPHY

Alltop, W. O. On the construction of block designs. *J. Comb. Theory*, **1**, 501–502 (1966).

Atiqullah, M. Some new solutions of symmetrical balanced incomplete block designs with $k = 9$ and $\lambda = 2$. *Bull. Calcutta Math. Soc.*, **50**, 23–28 (1958).

Assmus, E. F., and Mattson, H. F. Disjoint Steiner systems associated with the Mathieu groups. *Bull. Am. Math. Soc.*, **72**, 843–845 (1966).

Bays, S., and Belhote, G. Sur les systemes cyliques des triples de Steiner differents pour N premier de la forme $6n + 1$. *Comm. Math. Helv.*, **6**, 28–46 (1933).

Bays, S., and Deweck, E. Sur les systemes des quadrupules. *Comm. Math. Helv.*, **7**, 222–241 (1935).

Bhat, V. N. Some cyclic solutions of balanced incomplete block designs. *Sankhyā*, **A31**, 355–360.

Bhattacharya, K. N. A new solution in symmetrical balanced incomplete block designs ($v = b = 31$, $r = k = 10$, $\lambda = 3$). *Sankhyā*, **7**, 423–424 (1946).

Chowla, S. On difference sets. *Proc. Natl. Acad. Sci. U.S.*, **35**, 92–94 (1949).

Chowla, S., and Jones, B. W. A note on perfect difference sets. *Norske Vid Selsk. Forh. Trondheim*, **32**, 81–83 (1959).

Clatworthy, W. H. The subclass of balanced incomplete block designs with $r = 11$ replications. *Rev. Intern. Stat. Inst.*, **36**, 7–11 (1968).

Cole, F. N. Kirkman parades. *Bull. Am. Math. Soc.*, **28**, 435–437 (1922).

Cole, F. N., White, A. S., and Cummings, L. D., Jr. Complete classification of triad systems on fifteen elements. *Mem. Natl. Acad. Sci.*, **14**, Second Memoir 89 (1925).

Cochran, W. G., and Cox, G. M. *Experimental Designs*. J. Wiley, New York, 1957.

Cox, G. Enumeration and construction of balanced incomplete block designs. *Ann. Math. Stat.*, **11**, 72–85 (1940).

Dembowski, P. *Finite Geometries*. Springer Verlag, 1968.

Eckenstein, O. Bibliography of Kirkman's schoolgirl problem. *Mess. Math.*, **41**, 33–36 (1911).

Emch, A. Triple and multiple systems, their geometrical configurations and groups. *Trans. Am. Math. Soc.*, **31**, 25–42 (1929).

Evans, T. A., and Mann, H. B. On simple difference sets. *Sankhyā*, **11**, 357–364 (1951).

Gassner, B. J. Equal difference BIB designs. *Proc. Am. Math. Soc.*, **16**, 378–380 (1965).

Gordon, B., Mills, W. H., and Welch, L. R. Some new difference sets. *Can. J. Math.*, **14**, 614–625 (1962).

Guerin, R. Vue d'ensemble sur les plans en blocs incomplets equilibres et partiellement equilibres. *Rev. Intern. Stat. Inst.*, **33**, 24–58 (1965).

Halberstam, H., and Laxton, R. R. On perfect difference sets. *Quart. J. Math.*, **14**, 86–90 (1963).

Halberstam, H., and Laxton, R. R. Perfect difference sets. *Proc. Glasgow Math. Assn.*, **6**, 177–184 (1964).

Hall, M., Jr. A survey of difference sets. *Proc. Am. Math. Soc.*, **7**, 975–986 (1956).

Hall, M., Jr. Group theory and block designs. *Proc. Intern. Conf. Theory of Groups.* Gordon and Breach, New York, 1967.

Hall, M., and Swift, J. D. Determination of Steiner triple systems of order 15. *Math Tables Aids Comput.*, **9**, 146–155 (1955).

Hanani, H., and Schonheim, J. On Steiner systems. *Israel J. Math.*, **2**, 139–142 (1964).

Hughes, D. R. A note on difference sets. *Proc. Am. Math. Soc.*, **6**, 689–692 (1955).

Hughes, D. R. Partial difference sets. *Am. J. Math.*, **78**, 650–674 (1956).

Hussain, Q. M. On the totality of solutions for the symmetrical incomplete block designs: $\lambda = 2$, $k = 5$ or 6. *Sankhyā*, **7**, 204–208 (1945).

Hussain, Q. M. Symmetrical incomplete block designs with $\lambda = 2$, $k = 8$ or 9. *Bull. Calcutta Math. Soc.*, **37**, 115–123 (1945).

John, P. W. M. On obtaining balanced incomplete block designs from partially balanced association schemes. *Ann. Math. Stat.*, **38**, 618–619 (1967).

Kerawala, S. M. Note on symmetrical incomplete block designs: $\lambda = 2$, $k = 6$ or 7 *Bull. Calcutta Math. Soc.*, **38**, 190–192 (1946).

Lehmer, E. On residue difference sets. *Can. J. Math.*, **5**, 425–432 (1953).

Majindar, K. N. On some methods for construction of BIB designs. *Can. J. Math.*, **20**, 929–938 (1968).

Mann, H. B. Some theorems on difference sets. *Can. J. Math.*, **4**, 222–226 (1952).

Mann, H. B. Balanced incomplete block designs and abelian difference sets. *Illinois J. Math.*, **8**, 252–261 (1964).

Mann, H. B. Difference sets in elementary Abelian groups. *Illinois J. Math.*, **9**, 212–219 (1965).

Mann, H. B. A note on balanced incomplete block designs. *Ann. Math. Stat.*, **40**, 679–680 (1969).

Menon, P. K. Difference sets in Abelian groups. *Proc. Am. Math. Soc.*, **11**, 368–376 (1960).

Menon, P. K. On certain sums connected with Galois fields and their applications to difference sets. *Math. Ann.*, **154**, 341–364 (1964).

Muller, E. R. A method of constructing balanced incomplete block designs. *Biometrika*, **52**, 285–288 (1965).

Mullin, R. C., and Stanton, R. G. Classification and embedding of BIBD's. *Sankhyā*, **A30**, 91–100 (1968).

Mullin, R. C., and Stanton, R. G. Ring generated difference blocks. *Sankhyā*, **A30**, 101–106 (1968).

Netto, E. Zur Theorie der tripel Systeme. *Math. Ann.*, **42**, 143–152 (1893).

Newman, M. Multipliers of difference sets. *Can. J. Math.*, **15**, 121–124 (1963).

Preece, D. A. Incomplete block designs with $v = 2k$. *Sankhyā*, **A29**, 305–316 (1967).

Primrose, E. J. F. Resolvable balanced incomplete block designs. *Sankhyā*, **12**, 137–140 (1952).

Ramanujacharyulu, C. A new general series of balanced incomplete block designs. *Proc. Am. Math. Soc.*, **17**, 1064–1068 (1966).

Rankin, R. A. Difference sets. *Acta. Arithm.*, **9**, 161–168 (1964).

Roy, P. M. A note on the unreduced balanced incomplete block designs. *Sankhyā*, **13**, 11–16 (1953).

Ryser, H. J. *Combinatorial Mathematics.* Wiley, New York, 1963.

Savur, S. R. A note on the arrangement of incomplete blocks when $k = 3$ and $\lambda = 1$. *Ann. Eugen.*, **9**, 45–49 (1939).

Sprott, D. A. A note on balanced incomplete block designs. *Can. J. Math.*, **6**, 341–346 (1954).

Sprott, D. A. Balanced incomplete block designs and tactical configurations. *Ann. Math. Stat.*, **26**, 752–758 (1955).

Sprott, D. A. Some series of balanced incomplete block designs. *Sankhyā*, **17**, 185–192 (1956).

Stanton, R. G., and Sprott, D. A. Block intersections in balanced incomplete block designs. *Can. Math. Bull.*, **7**, 539–548 (1964).

Szamkolowicz, L. Sur une classification des triplets de Steiner. *Atti Accad. Naz. Lincei Rendic*, **36**, 125–128 (1964).

Szekeres, G. A new class of symmetric block designs. *J. Comb. Theory*, **6**, 219–221 (1969).

Takeuchi, K. A table of difference sets generating balanced incomplete block designs. *Rev. Intern. Inst. Stat.*, **30**, 361–366 (1962).

Takeuchi, K. On the construction of a series of BIB designs. *Rep. Stat. Appl. Res. Un. Japan Sci. Engrs.*, **10**, 226 (1963).

Tits, J. Sur les systemes de Steiner associes aux trois 'grands' groupes de Mathieu. *Rendic. Mat.*, **23**, 166–184. (1964).

Turyn, R. J. The multiplier theorem for difference sets. *Can. J. Math.*, **16**, 386–388 (1964).

Turyn, R. J. Character sums and difference sets. *Pacific J. Math.*, **15**, 319–346 (1965).

Whiteman, A. L. A family of difference sets. *Illinois J. Math.*, **6**, 107–121 (1962).

Zaidi, N. H. Symmetrical balanced incomplete block designs with $\lambda = 2$ and $k = 9$. *Bull. Calcutta Math. Soc.*, **55**, 163–167 (1963).

CHAPTER 6

Systems of Distinct Representatives and Youden Squares

6.1. INTRODUCTION

A certain tennis tournament is attended by n ladies and n gentlemen. Each man knows exactly k ladies and each woman knows exactly k gentlemen. No one wants to make any further introduction to make partners for a mixed-doubles tournament. Under these assumptions, will it be always possible for the men and women to be paired with each other to form the n teams for the tournament?

Let us number the ladies and gentlemen from 1 to n and let T_i be the set of the k ladies the ith gentleman knows ($i = 1, 2, \cdots, n$). Then if we can select the ladies in the order $(a_1, a_2, \cdots, a_n)$, which is a permutation of $(1, 2, \cdots, n)$ such that $a_i \in T_i$, we can form the partnership of the ith man with the lady designated a_i and the problem will be solved.

The problem considered above is of great combinatorial interest. This and related problems can be very neatly solved with the help of systems of distinct representatives.

6.2. SYSTEMS OF DISTINCT REPRESENTATIVES

Let a set S have v elements $1, 2, \cdots, v$ and let the nonvoid sets $T_1, T_2, \cdots, T_n$ be subsets of S that are not necessarily disjoint. We define

Definition 6.2.1. A set $(a_1, a_2, \cdots, a_n)$, where $a_i \in T_i$ for every i and $a_i \neq a_j$ for every $i \neq j$, is defined to be a system of distinct representatives (SDR) for the sets $T_1, T_2, \cdots, T_n$.

One will be interested to know the necessary and sufficient conditions for the existence of SDR for the sets $T_1, T_2, \cdots, T_n$. We state the result of P. Hall (1935) in the following theorem:

101

Theorem 6.2.1. A necessary and sufficient condition for the existence of an SDR for the sets $T_1, T_2, \cdots, T_n$ is that for every integral $k, i_1, i_2, \cdots, i_k$, satisfying $1 \leq k \leq n$ and $1 \leq i_1 < i_2 \cdots < i_k \leq n$, the condition

$$(6.2.1) \qquad\qquad |T_{i_1} \cup T_{i_2} \cup \cdots \cup T_{i_k}| \geq k$$

hold, where $|T|$ denotes the cardinality of the set T.

Proof. The necessity part of the theorem is obvious, and to prove the sufficiency part we proceed by induction on n, distinguishing two cases: (a) whenever $1 \leq k \leq n$ and $1 \leq i_1 < i_2 < \cdots < i_k \leq n$, then

$$|T_{i_1} \cup T_{i_2} \cup \cdots \cup T_{i_k}| \geq k + 1;$$

(b) for some $1 \leq k \leq n$ there are sets $T_{i_1}, T_{i_2}, \cdots, T_{i_k}$ such that

$$|T_{i_1} \cup T_{i_2} \cup \cdots \cup T_{i_k}| = k;$$

this completes all the possible cases for establishing the theorem.

The result is true for $n = 1$, and hence let us assume that it is true for every $m < n$.

Case a. Choose any element a_1 from T_1 and form the sets

$$(6.2.2) \qquad\qquad T_i^* = T_i - \{a_1\} \qquad i = 2, 3, \cdots, n.$$

Because of the assumptions of case a, whenever $1 \leq k \leq n - 1$ and $2 \leq v_1 < v_2 < v_3 \cdots < v_k \leq n$, then $|T_{v_1}^* \cup T_{v_2}^* \cup T_{v_3}^* \cup \cdots \cup T_{v_k}^*| \geq k$, and, from the induction hypothesis, there exists an SDR $(b_2, b_3, \cdots, b_n)$ for the sets $T_2^*, T_3^*, \cdots, T_n^*$. Then clearly $(a_1, b_2, b_3, \cdots, b_n)$ is an SDR for $T_1, T_2, T_3, \cdots, T_n$, thus completing the proof in this case.

Case b. Without loss of generality, let us assume that the sets $T_{i_1}, T_{i_2}, \cdots, T_{i_k}$ satisfying $|T_{i_1} \cup T_{i_2} \cup \cdots \cup T_{i_k}| = k$ are the first k sets. Then, because of the sufficiency condition and the induction hypothesis, there exists an SDR $(c_1, c_2, \cdots, c_k)$ for the sets $T_1, T_2, \cdots, T_k$. Now form the sets

$$(6.2.3) \qquad T_i' = T_i - \{c_1, c_2, \cdots, c_k\} \qquad i = k + 1, k + 2, \cdots, n.$$

Whenever $1 \leq h \leq n - k$ and $k < j_1 < j_2 \cdots < j_h \leq n$, then

$$|T_{j_1}' \cup T_{j_2}' \cup \cdots \cup T_{j_h}'| \geq h,$$

for otherwise

$$|T_1 \cup T_2 \cup \cdots \cup T_k \cup T_{j_1} \cup T_{j_2} \cup \cdots \cup T_{j_h}| < k + h,$$

contradicting the sufficiency condition. Thus again by the induction hypothesis an SDR $(d_{k+1}, d_{k+2}, \cdots, d_n)$ exists for the sets $T_{k+1}', T_{k+2}', \cdots, T_n'$. We can now easily verify that $(c_1, c_2, \cdots, c_k, d_{k+1}, d_{k+2}, \cdots, d_n)$ is an SDR for the sets $T_1, T_2, \cdots, T_n$, completing the proof for this case and hence completing the proof for the whole theorem.

The above theorem can be generalized to infinitely many sets T_i, although every set T_i is finite. In this connection we state the following result and refer to M. Hall, Jr. (1967) for its proof:

Theorem 6.2.2. A necessary and sufficient condition for the existence of an SDR for the sequence of finite sets $\{T_i\}$ is that for every integral k and distinct integral values $i_1, i_2, \cdots, i_k$, (6.2.1) hold.

M. Hall (1948) proved that, if $R_n(T_1, T_2, \cdots, T_n)$ denotes the set of SDRs for the sets $T_1, T_2, \cdots, T_n$, then the following obtains:

Theorem 6.2.3. If $|T_i| \geq s$ for every i, then

$$(6.2.4) \qquad \begin{aligned} |R_n(T_1, T_2, \cdots, T_n)| &\geq s! & &\text{if } s \leq n \\ &\geq s(s-1) \cdots (s-n+1) & &\text{if } s \geq n. \end{aligned}$$

Proof. The proof is again by induction on n. For $n = 1$ the result is trivial. Suppose that $n > 1$ and let the result be true for all $m < n$, distinguishing the two cases as in Theorem 6.2.1.

In case a the number a_1 can be selected in at least s ways, and the SDR $(b_2, b_3, \cdots, b_n)$ for the sets $T_2^*, T_3^*, \cdots, T_n^*$ can be selected in at least $(s-1)!$ or $(s-1)(s-2) \cdots (s-n+1)$ ways, depending on whether $s \leq n$ or $s \geq n$, respectively, from the induction hypothesis. Thus the SDR $(a_1, b_2, b_3, \cdots, b_n)$ for the sets $T_1, T_2, \cdots, T_n$ can be selected in at least $s!$ ways if $s \leq n$ or in at least $s(s-1) \cdots (s-n+1)$ ways if $s \geq n$.

In case b, since every set T_i has at least s elements, the number k, where the sets $T_1, T_2, \cdots, T_k$ have exactly k elements among them, satisfies $k \geq s$. Thus by the induction hypothesis the SDR $(c_1, c_2, \cdots, c_k)$ can be formed in at least $s!$ ways and the SDR $(c_1, c_2, \cdots, c_k, d_{k+1}, d_{k+2}, \cdots, d_n)$ can then be selected for the sets $T_1, T_2, \cdots, T_n$ in at least $s!$ ways; since $n \geq k \geq s$, the theorem is true for n in case b.

The proof is thus complete.

If, without any loss of generality, we add to Theorem 6.2.3 the hypothesis $|T_1| \leq |T_2| \leq \cdots \leq |T_n|$, then the maximum value of s is $|T_1|$, and the assertion of Theorem 6.2.3 is equivalent to

$$(6.2.5) \qquad |R_n(T_1, T_2, \cdots, T_n)| \geq \prod_i (|T_1| - i + 1) \qquad i \leq \min(n, |T_1|).$$

We can, in fact, prove a result that is stronger then Theorem 6.2.3; due to Rado (1967), it is given by the following theorem:

Theorem 6.2.4. If $T_1, T_2, \cdots, T_n$ are n subsets of a set S, such that

$|T_1| \leq |T_2| \leq \cdots \leq |T_n|$, and $R_n(T_1, T_2, \cdots, T_n)$ is the set of SDRs for $T_1, T_2, \cdots, T_n$, then

$$(6.2.6) \qquad |R_n(T_1, T_2, \cdots, T_n)| \geq \prod_i (|T_i| - i + 1) \qquad i \leq \min(n, |T_1|).$$

However, we omit the proof of this theorem since it follows the same lines as those used in Theorem 6.2.3.

Let R be the intersection set of the SDRs, treated as subsets of S. Then we have the following:

Theorem 6.2.5. If R contains exactly the elements $a_1, a_2, \cdots, a_l$ $(a_i \in T_i, i = 1, 2, \cdots, l)$, then $|T_1 \cup T_2 \cup \cdots \cup T_l| = l$.

Proof. Let $|T_1 \cup T_2 \cup \cdots \cup T_l| = p$. Then from Theorem 6.2.1 we have, since SDRs exist for the sets $T_1, T_2, \cdots, T_n$,

$$(6.2.7) \qquad p \geq l.$$

We now show that $p > l$ leads to a contradiction. Let $(a_1, a_2, \cdots, a_l, b_{l+1}, \cdots, b_n)$ be an SDR for the sets $T_1, T_2, \cdots, T_l, T_{l+1}, \cdots, T_n$. Let the element a_{l+1} belong, without loss of generality, to the set T_1. Then $(a_{l+1}, a_2, \cdots, a_l, b_{l+1}, \cdots, b_n)$ is also an SDR for the sets $T_1, T_2, \cdots, T_n$, and the intersection set of the SDRs will not contain the element a_1, which is a contradiction. Thus $p \not> l$, and we have the required result $p = l$.

We now make the following definition:

Definition 6.2.2. A partition of size n of a set S is a collection of n nonvoid disjoint sets $A_1, A_2, \cdots, A_n$ such that $S = A_1 \cup A_2 \cup \cdots \cup A_n$.

Theorem 6.2.6. Let $A_1, A_2, \cdots, A_n$ be a partition of S of size n and let $B_1, B_2, \cdots, B_m$ $(m \leq n)$ be any nonempty subsets of S. Furthermore, let for each integral k, $1 \leq k \leq m$ and k of the partitions A contain at most k of the subsets B; then on suitably renumbering m of the partitions A, $A_i \cap B_i \neq \phi$ for $i = 1, 2, \cdots, m$.

Proof. Let T_i be the set of all the partitions A that have nonvoid intersections with B_i $(i = 1, 2, \cdots, m)$. Then from the hypothesis of the theorem any k sets of $T_1, T_2, \cdots, T_m$ for $1 \leq k \leq m$ contain among themselves at least k partitions A and hence an SDR exists for $T_1, T_2, \cdots, T_m$. Let the SDR be designated $A_1, A_2, \cdots, A_m$. Then we have $A_i \cap B_i \neq \phi$ for $i = 1, 2, \cdots, m$, and the result is established.

As a corollary of the above theorem we have the result of Konig (1950):

Corollary 6.2.6.1. If $A_1, A_2, \cdots, A_n$ and $B_1, B_2, \cdots, B_n$ are two different partitions of size n of S and each of the A and B sets has the same finite cardinal number, then the A and B sets can be so numbered that $A_i \cap B_i \neq \phi$.

6.3. GENERALIZATION OF SDR

We now consider the generalization of SDRs due to Agrawal (1966b).

Definition 6.3.1. If $T_1, T_2, \cdots, T_n$ are n nonvoid subsets, not necessarily disjoint, of a given finite set S, then $(0_1, 0_2, \cdots, 0_n)$ will be called an $(m_1, m_2, \cdots, m_n)$-SDR if

1. $0_i \subseteq T_i$ $i = 1, 2, \cdots, n.$
2. $|0_i| = m_i$ $i = 1, 2, \cdots, n.$
3. $0_i \cap 0_j = \phi$ $i, j = 1, 2, \cdots, n; i \neq j.$

If $m_1 = m_2 = \cdots = m_n$, the sets will be said to possess an m-ple SDR.

Analogously to Theorem 6.2.1 we can prove the following:

Theorem 6.3.1. A necessary and sufficient condition for the existence of an $(m_1, m_2, \cdots, m_n)$-SDR for the sets $T_1, T_2, \cdots, T_n$ is that for every integral $k, i_1, i_2, \cdots, i_k$ satisfying $1 \leq k \leq n$ and $1 \leq i_1 < i_2 < \cdots < i_k \leq n$ the condition

$$(6.3.1) \qquad |T_{i_1} \cup T_{i_2} \cup \cdots \cup T_{i_k}| \geq \sum_{j=1}^{k} m_{i_j}$$

hold.

6.4. APPLICATIONS OF SDR IN THE CONSTRUCTION OF YOUDEN SQUARES

Definition 6.4.1. A Youden-square arrangement [Youden (1937)] is an arrangement of v symbols in k rows and v columns such that (a) every symbol occurs exactly once in each row, and (b) the columns form a symmetrical BIB design.

For example, the seven symbols arranged in three rows and seven columns in (6.4.1) is a Youden square because the columns form a symmetrical BIB design with the parameters $v = b = 7$, $r = 3 = k$, $\lambda = 1$, and each symbol occurs exactly once in each row.

$$
\begin{array}{ccccccc}
1 & 2 & 3 & 4 & 5 & 6 & 7 \\
2 & 3 & 4 & 5 & 6 & 7 & 1 \\
4 & 5 & 6 & 7 & 1 & 2 & 3
\end{array}
$$

(6.4.1)

In experiments Youden-square designs may be used to remove the heterogeneity of experimental units in two directions when sufficient material is not available to use an equal number of replicates and treatments.

We now show that a Youden-square design can always be constructed from a symmetrical BIB design. In fact, given a symmetrical BIB design with the parameters $v = b$, $r = k$, $\lambda = r(r - 1)/(v - 1)$, let us write the sets of the BIB design as columns. Then any h columns $(1 \leq h \leq v)$ contain between themselves hr symbols of which at least h are distinct, since each symbol can occur at most r times in these h columns. Thus, from Theorem 6.2.1, an SDR exists for the v columns, and this SDR will be a permutation of the symbols $1, 2, \cdots, v$. Let us bring this SDR to the first row. Then each of the v symbols occurs exactly once in the first row. Deleting the first row, we find that each column now contains $k - 1$ symbols, and any h of these columns will again contain at least h distinct symbols; hence another SDR exists for these deleted columns. By bringing this SDR to the second row, we can have each of the symbols occur exactly once in the second row. Continuing similarly, we can prove that the k rows can be so arranged that each symbol occurs exactly once in each of the k rows. We state this result as follows:

Theorem 6.4.1. A Youden square can always be constructed from a symmetrical BIB design.

Following the lines of proof used in Theorem 6.4.1, we can easily show that the tournament problem in the introduction always has a solution.

We may be interested to know the number of Youden squares that can be constructed from the corresponding symmetrical BIB design. The lower bound of this number—a bound that is, however, very low—is set in the following:

Theorem 6.4.2. At least $k!(k - 1)! \cdots 1!$ Youden squares can be constructed from the corresponding symmetrical BIB design.

Proof. Using Theorem 6.2.3, we can see that the ith row can be constructed in at least $(k - i + 1)!$ ways $(i = 1, 2, \cdots, k)$; hence the total number of Youden squares that can be constructed from a given symmetrical BIB design is at least $\prod_{i=1}^{k} (k - i + 1)!$.

6.5. FURTHER APPLICATIONS OF SDR AND ITS GENERALIZATION IN EXPERIMENTAL DESIGNS

In the preceding section we discussed the uses of SDR in the construction of Youden squares. We now study some further applications of these concepts.

Definition 6.5.1. A matrix with a single entry of 1 in each row and in each column and with all other elements zero is called a permutation matrix.

A permutation matrix is obtained from the identity matrix by permuting rows or columns or both. We prove

Theorem 6.5.1. If N is the incidence matrix of a symmetrical design with $v = b$, $r = k$, then N can be expressed as the sum of k permutation matrices.

Proof. Let the ith row of N have ones in columns $i_1, i_2, \cdots, i_k$ and zeros elsewhere. Form the sets $S_i = \{i_1, i_2, \cdots, i_k\}$ for $i = 1, 2, \cdots, v$. Then we can easily verify that any h of the sets S_i contain between themselves at least h of the integers $1, 2, \cdots, v$, which are the column numbers of $N (1 \leq h \leq v)$. Then from Theorem 6.2.1 an SDR exists for these sets, which will lead to a permutation matrix P_1. Consider $N - P_1$. In this matrix every row and column has got exactly $k - 1$ ones and zeros elsewhere, and we can proceed similarly on it to get another permutation matrix P_2. Continuing similarly, we can express N as $P_1 + P_2 + \cdots + P_k$, where $P_i (i = 1, 2, \cdots, k)$ are permutation matrices.

Theorem 6.5.2. If N is the incidence matrix of a symmetrical design with $v = b$, $r = k$, the symbols and sets of the design can be so arranged that N has zeros in the v diagonal positions.

Proof. Let $1, 2, \cdots, v$ be the symbols and let S_i denote the set of set numbers of the design in which the ith symbol does not occur ($i = 1, 2, \cdots, v$). Then any h of the sets S_i contain between themselves at least h symbols, and hence an SDR exists for the S_i sets. Let $(b_1, b_2, \cdots, b_v)$ be an SDR for $S_1, S_2, \cdots, S_v$. Without loss of generality, we renumber the sets of the design $b_1, b_2, \cdots, b_v$ respectively as $1, 2, \cdots, v$; with this renumbering N possesses the desired property.

We now generalize the concept of Youden squares to obtain latin rectangles as given by the following:

Definition 6.5.2. An $m \times n$ rectangle formed by the symbols $1, 2, \cdots, v$. ($v \geqq m$; $v \geqq n$; both m and n are not equal to v) is called a latin rectangle if every symbol occurs at most once in each row and in each column.

Latin rectangles can be used as experimental designs to remove the heterogeneity in two directions. The analysis of a general latin rectangle will normally be complicated. However, if we consider an $m \times v$ latin rectangle formed by the symbols $1, 2, \cdots, v$ and if the columns form a BIB design, then it becomes a Youden square, whose analysis is given in standard books. If the columns of an $m \times v$ latin rectangle form a PBIB design, the analysis can be easily worked out on similar lines as those used with Youden squares. We now consider certain combinatorial properties of latin rectangles.

The following theorem was proved by Ryser (1951):

Theorem 6.5.3. Let L be an $m \times n$ latin rectangle formed by the symbols $1, 2, \cdots, v$. Let $N(i)$ denote the number of times the ith symbol occurs in

L $(i = 1, 2, \cdots, v)$. A necessary and sufficient condition for L to be extended to a $v \times v$ latin square is that for each $i = 1, 2, \cdots, v$, $N(i) \geqq m + n - v$.

Proof. Let T_i $(i = 1, 2, \cdots, m)$ be the set of symbols that do not occur in the ith row of L. Then $|T_i| = v - n$ for every i. Let $M(i)$ denote the number of times the ith symbol occurs in these T_i sets. Then clearly $N(i) + M(i) = m$. Furthermore, if L can be extended to a latin square, $M(i) \leqq v - n$ and thus $N(i) \geqq m + n - v$, and the necessary part is established.

To prove the sufficiency part, we observe that if $N(i) \geqq m + n - v$, then $M(i) \leqq v - n$. Since $|T_i| = v - n$, we can easily see that any k of the sets T_i contain between themselves at least k distinct elements and an SDR exists for the sets $T_1, T_2, \cdots, T_m$. We add this SDR as a column to L. Proceeding in the same way again and again, we can get an $m \times v$ latin rectangle. Now using a similar argument, we can augment rows to get a $v \times v$ latin square, establishing the sufficiency of the condition.

Analogously to Theorem 6.4.2, we can easily prove the following:

Theorem 6.5.4. There are at least $v!(v - 1)! \cdots (v - m + 1)!$, $m \times v$ latin rectangles and hence at least $v!(v - 1)! \cdots 1!$, $v \times v$ latin squares.

Erdos and Kaplansky (1946) showed that an $m \times v$ latin rectangle can be embedded into an $(m + 1) \times v$ latin rectangle in approximately $v!/e^m$ ways if $m < v^{1/3}$.

We generalize the result of Theorem 6.5.4 to the following:

Theorem 6.5.5. There are at least $\binom{v}{n}^m n!(n - 1)! \cdots (n - m + 1)!$, $m \times n$ latin rectangles in v symbols $(n \geqq m)$.

Proof. We can select each row for the latin rectangle in $\binom{v}{n}$ ways, and hence all the m rows for the latin rectangle can be selected in $\binom{v}{n}^m$ ways. Then any k of the rows clearly contain between themselves at least k elements. Hence an SDR can be selected in $n(n - 1) \cdots (n - m + 1)$ ways, and this will be written as the first column. The second column can be selected in $(n - 1)(n - 2) \cdots (n - m)$ ways. Similarly proceeding, we show that the ith column can be selected in $(n - i + 1)(n - i) \cdots (n - m - i + 2)$ ways for $i = 1, 2, \cdots, n - m$. Then the $(n - m + j)$th column can be selected in $j!$ ways $(j = 1, 2, \cdots, m)$. Hence for a given selection of m rows there will be at least $n!(n - 1)! \cdots (n - m + 1)!$ latin rectangles, and since the rows can be selected in $\binom{v}{n}^m$ ways, there are at least $\binom{v}{n}^m n!(n - 1)! \cdots (n - m + 1)!$ latin rectangles.

The lower bounds given in Theorems 6.5.4 and 6.5.5 on the total number of latin rectangles and squares are far smaller than the actual number of latin rectangles and squares.

We now study some applications of generalized SDRs. The following theorem is due to Agrawal (1966b):

Theorem 6.5.6. In every binary, equireplicated design (with columns as sets of the design) of constant set size k such that $bk = vr$ and $b = mv$ the symbols can be rearranged in the columns so that every symbol occurs in a row m times.

Proof. Form the sets $T_1, T_2, \cdots, T_v$, where T_i is the set of all set numbers of the design containing the symbol i. Now

(6.5.1)
$$|T_{i_1} \cup T_{i_2} \cup \cdots \cup T_{i_n}| \geq \frac{hr}{k} = \frac{hmk}{k} = hm,$$

$$1 \leq i_1 \leq \cdots \leq i_h \leq v, 1 \leq h \leq v,$$

and hence from Theorem 6.3.1 an m-ple SDR $(0_1, 0_2, \cdots, 0_v)$ exists for $T_1, T_2, \cdots, T_v$. Write the first row such that the ith symbol occurs in the set numbers by 0_i ($i = 1, 2, \cdots, v$). We shall continue this procedure on the next $k - 1$ rows repeatedly to get the design in the required form.

The designs obtained in the above theorem will be useful as experimental designs to remove the heterogeneity in two ways, as was shown by Shrikhande (1951).

Theorem 6.5.7. It is possible to arrange v symbols in a $v \times v$ square such that (a) diagonal cells are blank, (b) every symbol occurs at most once in any row or column, and (c) the ith symbol does not occur in the ith row or the ith column ($i = 1, 2, \cdots, v$).

Proof. Let us consider the v sets $T_1^*, T_2^*, \cdots, T_v^*$, where

$$T_i^* = \{1, 2, \cdots, i - 1, i + 1, \cdots, v\}$$

and let

$$T_i^{*\prime} = T_i^* - \{1\}, \qquad \text{for } i = 1, 2, \cdots, v.$$

Then we can easily verify that any k of the $T_i^{*\prime}$ contain between themselves at least $k - 1$ or k distinct symbols, depending on whether the set contains 0 or 1 symbol in the $(0, 1, \cdots, 1)$-SDR. Hence a $(0, 1, 1, \cdots, 1)$-SDR exists for the v sets $T_i^{*\prime}$. Let γ_i occur from $T_i^{*\prime}$ ($i = 2, 3, \cdots, v$) in this SDR. We form the first column $(0, \gamma_2, \gamma_3, \cdots, \gamma_v)'$. Now let $T_i^{**} = T_i^* - \{\gamma_i\}$ and $T_i^{**\prime} = T_i^{**} - \{2\}$. We can easily verify that a $(1, 0, 1, \cdots, 1)$-SDR exists for the sets $T_i^{**\prime}$. Let δ_i occur from $T_i^{**\prime}$ for $i = 1, 3, \cdots, v$. We form the second column $(\delta_1, 0, \delta_3, \cdots, \delta_v)'$. Continuing on similar lines, we construct the v columns satisfying the requirements of the theorem.

Agrawal (1966a) showed that designs satisfying the requirements of Theorem 6.5.7 are row-, column-, and treatment-balanced, generalized, two-way-elimination-of-heterogeneity designs. He further showed that these designs can be constructed for odd v by cyclically permuting the row $(1, 2, \cdots, v)$, v times and finally omitting the diagonal symbols after renumbering the symbols suitably. Trial-and-error solutions were given by him for $v = 4$ and 16. Theorem 6.5.7 and trial-and-error solutions for $v = 6, 8$, and 10 have been given by Raghavarao (1970). A general solution for even v is unknown to this author.

REFERENCES

1. Agrawal, H. (1966a). Some systematic methods of construction of designs for two way elimination of heterogeneity. *Calcutta Stat. Assn. Bull.*, **15**, 93–108.
2. Agrawal, H. (1966b). Some generalizations of distinct representatives with applications to statistical designs. *Ann. Math. Stat.*, **37**, 525–526.
3. Erdos, P., and Kaplansky, I. (1946). The asymptotic number of latin rectangles. *Am. J. Math.*, **68**, 230–236.
4. Hall, M. (1948). Distinct representatives of subsets. *Bull. Am. Math. Soc.*, **54**, 922–926.
5. Hall, M., Jr. (1967). *Combinatorial Theory.* Blaisdell Publishing Company.
6. Hall, P. (1935). On representatives of subsets. *J. London Math. Soc.*, **10**, 26–30.
7. Konig, D. (1950). *Theorie der endlichen und unendlichen Graphen.* Chelsea. New York, pp. 170–178.
8. Rado, R. (1967). On the number of systems of distinct representatives of sets, *J. London Math. Soc.*, **42**, 107–109.
9. Raghavarao, D. (1970). A note on some balanced generalized two-way elimination of heterogeneity designs. *J. Indian Soc. Agric. Stat.*, **22**, 49–52.
10. Ryser, H. J. (1951). A combinatorial theorem with an application to Latin rectangles. *Proc. Am. Math. Soc.*, **2**, 550–552.
11. Shrikhande, S. S. (1951). Designs for two-way elimination of heterogeneity. *Ann. Math. Stat.*, **22**, 235–247.
12. Youden, W. J. (1937). Use of incomplete block replications in estimating tobacco virus. *Contributions from Boyce Thompson Institute*, **9**, 317–326

BIBLIOGRAPHY

Chakrabarti, M. C. *Mathematics of Design and Analysis of Experiments.* Asia Publishing House, 1962.
Mann, H. B., and Ryser, H. J. Systems of distinct representatives. *Ann. Math. Monthly*, **60**, 397–401 (1953).

Mirsky, L., and Perfect, H. Systems of representatives. *J. Math. Anal. Appl.*, **3**, 520–568 (1966).

Ryser, H. J. *Combinatorial Mathematics.* Wiley, New York, 1963.

Smith, C. A. B., and Hartley, H. O. The construction of Youden squares. *J. Royal Stat. Soc.*, **B10**, 262–263 (1948).

CHAPTER 7

Tactical Configurations and
Doubly Balanced Designs

7.1. INTRODUCTION

In Chapter 4 we introduced BIB designs and studied them in detail in Chapter 5. We now try to specialize this definition to include a class of designs that are BIB designs and possess some additional properties.

Definition 7.1.1. Given a set E of v symbols, and given positive integers k, $l (l \leq k \leq v)$ and δ, a tactical configuration (or simply configuration) $C[k, l, \delta, v]$ is defined to be a system of subsets of E, having k symbols each, such that every subset of E having l symbols is contained in exactly δ sets of the system.

Tactical configurations have been discussed in detail by Carmichael (1956). Clearly the $C[k, 2, \lambda, v]$ configuration is a BIB design, and $C[k, 3, \delta, v]$ configurations are known as doubly balanced designs [cf. Calvin (1954)].

Theorem 7.1.1. A necessary condition for the existence of a $C[k, l, \delta, v]$ configuration is that

(7.1.1) $$\delta \frac{\binom{v - h}{l - h}}{\binom{k - h}{l - h}} = \text{an integer} \qquad \text{for } h = 0, 1, \cdots, l - 1.$$

Proof. Let x be the number of sets of the configuration $C[k, l, \delta, v]$ in which a given set of h fixed symbols of E occurs $(h = 0, 1, \cdots, l - 1)$. Then the total number of l-plets formed including the h fixed symbols is on one side

$$x \binom{k-h}{l-h}$$

and on the other side

$$\delta \binom{v-h}{l-h}.$$

Hence

(7.1.2)
$$x \binom{k-h}{l-h} = \delta \binom{v-h}{l-h},$$

which implies

$$\delta \frac{\binom{v-h}{l-h}}{\binom{k-h}{l-h}}$$

is an integer for $h = 0, 1, \cdots, l - 1$.

As a corollary of the above theorem, we have the following:

Corollary 7.1.1.1. A $C[k, l, \delta, v]$ configuration is also a

$$C\left[k, h, \delta \frac{\binom{v-h}{l-h}}{\binom{k-h}{l-h}}, v \right]$$

configuration, where $h < l$.

Proof. Since in a $C[k, l, \delta, v]$ configuration every given set of h symbols occurs in

$$\delta \frac{\binom{v-h}{l-h}}{\binom{k-h}{l-h}}$$

sets, it follows that the configuration is also a

$$
C\left[k, h, \delta \frac{\binom{v-h}{l-h}}{\binom{k-h}{l-h}}, v\right]
$$

configuration. The following corollary is evident from the definition.

Corollary 7.1.1.2. The total number of sets in a $C[k, l, \delta, v]$ configuration is

$$
\delta \frac{\binom{v}{l}}{\binom{k}{l}}.
$$

7.2. TACTICAL CONFIGURATIONS $C[4, 3, \delta, v]$

A tactical configuration $C[4, 3, \delta, v]$ is known as a quadruple system when $\delta = 1$. Clearly such a configuration is a doubly balanced design.

Let a quadruple system consisting of $v(v - 1)(v - 2)/24$ quadruples be given on v symbols $x_1, x_2, \cdots, x_v$. From this we shall generate a quadruple system on $2v$ symbols $x_1, x_2, \cdots, x_v, y_1, y_2, \cdots, y_v$ as follows: From each quadruple, say (x_a, x_b, x_c, x_d) on the v symbols, we form eight quadruples $(x_a, x_b, x_c, x_d), (y_a, y_b, y_c, y_d), (x_a, x_b, y_c, y_d), (x_a, y_b, x_c, y_d), (x_a, y_b, y_c, x_d),$ $(y_a, x_b, x_c, y_d), (y_a, x_b, y_c, x_d), (y_a, y_b, x_c, x_d)$. In this way we shall generate $v(v - 1)(v - 2)/3$ quadruples. We add $v(v - 1)/2$ more quadruples (x_a, x_b, y_a, y_b) $(a \neq b, a, b = 1, 2, \cdots, v)$. Thus in all we get $v(v - 1)(2v - 1)/6$ quadruples forming the quadruple system on $2v$ symbols.

Clearly (x_1, x_2, x_3, x_4) is a quadruple on four symbols. From this by the above technique we get the following configuration $C[4, 3, 1, 8]$:

$$
\begin{array}{lll}
(x_1, x_2, x_3, x_4), & (y_1, y_2, y_3, y_4), & (y_1, y_2, x_3, x_4), \\
(y_1, y_2, x_1, x_2), & (y_1, y_3, x_2, x_4), & (y_1, y_3, x_1, x_3), \\
(y_1, y_4, x_2, x_3), & (y_1, y_4, x_1, x_4), & (y_2, y_3, x_1, x_4), \\
(y_2, y_3, x_2, x_3), & (y_2, y_4, x_1, x_3), & (y_2, y_4, x_2, x_4), \\
(y_3, y_4, x_1, x_2), & (y_3, y_4, x_3, x_4).
\end{array}
$$

(7.2.1)

The conditions given in Theorem 7.1.1 are only necessary conditions. The necessary and sufficient conditions were given for the existence of these configurations by Hanani.

Theorem 7.2.1. [Hanani (1960)]. A quadruple system exists if and only if $v \equiv 2$ or $4 \pmod 6$.

Theorem 7.2.2. [Hanani (1963)]. The configuration $C[4, 3, \delta, v]$ exists if and only if

$$(7.2.2) \qquad \begin{aligned} \delta v &\equiv 0 \pmod 2, \qquad \delta(v - 1)(v - 2) \equiv 0 \pmod 3, \\ \delta v(v - 1)(v - 2) &\equiv 0 \pmod 8. \end{aligned}$$

The necessary parts of both these theorems follow from Theorem 7.1.1. The sufficiency part was established by Hanani by actually constructing the configurations, and in this connection we refer the reader to the original papers of Hanani.

7.3. AN INEQUALITY FOR $C[k, l, \delta, v]$ CONFIGURATIONS

Let b be the number of sets in a $C[k, l, \delta, v]$ configuration. When $k = v - 1$, clearly either $b = v$ or b is a multiple of v and there is nothing to prove. We call such designs trivial designs. The following inequality was obtained by Raghavarao (1970):

Theorem 7.3.1. The inequality

$$(7.3.1) \qquad b \geq (l - 1)(v - l + 2)$$

holds for nontrivial $C[k, l, \delta, v]$ configurations.

Proof. Let the v symbols of the set E be numbered $1, 2, \cdots, v$ and let the b sets be numbered $1, 2, \cdots, b$. Define a $(v - l + 2) \times b$ matrix $N_1 = (n_{ij})$, where

$$(7.3.2) \qquad \begin{aligned} n_{ij} &= 1 \qquad \text{if the } i\text{th element occurs in the } j\text{th set;} \\ &= 0 \qquad \text{otherwise;} \qquad i = l - 1, l, l + 1, \cdots, v; j = 1, 2, \cdots, b. \end{aligned}$$

Define $l - 2$ matrices each of order $(v - l + 2) \times b$ by

$$M_\alpha = [m_{(\alpha), i, j}] \qquad (\alpha = 1, 2, \cdots, l - 2),$$

where

$$(7.3.3) \qquad \begin{aligned} m_{(\alpha), i, j} &= 1 \qquad \text{if the } (\alpha + 1)\text{-plet } (1, 2, \cdots, \alpha, i) \text{ occurs in the } j\text{th set;} \\ &= 0 \qquad \text{otherwise;} \qquad i = l - 1, l, l + 1, \cdots, v; \\ & \qquad\qquad\qquad\qquad j = 1, 2, \cdots, b. \end{aligned}$$

Consider

$$(7.3.4) \qquad N^* = \begin{bmatrix} N_1 \\ M_1 \\ M_2 \\ \vdots \\ M_{l-2} \end{bmatrix}.$$

Then we can easily verify that

$$(7.3.5) \qquad N^*N^{*\prime} = \begin{bmatrix} A_1 & A_2 & A_3 & A_4 & \cdots & A_{l-1} \\ A_2 & A_2 & A_3 & A_4 & \cdots & A_{l-1} \\ A_3 & A_3 & A_3 & A_4 & \cdots & A_{l-1} \\ A_4 & A_4 & A_4 & A_4 & \cdots & A_{l-1} \\ \cdot & \cdot & \cdot & \cdot & \cdots & \cdot \\ A_{l-1} & A_{l-1} & A_{l-1} & A_{l-1} & \cdots & A_{l-1} \end{bmatrix},$$

where

$$(7.3.6) \qquad A_i = (\lambda_i - \lambda_{i+1})I_{v-l+2} + \lambda_{i+1}E_{v-l+2,\,v-l+2};$$

$$\lambda_i = \delta \frac{\dbinom{v-i}{l-i}}{\dbinom{k-i}{l-i}}$$

is the number of sets in which every i-plet occurs ($i \leq l$, $i = 1, 2, \cdots, l$). Of course, $\lambda_l = \delta$. Performing row operations to reduce $N^*N^{*\prime}$ into a lower triangular matrix and taking determinants, we get

$$(7.3.7) \quad |N^*N^{*\prime}| = |A_{l-1}|\,|A_{l-2} - A_{l-1}|\,|A_{l-3} - A_{l-2}| \cdots |A_1 - A_2|.$$

Since $\lambda_1 > \lambda_2 > \lambda_3 > \cdots > \lambda_l$, we see that each factor on the right-hand side of (7.3.7) is nonzero when $v \neq k + 1$, and in such cases $|N^*N^{*\prime}| \neq 0$. Thus

$$(7.3.8) \qquad \begin{aligned} (l-1)(v-l+2) &= \mathrm{rank}\,(N^*N^{*\prime}) \\ &= \mathrm{rank}\,(N^*) \leq b \end{aligned}$$

for all nontrivial designs, and the proof is complete.

Considering the particular cases of the above theorem when $l = 2$ and 3, we get the following:

Corollary 7.3.1.1. For a BIB design the inequality $b \geq v$ holds.

Corollary 7.3.1.2. For a doubly balanced design $b \geq 2(v - 1)$ holds.

The result of Corollary 7.3.1.1 was already discussed in Chapter 5. The inequality $b \geq 2(v - 1)$ for doubly balanced designs was obtained by Raghavarao and Tharthare (1967). The lower bound for b obtained for a doubly balanced design in the corollary cannot be improved on. In a doubly balanced design let λ be the number of sets in which a given pair of symbols occurs together and let r denote the number of sets in which any particular symbol occurs. The terms v, b, r, k, λ, and δ are known as the parameters of the doubly balanced incomplete block design and they satisfy

$$(7.3.9) \qquad vr = bk,$$

$$(7.3.10) \qquad r(k - 1) = \lambda(v - 1),$$

and

$$(7.3.11) \qquad \lambda(k - 2) = \delta(v - 2).$$

When $b = 2(v - 1)$, since $vr = bk$, we must have $v = 2k$ and $r = v - 1$. Then clearly $\lambda = k - 1$. Now, from (7.3.11), $\delta = (k - 2)/2$, and since δ is an integer, k must be even, say $2t$. Thus the series of designs having the lower bound for b has the parameters

$$(7.3.12) \qquad \begin{array}{llll} v = 4t, & b = 2(4t - 1), & r = 4t - 1, & k = 2t, \\ \lambda = 2t - 1, & \delta = t - 1, & & \end{array}$$

and we shall actually construct this series in the next section.

7.4. CONSTRUCTION OF A SERIES OF DOUBLY BALANCED DESIGNS

To enable us to construct the doubly balanced designs with the parameters given in (7.3.12), we first prove the following:

Theorem 7.4.1. If a BIB design D exists with the parameters v, b, r, k, and λ and if D^* is the complement of D, then by considering the sets in which a particular symbol (say θ) occurs in D and D^* and omitting θ, we get a pairwise balanced design $[v - 1; k - 1, v - k - 1]$ of index $b - 3r + 3\lambda$.

Proof. Let N be the incidence matrix of the BIB design D, which without loss of generality can be assumed to be

$$(7.4.1) \qquad N = \begin{bmatrix} E_{1,r} & 0_{1,b-r} \\ N_1 & N_2 \end{bmatrix}.$$

Then $\overline{N}$, the incidence matrix of D^*, is given by

$$(7.4.2) \qquad \overline{N} = E_{v,b} - N = \begin{bmatrix} 0_{1,r} & E_{1,b-r} \\ E_{v-1,r} - N_1 & E_{v-1,b-r} - N_2 \end{bmatrix}.$$

Now, if we show

$$(7.4.3) \qquad M = [N_1 \quad E_{v-1,b-r} - N_2]$$

to be the incidence matrix of a pairwise balanced design $[v - 1; k - 1, v - k - 1]$ of index $b - 3r + 3\lambda$, we are through with the proof. Clearly the design with incidence matrix M has $v - 1$ symbols arranged in r sets containing $k - 1$ symbols each and $b - r$ sets containing $v - k - 1$ symbols each. Thus the first part is established. By using the relation of the BIB design to its incidence matrix N, we easily verify

$$(7.4.4) \qquad MM' = (r - \lambda)I_{v-1} + (b - 3r + 3\lambda)E_{v-1,v-1},$$

which shows that the index of the pairwise balanced design is $b - 3r + 3\lambda$. Thus the proof is complete.

The symmetrical BIB design with the parameters

$$(7.4.5) \qquad v^* = 4t - 1 = b^*, \qquad r^* = 2t - 1 = k^*, \qquad \lambda^* = t - 1$$

is constructed for many values of t in Chapters 5 and 17.

Theorem 7.4.2 [Raghavarao and Tharthare (1970)]. To each set of the BIB design with the parameters given in (7.4.5) add a new symbol ϕ. Then these sets and the sets of the complement design of (7.4.5) constitute a doubly balanced design with the parameters given in (7.3.12).

Proof. The parameters $v, b, r, k,$ and λ need no explanation. If we consider the sets in which ϕ occurs, every pair other than ϕ occurs together in $\delta = t - 1$ sets. If we consider the sets in which any symbol $\alpha \neq \phi$ occurs, then from Theorem 7.4.1 every pair of symbols other than α occurs together in $b^* - 3r^* + 3\lambda^* = t - 1 = \delta$ sets. Thus we show that all possible triples occur together in δ sets, completing the proof of the theorem.

7.5. $C[k, l, \delta, v]$ CONFIGURATIONS AS PARTIALLY BALANCED ARRAYS

Chakravarti (1961) studied the problem of constructing partially balanced arrays from general $C[k, l, \delta, v]$ tactical configurations. Since infinitely many configurations exist with $l = 2, 3$, whereas very little is known about configurations with $l > 3$, we give in this section the series of partially balanced arrays that can be constructed from BIB and doubly balanced designs.

Theorem 7.5.1. The incidence matrix N of a BIB design with the parameters v, b, r, k, and λ is a partially balanced array with v constraints, b assemblies, in two symbols 0 and 1, of strength 2, with the λ-parameters given by

(7.5.1)
$$\begin{aligned}
\lambda(x_1, x_2) &= \lambda && \text{if } x_1 = 1, x_2 = 1; \\
&= r - \lambda && \text{if } x_1 = 1, x_2 = 0 \text{ or } x_1 = 0, x_2 = 1; \\
&= b - 2r + \lambda && \text{if } x_1 = 0, x_2 = 0.
\end{aligned}$$

Theorem 7.5.2. The incidence matrix N^* of a doubly balanced design with the parameters v, b, r, k, λ, and δ is a partially balanced array with v constraints, b assemblies, in two symbols 0 and 1, of strength 3, with the λ-parameters given by

(7.5.2)
$$\begin{aligned}
\lambda(x_1, x_2, x_3) &= \delta && \text{if } x_1 = x_2 = x_3 = 1; \\
&= \lambda - \delta && \text{if two } x \text{ symbols are unity}; \\
&= r - 2\lambda + \delta && \text{if one } x \text{ is unity}; \\
&= b - 3r + 3\lambda - \delta && \text{if } x_1 = x_2 = x_3 = 0.
\end{aligned}$$

The proofs are straightforward and hence we omit them.

All the BIB designs studied in Chapter 5 give many series of partially balanced arrays of strength 2. Since the quadruple systems exist for every $v \equiv 2$ or 4 (mod 6), we have the following:

Corollary 7.5.2.1. If $v \equiv 2$ or 4 (mod 6), then a partially balanced array exists with v constraints, $v(v-1)(v-2)/24$ assemblies, in two symbols 0 and 1, of strength 3, with the λ-parameters given by

(7.5.3)
$$\begin{aligned}
\lambda(x_1, x_2, x_3) &= 1 && \text{if } x_1 = x_2 = x_3 = 1; \\
&= \frac{v-4}{2} && \text{if two } x \text{ symbols are unity}; \\
&= \frac{(v-4)(v-5)}{6} && \text{if one } x \text{ is unity}; \\
&= \frac{v^3 - 15v^2 + 38v - 120}{24} && \text{if } x_1 = x_2 = x_3 = 0.
\end{aligned}$$

Since the doubly balanced designs exist with the parameters given in (7.3.12) for infinitely many values of t, we have

Corollary 7.5.2.2. A partially balanced array exists with $4t$ constraints,

$2(4t - 1)$ assemblies, in two symbols 0 and 1, of strength 3, with the λ-parameters given by

$$\lambda(x_1, x_2, x_3) = t - 1 \quad \text{if } x_1 = x_2 = x_3 = 1,$$

(7.5.4)
$$= t \quad \text{if two } x \text{ symbols are unity,}$$
$$= t \quad \text{if one } x \text{ is unity,}$$
$$= t - 1 \quad \text{if } x_1 = x_2 = x_3 = 0$$

for infinitely many t.

REFERENCES

1. Carmichael, R. D. (1956). *Introduction to the Theory of Groups of Finite Order.* Dover Publications.
2. Calvin, L. D. (1954). Doubly balanced incomplete block designs for experiments in which the treatment effects are correlated. *Biometrics,* **10,** 61–88.
3. Chakravarti, I. M. (1961). On some methods of construction of partially balanced arrays. *Ann. Math. Stat.,* **32,** 1181–1185.
4. Hanani, H. (1960). On quadruple systems. *Can. J. Math.,* **12,** 145–157.
5. Hanani, H. (1963). On some tactical configurations. *Can. J. Math.,* **15,** 702–722.
6. Raghavarao, D. (1970). Some results on tactical configurations and nonexistence of difference set solutions for certain symmetrical PBIB designs. *Ann. Inst. Stat. Math.,* **22,** 501–506.
7. Raghavarao, D., and Tharthare, S. K. (1967). An inequality for doubly balanced incomplete block designs. *Calcutta Stat. Assn. Bull.,* **16,** 37–39.
8. Raghavarao, D., and Tharthare, S. K. (1970). A new series of doubly balanced designs. *Calcutta Stat. Assn. Bull,* **19,** 95–96.

BIBLIOGRAPHY

Netto, E. *Lehrbuch der Combinatorik,* 2nd edition. Teubner, Leipzig, 1927. Reprint Chelsea, New York.

CHAPTER 8

Partially Balanced Incomplete Block Designs

8.1. INTRODUCTION AND DEFINITION

We have shown in Section 4.5 that among the class of all connected incomplete block designs the balanced design is the most efficient design. Among equiblock-sized and equireplicated designs the BIB design is the only balanced, and hence the most efficient, design. Unfortunately, however, BIB designs exist in a limited number of cases, and hence it was necessary to introduce new designs. Bose and Nair (1939) defined partially balanced incomplete block (PBIB) designs in which all the elementary contrasts are not estimated with the same variance.

For defining PBIB designs, we need the concept of the association scheme for the v symbols as given below:

Definition 8.1.1. Given v symbols $1, 2, \cdots, v$, a relation satisfying the following conditions is said to be an association scheme with m classes:

1. Any two symbols are either 1st, 2nd, $\cdots$, or mth associates, the relation of association being symmetrical; that is, if the symbol α is the ith associate of the symbol β, then β is the ith associate of α.
2. Each symbol α has n_i ith associates, the number n_i being independent of α.
3. If any two symbols α and β are ith associates, then the number of symbols that are jth associates of α, and kth associates of β, is p_{jk}^i and is independent of the pair of ith associates α and β.

The numbers v, n_i $(i = 1, 2, \cdots, m)$ and p_{jk}^i $(i, j, k = 1, 2, \cdots, m)$ are called the parameters of the association scheme.

Given an association scheme for the v symbols, we define a PBIB design as follows:

Definition 8.1.2. If we have an association scheme with m classes and given parameters, we get a PBIB design with m associate classes if the v symbols are arranged into b sets of size k $(< v)$ such that

1. Every symbol occurs at most once in a set.
2. Every symbol occurs in exactly r sets.
3. If two symbols α and β are ith associates, then they occur together in λ_i sets, the number λ_i being independent of the particular pair of ith associates α and β.

The numbers v, b, r, k, λ_i $(i = 1, 2, \cdots, m)$ are called the parameters of the design. It is to be remarked here that the concept of the association scheme plays a fundamental role in the analysis and classification of PBIB designs. This concept, though inherent in Bose and Nair's definition, was explicitly introduced by Bose and Shimamoto (1952). In the original definition of Bose and Nair it was assumed that all λ_i parameters are distinct, but this restriction was removed by Nair and Rao (1942). In the earlier papers it was assumed that $p_{jk}{}^i = p_{kj}{}^i$, but Bose and Mesner (1959) showed that this condition is redundant and is a consequence of Definition 8.1.1.

Bose and Clatworthy (1955) showed that it is unnecessary to assume the constancy of all the parameters $p_{jk}{}^i$ in the case of two-associate-class schemes. If we assume that $n_1, n_2, p_{11}{}^1$, and $p_{11}{}^2$ are constant, all other conditions automatically follow.

8.2. RELATIONS BETWEEN THE PARAMETERS OF PBIB DESIGNS

We can easily verify that

$$(8.2.1) \qquad\qquad vr = bk.$$

With respect to every symbol, since each of the $v - 1$ other symbols is classified as the 1st, 2nd, $\cdots$, or mth associate and as every symbol has n_i ith associates, we have

$$(8.2.2) \qquad\qquad \sum_{i=1}^{m} n_i = v - 1.$$

Let us consider the r sets in which a particular symbol α occurs. From these sets we can form $r(k - 1)$ pairs of symbols, keeping α as one of the symbols. Among these pairs the ith associates of α must occur λ_i times, and there are n_i ith associates of α $(i = 1, 2, \cdots, m)$. Hence

$$(8.2.3) \qquad\qquad \sum_{i=1}^{m} n_i \lambda_i = r(k - 1).$$

Let α and β be ith associates. In this case the kth associates of α

$(k = 1, 2, \cdots, m)$ should cover all the n_j jth associates of β $(j \neq i)$. Thus $\sum_{k=1}^{m} p_{jk}^{i} = n_j$ for $j \neq i$. When $i = j$, α itself will be one of the jth associates of β. Hence the kth associates of α $(k = 1, 2, \cdots, m)$ should cover all the $n_j - 1$ jth associates of β. Thus $\sum_{k=1}^{m} p_{jk}^{i} = n_j - 1$. These two relations can be combined to form

$$(8.2.4) \qquad \sum_{k=1}^{m} p_{jk}^{i} = n_j - \delta_{ij},$$

where δ_{ij} is the Kronecker delta, taking the value 1 if $i = j$ and 0 otherwise.

Let α be any symbol, and let G_i be the set of the ith associates of α and G_j the set of the jth associates of α. Any symbol in G_i has p_{jk}^{i} kth associates in G_j, and any symbol in G_j has p_{ik}^{j} kth associates in G_i. If we form kth associate pairs from G_i and G_j, on one side we get $n_i p_{jk}^{i}$ and on the other side $n_j p_{ik}^{j}$. Hence

$$(8.2.5) \qquad n_i p_{jk}^{i} = n_j p_{ik}^{j}.$$

The parameters p_{jk}^{i} can be written in m symmetrical matrices $P_i = (p_{jk}^{i})$ for $i = 1, 2, \cdots, m$, where p_{jk}^{i} is the (j, k)th term of the matrix P_i. By making use of (8.2.4) and (8.2.5), we can show that there are $m(m^2 - 1)/6$ independent p_{jk}^{i} parameters.

8.3. ASSOCIATION MATRICES—THEIR ALGEBRAIC PROPERTIES AND APPLICATIONS

Let us define m, $v \times v$ matrices $B_i = (b_{\alpha\beta}^{i})$ for $i = 1, 2, \cdots, m$, where

$$(8.3.1) \qquad \begin{aligned} b_{\alpha\beta}^{i} &= 1 \qquad \text{if } \alpha \text{ and } \beta \text{ are } i\text{th associates}; \\ &= 0 \qquad \text{otherwise,} \end{aligned}$$

$b_{\alpha\beta}^{i}$ being the (α, β)th element of B_i. We observe that the B_i matrices $(i = 1, 2, \cdots, m)$ are symmetrical matrices, with row and column totals each equal to n_i.

In addition, let every symbol be the 0th associate of itself and of no other symbol. Then we see that

$$(8.3.2) \quad B_0 = I_v, \qquad n_0 = 1, \qquad p_{ij}^{0} = n_i \delta_{ij}, \qquad p_{0k}^{i} = \delta_{ik}, \qquad \lambda_0 = r.$$

The four parametric relations (8.2.2), (8.2.3), (8.2.4), and (8.2.5) proved in Section 8.2 can now be written as

$$(8.3.3) \qquad \sum_{i=0}^{m} n_i = v,$$

$$(8.3.4) \qquad \sum_{i=0}^{m} n_i \lambda_i = rk,$$

(8.3.5)
$$\sum_{k=0}^{m} p_{jk}{}^{i} = n_j,$$

(8.3.6)
$$n_i p_{jk}{}^{i} = n_j p_{ik}{}^{j},$$

respectively.

The matrices $B_0, B_1, B_2, \cdots, B_m$ are said to be the association matrices of the association scheme of the symbols.

Given two symbols α and β, they are either 0th, 1st, $\cdots$, or mth associates, and hence only one of the elements $b_{\alpha\beta}{}^{0}, b_{\alpha\beta}{}^{1}, \cdots, b_{\alpha\beta}{}^{m}$ is unity. Hence

(8.3.7)
$$\sum_{i=0}^{m} B_i = E_{v,v}.$$

From the same consideration, it follows that

(8.3.8)
$$\sum_{i=0}^{m} c_i B_i = 0_{v,v}$$

holds if and only if

(8.3.9)
$$c_0 = c_1 = \cdots = c_m = 0;$$

hence the linear functions of $B_0, B_1, \cdots, B_m$ form a vector space of $(m+1)$-dimensionality with basis $B_0, B_1, \cdots, B_m$.

Since the (α, β)th element of $B_j B_k$ can be interpreted as the number of symbols common to the jth associates of α and the kth associates of β, we have

(8.3.10)
$$B_j B_k = \sum_{i=0}^{m} p_{jk}{}^{i} B_i \qquad j, k = 0, 1, \cdots, m;$$

and hence the multiplication is closed in the set of linear functions of $B_0, B_1, \cdots, B_m$. Clearly this set of linear functions of $B_0, B_1, \cdots, B_m$ forms a commutative group. Thus it forms a ring with unit element, which will be a linear associative algebra if the coefficients in the linear functions range over a field. Multiplication of the B-matrices is commutative. In fact

(8.3.11)
$$B_k B_j = B_k' B_j' = (B_j B_k)' = \left(\sum_{i=0}^{m} p_{jk}{}^{i} B_i \right)'$$
$$= \sum_{i=0}^{m} p_{jk}{}^{i} B_i' = \sum_{i=0}^{m} p_{jk}{}^{i} B_i = B_j B_k.$$

As a consequence to (8.3.11), we have

(8.3.12)
$$p_{jk}{}^{i} = p_{kj}{}^{i}.$$

Now we shall provide a regular representation in $(m+1) \times (m+1)$

matrices of the algebra given by the B-matrices. Since matrix multiplication is associative,

$$(8.3.13) \qquad B_i(B_j B_k) = \sum_{u,t} p_{jk}{}^u p_{iu}{}^t B_t = (B_i B_j) B_k = \sum_{u,t} p_{ij}{}^u p_{uk}{}^t B_t .$$

The independence of B_0, B_1, $\cdots$, B_m implies that

$$(8.3.14) \qquad \sum_u p_{jk}{}^u p_{iu}{}^t = \sum_u p_{ij}{}^u p_{uk}{}^t .$$

Now let us define $\mathscr{P}_i$-matrices by

$$(8.3.15) \qquad \mathscr{P}_i = (p_{ji}{}^k) = \begin{bmatrix} p_{0i}{}^0 & p_{0i}{}^1 & \cdots & p_{0i}{}^m \\ p_{1i}{}^0 & p_{1i}{}^1 & \cdots & p_{1i}{}^m \\ \cdot & \cdot & \cdots & \cdot \\ p_{mi}{}^0 & p_{mi}{}^1 & \cdots & p_{mi}{}^m \end{bmatrix}.$$

We remark that the $\mathscr{P}_i$-matrices of (8.3.15) are different from the P_i-matrices of Section 8.2. In terms of $\mathscr{P}_i$-matrices, (8.3.14) implies that

$$(8.3.16) \qquad \mathscr{P}_j \mathscr{P}_k = \sum_{i=0}^{m} p_{jk}{}^i \mathscr{P}_i .$$

Thus the $\mathscr{P}$-matrices multiply in the same manner as the B-matrices. We can easily verify that $\mathscr{P}_0$, $\mathscr{P}_1$, $\cdots$, $\mathscr{P}_m$ are linearly independent. They thus form the basis for a vector space and combine in the same way as the B-matrices in addition as well as in multiplication. They provide a regular representation in $(m + 1) \times (m + 1)$ matrices of the algebra given by the B-matrices, which are $v \times v$ matrices.

We now prove the following:

Theorem 8.3.1. The distinct characteristic roots of

$$(8.3.17) \qquad B = \sum_{i=0}^{m} c_i B_i$$

and the distinct characteristic roots of

$$(8.3.18) \qquad \mathscr{P} = \sum_{i=0}^{m} c_i \mathscr{P}_i$$

are the same.

Proof. Let $f(\lambda)$ and $g(\lambda)$ be the minimal polynomials of B and $\mathscr{P}$, respectively. They are monic polynomials. The term $f(B)$ can be expressed as

$$(8.3.19) \qquad f(B) = \sum_{i=0}^{m} d_i B_i ,$$

whose representation in $(m + 1) \times (m + 1)$ matrices will be

$$(8.3.20) \qquad f(\mathscr{P}) = \sum_{i=0}^{m} d_i \mathscr{P}_i.$$

Since $f(\lambda)$ is the minimal polynomial of B, we have $f(B) = 0_{v,v}$, which implies that $d_0 = d_1 = \cdots = d_m = 0$ and hence $f(\mathscr{P}) = 0_{v,v}$. Since $f(\mathscr{P}) = 0$ and since $g(\lambda)$ is the minimal polynomial of $\mathscr{P}$, we have the result that $g(\lambda)$ divides $f(\lambda)$. We can similarly show that $f(\lambda)$ divides $g(\lambda)$. Since both $f(\lambda)$ and $g(\lambda)$ are monic polynomials, $f(\lambda) = g(\lambda)$. Thus B and $\mathscr{P}$ have the same distinct characteristic roots. This completes the proof of the theorem.

As an immediate corollary to the above theorem we have

Corollary 8.3.1.1. If N is the incidence matrix of a PBIB design, then the distinct characteristic roots of

$$(8.3.21) \qquad NN' = rB_0 + \lambda_1 B_1 + \cdots + \lambda_m B_m$$

and

$$(8.3.22) \qquad \mathscr{P} = r\mathscr{P}_0 + \lambda_1 \mathscr{P}_1 + \cdots + \lambda_m \mathscr{P}_m$$

are the same.

This result was first obtained by Connor and Clatworthy (1954) by a longer method. The present approach is due to Bose and Mesner (1959). Connor and Clatworthy also determined the roots and their multiplicities of NN' for $m = 2$, the three distinct roots of NN' being

$$(8.3.23) \quad \theta_0 = rk, \qquad \theta_i = r - \tfrac{1}{2}\{(\lambda_1 - \lambda_2)[-\gamma + (-1)^i \sqrt{\Delta}] + (\lambda_1 + \lambda_2)\}$$
$$i = 1, 2,$$

where

$$(8.3.24) \quad \gamma = p_{12}^{\;2} - p_{12}^{\;1}, \qquad \beta = p_{12}^{\;1} + p_{12}^{\;2}, \qquad \Delta = \gamma^2 + 2\beta + 1.$$

Let α_0, α_1, α_2 be the multiplicities of the roots θ_0, θ_1, and θ_2, respectively, of NN'. If the design is connected, then θ_0 is a simple root and hence $\alpha_0 = 1$ and the other multiplicities satisfy

$$(8.3.25) \qquad \alpha_1 + \alpha_2 = v - 1$$

and

$$(8.3.26) \qquad \text{tr}\,(NN') = rk + \alpha_1 \theta_1 + \alpha_2 \theta_2 = vr,$$

where tr stands for the trace of the matrix. Solving the above two equations, we get

$$(8.3.27) \qquad \alpha_i = \frac{n_1 + n_2}{2} + (-1)^i \left[\frac{(n_1 - n_2) + \gamma(n_1 + n_2)}{2\sqrt{\Delta}} \right].$$

It is to be noted that the multiplicities depend only on the parameters of the association scheme with m classes, but not on the parameters of the design for any m. Furthermore, for an association scheme to exist the multiplicities must be integral. This imposes a necessary condition for the existence of any m-associate-class association scheme.

The association matrices can be applied in establishing the parametric relations of (8.3.3), (8.3.4), (8.3.5), and (8.3.6), which we leave as an exercise to the reader.

8.4. CLASSIFICATION OF TWO-ASSOCIATE-CLASS PBIB DESIGNS

The then known two-associate-class PBIB designs were classified by Bose and Shimamoto (1952) into the following five types depending on the association scheme:

1. Group divisible (GD).
2. Simple (SI).
3. Triangular (T).
4. Latin-square type (L_i).
5. Cyclic (C).

Definition 8.4.1. A PBIB design with two associate classes is said to be group divisible† if there are $v = mn$ symbols and the symbols can be divided into m groups of n symbols each, such that any two symbols of the same group are first associates and two symbols from different groups are second associates.

For a group-divisible design we can easily verify that the parameters of the association scheme are

$$n_1 = n - 1, \qquad n_2 = n(m - 1),$$

$$\text{(8.4.1)} \qquad P_1 = \begin{bmatrix} n - 2 & 0 \\ 0 & n(m - 1) \end{bmatrix}, \qquad P_2 = \begin{bmatrix} 0 & n - 1 \\ n - 1 & n(m - 2) \end{bmatrix}.$$

By applying the results of Section 8.3, we can easily see that the characteristic roots of NN' other than rk are $r - \lambda_1$ and $rk - v\lambda_2$, with the respective

---

† In the definition there is no group notion in the mathematical sense. However, the term "group-divisible designs" is now taken as a standard definition by statisticians. Dembowski (1968) calls them divisible partial designs.

multiplicities $m(n-1)$ and $m-1$. Depending on the values of the characteristic roots, the group-divisible designs are further subdivided into three classes by Bose and Connor (1952):

1. Singular (S) if $r - \lambda_1 = 0$.
2. Semiregular (SR) if $r - \lambda_1 > 0$ and $rk - v\lambda_2 = 0$.
3. Regular (R) if $r - \lambda_1 > 0$ and $rk - v\lambda_2 > 0$.

We shall study the group-divisible designs in detail in Sections 8.5 and 8.6.

Definition 8.4.2. A PBIB design with two associate classes is said to be simple if either (a) $\lambda_1 \neq 0$, $\lambda_2 = 0$ or (b) $\lambda_1 = 0$, $\lambda_2 \neq 0$.

Since case b can be changed to case a by interchanging the designation of first and second associates, case a is taken as the standard. It is also to be noted here that a design of the simple type may also belong to one of the other types—namely, group divisible, triangular, latin square, or cyclic. Simple designs with $\lambda_1 = 1$, $\lambda_2 = 0$ have a close relationship with partial geometries, and we study them in Chapter 9 in detail.

Definition 8.4.3. A PBIB design with two associate classes is said to be triangular if the number of symbols $v = n(n-1)/2$ and the association scheme is an array of n rows and n columns with the following properties:

1. The positions in the principal diagonal (running from the upper left-hand to the lower right-hand corner) are left blank.
2. The $n(n-1)/2$ positions above the principal diagonal are filled by the numbers $1, 2, \cdots, n(n-1)/2$ corresponding to the symbols.
3. The $n(n-1)/2$ positions below the principal diagonal are filled so that the array is symmetrical about the principal diagonal.
4. For any symbol i the first associates are exactly those that occur in the same row (or in the same column) as i.

The following relations clearly hold:

$$n_1 = 2(n-2), \qquad n_2 = \frac{(n-2)(n-3)}{2},$$

(8.4.2)
$$P_1 = \begin{bmatrix} n-2 & n-3 \\ n-3 & \dfrac{(n-3)(n-4)}{2} \end{bmatrix}, \qquad P_2 = \begin{bmatrix} 4 & 2n-8 \\ 2n-8 & \dfrac{(n-4)(n-5)}{2} \end{bmatrix}.$$

By applying the results of the last section we can see that the characteristic roots of NN' other than rk are $r + (n-4)\lambda_1 - (n-3)\lambda_2$ and $r - 2\lambda_1 + \lambda_2$, with the respective multiplicities $n-1$ and $n(n-3)/2$. We shall study these designs in detail in Sections 8.7 and 8.8.

Definition 8.4.4. A PBIB design with two associate classes is said to be a latin-square-type design with i constraints if the number of symbols $v = s^2$ satisfies the following association scheme: the s^2 symbols are arranged in an $s \times s$ square array and $i - 2$ MOLS are superimposed. Two symbols are first associates if and only if they occur in the same row or column of the array or in positions occupied by the same letter in any of the latin squares.

It can be checked in this case that

$$n_1 = i(s - 1), \qquad n_2 = (s - i + 1)(s - 1),$$

(8.4.3)
$$P_1 = \begin{bmatrix} (i-1)(i-2) + s - 2 & (s-i+1)(i-1) \\ (s-i+1)(i-1) & (s-i+1)(s-i) \end{bmatrix},$$

$$P_2 = \begin{bmatrix} i(i-1) & i(s-i) \\ i(s-i) & (s-i)(s-i-1) + s - 2 \end{bmatrix}.$$

In this case we can also see that the characteristic roots of NN' of L_i designs other than rk are $r + (s - i)\lambda_1 - (s - i + 1)\lambda_2$ and $r - i\lambda_1 + (i - 1)\lambda_2$, with the respective multiplicities $i(s - 1)$ and $(s - i + 1)(s - 1)$. We shall study these designs in detail in Sections 8.9 and 8.10.

Definition 8.4.5. A non-group-divisible PBIB design with two associate classes is called cyclic if the set of first associates of the ith symbol is $(i + d_1, i + d_2, \cdots, i + d_{n_1})$ mod v, where the d elements satisfy the following conditions:

1. The d elements are all different and $0 < d_j < v$ for $j = 1, 2, \cdots, n_1$.

2. Among the $n_1(n_1 - 1)$ differences $d_j - d_{j'}$ each of the $d_1, d_2, \cdots, d_{n_1}$ elements occurs $p_{11}{}^1$ times and each of the $e_1, e_2, \cdots, e_{n_2}$ elements occurs $p_{11}{}^2$ times, where $d_1, d_2, \cdots, d_{n_1}$; $e_1, e_2, \cdots, e_{n_2}$ are all distinct nonzero elements of the module M of v elements $0, 1, 2, \cdots, v - 1$.

3. The set $D = (d_1, d_2, \cdots, d_{n_1})$ is such that $D = (-d_1, -d_2, \cdots, -d_{n_1})$.

All the known cyclic association schemes have the parameters

$$v = 4t + 1, \qquad n_1 = n_2 = 2t,$$

(8.4.4)
$$P_1 = \begin{bmatrix} t - 1 & t \\ t & t \end{bmatrix}, \qquad P_2 = \begin{bmatrix} t & t \\ t & t - 1 \end{bmatrix}.$$

Conditions 1 and 2 were only included in the definition of the cyclic association scheme by Bose and Shimamoto. Nandi and Adhikary (1966) realized that conditions 1 and 2 are enough to make this an association scheme with two clases if $p_{11}{}^1 = p_{11}{}^2$, but not when $p_{11}{}^1 \neq p_{11}{}^2$. They imposed condition 3, which will be automatically satisfied if $p_{11}{}^1 = p_{11}{}^2$, to define an association scheme even when $p_{11}{}^1 \neq p_{11}{}^2$. It may be added that all examples given by Bose and Shimamoto satisfy $p_{11}{}^1 = p_{11}{}^2$, and hence condition 3 was automatically satisfied.

Tables of two-associate PBIB designs were prepared by Bose, Clatworthy, and Shrikhande (1954) and were later extended by Clatworthy (1956). In the 1954 tables the parameters of 124 singular group-divisible designs, 91 semi-regular group-divisible designs, 68 regular group-divisible designs, 27 simple designs, 36 triangular designs, 20 latin-square-type designs, and 10 cyclic designs were listed together with their plans. The symbols S_n, SR_n, R_n, Sl_n, T_n, LS_n, and C_n refer to the nth serial number design of singular group-divisible, semiregular group-divisible, regular group-divisible, triangular, latin-square-type, and cyclic designs, respectively. We use these tables in Chapter 11 in the construction of SUB arrangements.

The two-associate-class association schemes that are not covered by Bose and Shimamoto fall into two categories. The first category contains pseudo-triangular, pseudo-latin-square type, and pseudocyclic association schemes. These are the association schemes with the parameters given in (8.4.2), (8.4.3), and (8.4.4), whatever their combinatorial structure. We shall study the pseudo-cyclic association scheme in detail in Section 8.11.

The other category of two-associate-class designs that do not fit the Bose–Shimamoto classification differs in association-scheme parameters from group-divisible, triangular, latin-square-type, or cyclic, designs. In general these designs do not satisfy $\lambda_1 = 0$ or $\lambda_2 = 0$ and hence are not simple. The NL_j family of designs with the parameters given in (8.4.3), where s and i are replaced by $-t$ and $-j$, belongs to this class and was studied by Mesner (1967). Other examples of association schemes of this type were given for $v = 15$ by Clatworthy (1955) and for $v = 50$ by Hoffman and Singleton (1960).

8.5. COMBINATORIAL PROBLEMS OF GROUP-DIVISIBLE DESIGNS

8.5.1. Uniqueness of the Association Scheme

In this section we show that the parameters given in (8.4.1) uniquely represent the group-divisible association scheme. Let θ be any symbol and let $\theta_1, \theta_2, \cdots, \theta_{n_1}$ be its first associates. Consider the symbols θ and θ_1. Since $n_1 = n - 1$ and $p_{11}{}^1 = n - 2$, the first associates of θ_1 except θ are the same as the first associates of θ except θ_1. Also, as $p_{12}{}^1 = 0$, we can divide the symbols into m groups of n symbols each such that symbols in the same group are first associates and symbols in different groups are second associates. Hence the symbols have the group-divisible association scheme, which is uniquely determined by its parameters.

8.5.2. Singular Group-Divisible Designs

We observed in Section 8.4 that singular group-divisible designs are characterized by the property $r = \lambda_1$. In this case we can easily establish the following:

Theorem 8.5.1. [Bose and Connor (1952)]. The existence of a BIB design with the parameters v^*, b^*, r^*, k^*, λ^* is equivalent to the existence of singular group-divisible designs with the parameters

$$(8.5.1) \quad v = v^*n, \quad b = b^*, \quad r = r^*, \quad k = k^*n, \quad \lambda_1 = r^*, \quad \lambda_2 = \lambda^*,$$

for every n.

The proof of this theorem follows by noting that each symbol of the BIB design with the parameters v^*, b^*, r^*, k^*, λ^* can be replaced by a group of n symbols to get the singular group-divisible design with the parameters given in (8.5.1). Conversely, every group-divisible design with the parameters given in (8.5.1) must be related in this way to a BIB design with the parameters v^*, b^*, r^*, k^*, λ^*. This result, along with the known results of BIB designs, gives us the following theorems [cf. Kapadia (1966)]:

Theorem 8.5.2. A given set of a singular group-divisible design cannot have more than

$$(8.5.2) \quad b - 1 - \left[\frac{k(\lambda_1 - 1)^2}{(\lambda_1 - 1)n + (k - n)(\lambda_2 - 1)} \right]$$

disjoint sets with it. If some set has that many disjoint sets, then $n + [(k - n)(\lambda_2 - 1)]/(\lambda_1 - 1)$ is an integer, and each nondisjoint set has $n + [(k - n)(\lambda_2 - 1)]/(\lambda_1 - 1)$ symbols in common with that given set.

Theorem 8.5.3. The necessary and sufficient condition for a set of a singular group-divisible design to have the same number of symbols in common with each of the remaining sets is that $b = m$.

Theorem 8.5.4. For a resolvable, singular, group-divisible design $b \geq r + m - 1$. A resolvable, singular, group-divisible design will be affine resolvable if and only if $b = r + m - 1$.

8.5.3. Semiregular Group-Divisible Designs

Semiregular group-divisible designs are characterized by the property $rk - v\lambda_2 = 0$. Thus we have

$$(8.5.3) \quad v - m + 1 = \text{rank } (NN') = \text{rank } (N) \leq \min (v, b) \leq b.$$

If the design is resolvable, then

$$(8.5.4) \quad v - m + 1 \leq b - r + 1.$$

This result can be summarized as follows:

Theorem 8.5.5. For a semiregular group-divisible design $b \geq v - m + 1$. Furthermore, if the design is resolvable, then $b \geq v - m + r$.

We prove the following important property of the set structure of the design due to Bose and Connor (1952):

Theorem 8.5.6. For a semiregular group-divisible design k is divisible by m. If $k = cm$, then every set contains c symbols from each group.

Proof. Let x_{ij} be the number of symbols occurring in the jth set of the design from the ith group of the association scheme ($i = 1, 2, \cdots, m$; $j = 1, 2, \cdots, b$). Then

$$(8.5.5) \qquad \sum_{j=1}^{b} x_{ij} = nr$$

and

$$(8.5.6) \qquad \sum_{j=1}^{b} x_{ij}(x_{ij} - 1) = n(n - 1)\lambda_1.$$

Let us define $\bar{x}_i = \sum_{j=1}^{b} x_{ij}/b = nr/b = k/m = c$, say. Now

$$\sum_{j=1}^{b} (x_{ij} - \bar{x}_i)^2 = \sum_{j=1}^{b} x_{ij}^2 - b(\bar{x}_i)^2$$

$$= n[r + (n - 1)\lambda_1] - \frac{bk^2}{m^2}$$

$$(8.5.7) \qquad = n(n\lambda_2) - \frac{nrk}{m}$$

$$= \frac{n(v\lambda_2 - rk)}{m}$$

$$= 0,$$

using the property that $rk - v\lambda_2 = 0$ and $r(k - 1) = \lambda_1(n - 1) + \lambda_2(v - n)$. Hence $x_{i1} = x_{i2} = \cdots = x_{ib} = c$, and every set has c symbols from the ith group of the association scheme. Since this result holds for every i, we have the stated result.

The semiregular group-divisible designs with $\lambda_1 = 0$ are closely related to orthogonal arrays [cf. Bose, Shrikhande, and Bhattacharya (1953)]. We prove the theorem:

Theorem 8.5.7. The existence of a semiregular group-divisible design with the parameters

(8.5.8)
$$v = mn, \quad b = n^2 \lambda_2, \quad r = n\lambda_2, \quad k = m,$$
$$\lambda_1 = 0, \quad \lambda_2, \quad m, \quad n$$

implies and is implied by the existence of an orthogonal array

$$A = (\lambda_2 n^2, m, n, 2)$$

of strength 2.

Proof. Let the symbols of the ith group be numbered $(i-1)n + 0$, $(i-1)n + 1, \cdots, (i-1)n + (n-1)$, where $i = 1, 2, \cdots, m$. In view of Theorem 8.5.6, every set contains exactly one symbol from each of the m groups. Let us exhibit the sets of the design as the columns of a rectangular scheme in which the symbols of the ith group occupy the ith row $(i = 1, 2, \cdots, m)$. Replacing the symbols $(i-1)n + \theta$ of the ith group by θ $(\theta = 0, 1, \cdots, n-1)$, we get an orthogonal array A of size $\lambda_2 n^2$, m constraints, n levels, and strength 2.

The converse can be proved by tracing back the above steps.
The following corollary to the above theorem is obvious:

Corollary 8.5.7.1. The existence of a semiregular group-divisible design with the parameters given in (8.5.8) implies the existence of the semiregular group-divisible design with the parameters

(8.5.9)
$$v = m_1 n, \quad b = n^2 \lambda_2, \quad r = n\lambda_2, \quad k = m_1,$$
$$m_1, \quad n, \quad \lambda_1 = 0, \quad \lambda_2,$$

where m_1 can be any integer satisfying $0 < m_1 < m$.

We studied affine resolvable BIB designs in Chapter 5. By noting the definition of an affine resolvable BIB design, we get

Theorem 8.5.8. By interchanging the role of sets and symbols in an affine resolvable BIB design with the parameters

(8.5.10)
$$v' = n^2[(n-1)t + 1], \quad b' = n(n^2 t + n + 1),$$
$$r' = n^2 t + n + 1, \quad k' = n[(n-1)t + 1], \quad \lambda' = nt + 1,$$

we get a semiregular group-divisible design with the parameters

(8.5.11)
$$v = n(n^2 t + n + 1), \quad m = n^2 t + n + 1, \quad n,$$
$$b = n^2[(n-1)t + 1], \quad r = n[(n-1)t + 1],$$
$$k = n^2 t + n + 1, \quad \lambda_1 = 0, \quad \lambda_2 = (n-1)t + 1.$$

The procedure involved in the above theorem is known as dualization, and we shall study it at greater length in Chapter 10.

We now obtain bounds for the number of symbols in common between the jth and the uth sets, denoted by S_{ju}, given by Agrawal (1964). We prove the following theorem:

Theorem 8.5.9. For a semiregular group-divisible design S_{ju} satisfies the inequality

$$\max[0, 2k - v, -r + \lambda_1 + k] \leq S_{ju}$$

$$(8.5.12) \qquad \leq \min\left[k, \frac{2r(k - 1) + \lambda_1}{b} - k + r - \lambda_1\right].$$

Proof. Let us renumber the sets in such a manner that the jth and the uth sets occupy the first and second positions. Since the characteristic roots of $N'N$ are rk, 0, and $r - \lambda_1$, the matrix

$$A = N'N - \frac{r(k - 1) + \lambda_1}{b} E_{b,b}$$

has the roots 0 and $r - \lambda_1$. If $\mathbf{y}$ is any $b \times 1$ vector, then

$$(8.5.13) \qquad 0 \leq \mathbf{y}'N'N\mathbf{y} - \frac{r(k - 1) + \lambda_1}{b} \mathbf{y}'E_{b,b}\mathbf{y} \leq (r - \lambda_1)\mathbf{y}'\mathbf{y}.$$

Choosing for $\mathbf{y}$ the column vectors $(1/\sqrt{2}, -1/\sqrt{2}, 0, \cdots, 0)'$ and $(1/\sqrt{2}, +1/\sqrt{2}, 0, \cdots, 0)'$, respectively, and substituting in (8.5.13), we have

$$(8.5.14) \qquad k - r + \lambda_1 \leq S_{12} \leq r - \lambda_1 + \frac{2r(k - 1) + \lambda_1}{b} - k.$$

If $N^* = E_{v,b} - N$, then $N^{*'}N^* = N'N + (v - 2k)E_{b,b}$, the elements of which are nonnegative, and hence

$$(8.5.15) \qquad\qquad\qquad S_{12} - 2k + v \geq 0.$$

Combining (8.5.14) and (8.5.15) and remembering that $0 \leq S_{12} \leq k$, we get the required result.

The upper bound for the number of disjoint sets in a semiregular group-divisble design was obtained by S. M. Shah (1964).

Theorem 8.5.10. A given set of a semiregular group-divisible design cannot have more than

$$(8.5.16) \qquad b - 1 - \frac{v(v - m)(r - 1)^2}{(v - k)(b - r) - (v - rk)(v - m)}$$

disjoint sets with it. If some set has that many disjoint sets, then

$$\frac{k[(v - k)(b - r) - (v - rk)(v - m)]}{v(v - m)(r - 1)}$$

is an integer and each nondisjoint set has

(8.5.17)
$$\frac{k[(v - k)(b - r) - (v - rk)(v - m)]}{v(v - m)(r - 1)}$$

symbols in common with that given set.

Proof. Without loss of generality, let us assume that the first set is the given set and has d disjoint sets. Let the first set have x_i symbols in common with the ith set, $i = d + 2, d + 3, \cdots, b$. Clearly

(8.5.18)
$$\sum_{i=d+2}^{b} x_i = k(r - 1)$$

and

(8.5.19)
$$\sum_{i=d+2}^{b} x_i(x_i - 1) = \frac{k[(k - m)\lambda_1 + k(m - 1)\lambda_2 - m(k - 1)]}{m}.$$

Define

(8.5.20)
$$\bar{x} = \sum_{i=d+2}^{b} \frac{x_i}{(b - d - 1)}.$$

Then

(8.5.21)
$$\sum_{i=d+2}^{b} (x_i - \bar{x})^2 = \frac{k^2[(v - k)(b - r) - (v - rk)(v - m)]}{v(v - m)} - \frac{k^2(r - 1)^2}{b - d - 1}$$
$$\geqq 0$$

gives d to be less than or equal to (8.5.16).

If d equals (8.5.16), then

(8.5.22)
$$x_i = \frac{k[(v - k)(b - r) - (v - rk)(v - m)]}{v(v - m)(r - 1)} \qquad i = d + 2, d + 3, \cdots, b$$

is an integer, and the latter conclusion of the theorem follows.

As a consequence of the above theorem we have

Corollary 8.5.10.1. A resolvable semiregular group-divisible design is affine resolvable if and only if (a) $b = v - m + r$ and (b) k^2/v is an integer.

Corollary 8.5.10.2. The necessary and sufficient conditions for every pair of sets to have the same number of symbols in common is that (a) $b = v - m + 1$ and (b) $k(r - 1)/(v - m)$ be an integer.

The designs possessing the property introduced in the last corollary are called linked-block designs, and we shall discuss them in Chapter 10.

8.5.4. Regular Group-Divisible Designs

By considering the rank of NN', we have the following theorem:

Theorem 8.5.11. For a regular group-divisible design $b \geqq v$. Furthermore, if the design is resolvable, then $b \geqq v + r - 1$.

Consider any $t \leqq b$ sets of the design. Let the submatrix of N that corresponds to these t sets be denoted by N_0. Let S_{ju} be the number of symbols in common between the jth and the uth chosen sets ($j, u = 1, 2, \cdots, t$). Then the $t \times t$ symmetrical matrix

$$(8.5.23) \qquad S_t^I = N_0' N_0 = (S_{ju})$$

is said to be the intersection structural matrix of the t chosen sets.

We shall consider another sort of structural matrix. Let $S_{ju}{}^w$ denote the number of symbols from the wth group that are in common between the jth and the uth chosen sets. Then we easily verify that

$$(8.5.24) \qquad \sum_{w=1}^{m} S_{ju}{}^w = S_{ju},$$

$$(8.5.25) \qquad \sum_{w=1}^{m} S_{jj}{}^w = k.$$

The matrix

$$(8.5.26) \qquad S_t^G = G_t' G_t,$$

where G_t has as its (α, β)th element $S_{\beta\beta}{}^\alpha$, is said to be the group structural matrix of the t chosen sets.

Let us rearrange the sets of the group-divisible design in such a way that the t chosen sets occupy the first positions. Define

$$(8.5.27) \qquad N_1 = \begin{bmatrix} N \\ I_t & 0_{t, b-t} \end{bmatrix}.$$

Then

$$(8.5.28) \qquad N_1 N_1' = \begin{bmatrix} NN' & N_0 \\ N_0' & I_t \end{bmatrix}$$

and

(8.5.29) $|N_1 N_1'| = (rk)^{-t+1}(r - \lambda_1)^{v-t-m}(rk - v\lambda_2)^{m-t-1}|C_t|,$

where the typical element of C_t is

(8.5.30) $c_{ju} = (rk - v\lambda_2)(rk\Delta_{ju} + \lambda_2 k^2)$

$$+ (\lambda_1 - \lambda_2)\left(rk \sum_{w=1}^{m} S_{jj}{}^w S_{uu}{}^w - n\lambda_2 k^2\right),$$

where $\Delta_{ju} = r - \lambda_1 - k$ or $-S_{ju}$, depending on whether $j = u$ or $j \neq u$. The matrix C_t is said to be the characteristic matrix of the t chosen sets.

We can easily prove

Theorem 8.5.12. For a regular group-divisible design

(8.5.31) $C_t = rk(rk - v\lambda_2)\{(r - \lambda_1)I_t - S_t'\}$
$$+ rk(\lambda_1 - \lambda_2)S_t{}^G + \lambda_2 k^2(r - \lambda_1)E_{t,t}.$$

By consideration of the value of $|N_1 N_1'|$, the following theorem can be proved, the proof being left as an exercise:

Theorem 8.5.13. If C_t is the characteristic matrix of any t sets chosen from a regular group-divisible design with the parameters v, b, r, k, m, n, λ_1, and λ_2, then the following hold:

1. $|C_t| \geqq 0$ if $t < b - v$.
2. $|C_t| = 0$ if $t > b - v$.
3. $r^{-2(t-1)}(r - \lambda_1)^{v-t-m}(r^2 - v\lambda_2)^{m-t-1}|C_t|$ is a perfect integral square if $t = b - v$.

We now prove

Theorem 8.5.14. For a regular symmetrical group-divisible design the number of symbols S_{ju} in common between the jth and the uth sets satisfies the inequalities

(8.5.32) $\dfrac{\lambda_2(r - \lambda_1)}{(r^2 - v\lambda_2)} \leqq S_{ju} \leqq \lambda_1$

when $\lambda_1 > \lambda_2$. The inequalities are reversed when $\lambda_1 < \lambda_2$.

Proof. Taking $t = 1$, from statement 2 of Theorem 8.5.13, we have $c_{11} = 0$, which implies that

(8.5.33) $\displaystyle\sum_{w=1}^{m} (S_{11}{}^w)^2 = r^2 - v\lambda_2 + n\lambda_2.$

Now let $t = 2$. Since $c_{11} = 0 = c_{22}$ and since $|C_2| = 0$, it is necessary that $c_{12} = c_{21} = 0$. Hence

$$(8.5.34) \qquad S_{12} = \lambda_2 + \frac{e(\lambda_1 - \lambda_2)}{r^2 - v\lambda_2},$$

where

$$(8.5.35) \qquad e = \sum_{w=1}^{m} S_{11}{}^w S_{22}{}^w - n\lambda_2.$$

Since $S_{jj}{}^w \geqq 0$ and from (8.5.33) we have

$$(8.5.36) \qquad -n\lambda_2 \leqq e \leqq r^2 - v\lambda_2.$$

Making use of (8.5.36) in (8.5.34), we get the required inequality.

Connor (1952) introduced the concepts of intersection, group-structural, and characteristic matrices and proved Theorem 8.5.14.

8.6. METHODS OF CONSTRUCTING GROUP-DIVISIBLE DESIGNS

8.6.1. Construction of Group-Divisible Designs from known BIB Designs

Many of the methods of constructing group-divisible designs that we describe in this section are given by Bose, Shrikhande, and Bhattacharya (1953).

Theorems 8.5.1 and 8.5.8 can be used to obtain many singular and semi-regular group-divisible designs from known BIB designs.

We saw in Chapter 5 that the affine resolvable $0S1$ series of BIB designs with the parameters

$$(8.6.1) \quad v^* = s^2, \qquad b^* = s(s + 1), \qquad r^* = s + 1, \qquad k^* = s, \qquad \lambda^* = 1$$

exists when s is a prime or a prime power. Making use of Theorem 8.5.8, we have

Theorem 8.6.1. If s is a prime or a prime power, a group-divisible design with the parameters

$$(8.6.2) \quad \begin{aligned} v &= s(s + 1), \qquad m = s + 1, \qquad n = s, \qquad b = s^2, \\ r &= s, \qquad k = s + 1, \qquad \lambda_1 = 0, \qquad \lambda_2 = 1 \end{aligned}$$

always exists.

Theorem 8.6.2. By omitting the sets in which a symbol θ occurs from a BIB design with the parameters $v^*, b^*, r^*, k^*, \lambda^* = 1$, we obtain a group-divisible design with the parameters

$$(8.6.3) \quad \begin{aligned} v &= v^* - 1, \qquad b = b^* - r^*, \qquad r = r^* - 1, \qquad k = k^*, \\ m &= r^*, \qquad n = k^* - 1, \qquad \lambda_1 = 0, \qquad \lambda_2 = 1, \end{aligned}$$

where two symbols belong to the same group if they occur together in the same set as θ in the solution of the BIB design.

Proof. In the r^* sets where θ occurs, on omitting θ, we find that they are disjoint and the remaining $v^* - 1 = r^*(k^* - 1)$ symbols form r^* groups of $k^* - 1$ symbols each. We can easily verify that the sets in which θ does not occur form a group-divisible design.

In view of the existence of the $OS1$ series of designs when s is a prime or a prime power, we have the following:

Corollary 8.6.2.1. If s is a prime or a prime power, the series of group-divisible designs with the parameters

(8.6.4)
$$v = b = s^2 - 1, \quad r = k = s, \quad m = s + 1,$$
$$n = s - 1, \quad \lambda_1 = 0, \quad \lambda_2 = 1$$

always exists.

Theorem 8.6.3. If a resolvable solution or at least a solution with one complete replication exists for a BIB design with the parameters v^*, b^*, r^*, k^*, λ^* in which v^* is divisible by k^*, then a group-divisible design with the parameters

(8.6.5)
$$v = v^*, \quad b = tb^* + a\,\frac{v^*}{k^*}, \quad r = tr^* + a, \quad k = k^*,$$
$$m = \frac{v^*}{k^*}, \quad n = k^*, \quad \lambda_1 = t\lambda^* + a, \quad \lambda_2 = t\lambda^*,$$

where a and t are integers ($t > 0$, $a \geqq -t\lambda^*$), always exists.

Proof. Let N be the incidence matrix of a BIB design and let N_1 be a submatrix of N corresponding to v^*/k^* sets, giving a complete replicate. Let us renumber the symbols, if necessary, so that the v^*/k^* groups of k^* symbols correspond to the symbols in the sets of N_1. Then

(8.6.6)
$$M = [E_{1,t} \times N \,|\, E_{1,a} \times N_1],$$

where $\times$ denotes the Kronecker product of matrices, can easily be verified to be the incidence matrix of a group-divisible design with the parameters given in (8.6.5).

For the $OS1$ series we have the following corollary:

Corollary 8.6.3.1. If s is a prime or a prime power, a group-divisible design with the parameters

(8.6.7)
$$v = s^2, \quad b = t(s^2 + s) + as, \quad r = t(s + 1) + a, \quad k = s,$$
$$m = s, \quad n = s, \quad \lambda_1 = t + a, \quad \lambda_2 = t,$$

where a and t are integers ($t > 0$, $a \geqq -t$), always exists.

A particular series of interest will be with $t = 1$, $a = -1$.

When s is a prime or a prime power, by considering EG(3, s) geometry and considering the points of the geometry as symbols and planes as sets, we get a BIB design with the parameters

$$(8.6.8) \quad v = s^3, \quad b = \frac{s(s^3 - 1)}{s - 1}, \quad r = \frac{s^3 - 1}{s - 1}, \quad k = s^2, \quad \lambda = s,$$

and the design can easily be seen to be affine resolvable. Applying the above theorem for this series, we get the following corollary:

Corollary 8.6.3.2. If s is a prime or a prime power, a group-divisible design with the parameters

$$(8.6.9) \quad \begin{array}{llll} v = s^3, & b = s^2(s + 1), & r = s(s + 1), & k = s^2, \\ m = s, & n = s^2, & \lambda_1 = s, & \lambda_2 = s + 1 \end{array}$$

always exists.

The group-divisible design with the parameters given in (8.6.9) was obtained by Raghavarao (1959) through hypergraecolatin cubes of the first order.

We also note that Corollary 8.5.7.1 can be used for constructing various group-divisible designs.

8.6.2. Construction of Group-Divisible Designs by the Method of Difference

With the same notation as in Section 5.7, the following theorem can be proved along lines similar to those used in Theorems 5.7.1 and 5.7.2:

Theorem 8.6.4. Let M be a module with m elements and to each element of M let there correspond n symbols. Let it be possible to find t initial sets $S_1, S_2, \cdots, S_t$, each containing k symbols, and an initial group G containing n symbols such that (a) the $n(n - 1)$ differences arising from G are all different and (b) among the $k(k - 1)t$ differences arising from the initial sets each difference occurs λ_2 times, except those that arise from G, each of which occurs λ_1 times. Then by developing the initial sets $S_1, S_2, \cdots, S_t$ we get a group-divisible design with the parameters $v = mn$, $b = mt$, $r = kt/n$, k, m, n, λ_1, λ_2, the groups of the association scheme being obtained by developing the initial group G.

Customarily M will be taken to be the module of residue classes mod m, and the initial group G consists of the symbols $0_1, 0_2, \cdots, 0_n$. We illustrate the above theorem as follows:

Illustration 8.6.4.1. Let M consist of 0, 1, 2, 3, and 4 and let to each element of the module correspond two symbols denoted by the subscripts. For the initial sets

$$(8.6.10) \quad \begin{matrix} (0_1, 1_2, 2_2, 4_2), & (0_2, 1_1, 2_1, 4_1), \\ (0_1, 2_2, 3_2, 4_2), & (0_2, 2_1, 3_1, 4_1) \end{matrix}$$

we can easily verify that the nonzero pure and mixed differences are repeated exactly three times and the zero mixed differences occur zero times. Hence, on development, these initial sets will give a group-divisible design with the parameters

$$(8.6.11) \quad \begin{matrix} v = 10, & b = 20, & r = 8, & k = 4, \\ m = 5, & n = 2, & \lambda_1 = 0, & \lambda_2 = 3, \end{matrix}$$

with the groups obtained by developing the initial set $(0_1, 0_2)$. We can easily see that the complete solution of (8.6.11) is

0_1	1_1	2_1	3_1	4_1	0_2	1_2	2_2	3_2	4_2	0_1	1_1	2_1	3_1	4_1	0_2	1_2	2_2	3_2	4_2
1_2	2_2	3_2	4_2	0_2	1_1	2_1	3_1	4_1	0_1	2_2	3_2	4_2	0_2	1_2	2_1	3_1	4_1	0_1	1_1
2_2	3_2	4_2	0_2	1_2	2_1	3_1	4_1	0_1	1_1	3_2	4_2	0_2	1_2	2_2	3_1	4_1	0_1	1_1	2_1
4_2	0_2	1_2	2_2	3_2	4_1	0_1	1_1	2_1	3_1	4_2	0_2	1_2	2_2	3_2	4_1	0_1	1_1	2_1	3_1

where the sets are written as columns, for which the groups are the five columns

$$\begin{matrix} 0_1 & 1_1 & 2_1 & 3_1 & 4_1 \\ 0_2 & 1_2 & 2_2 & 3_2 & 4_2 \end{matrix}$$

The scope of the method of differences can be further extended by using the concepts of the partial cycle of C. R. Rao (1946) and the double module, and the reader may refer to Bose, Shrikhande, and Bhattacharya (1953) in this connection.

8.6.3. Construction of Group-Divisible Designs through Finite Geometries

Theorem 8.6.5. If s is a prime or a prime power, a semiregular group-divisible design with the parameters

$$(8.6.12) \quad \begin{matrix} v = ms, & b = s^3, & r = s^2, & k = m, \\ m, & n = s, & \lambda_1 = 0, & \lambda_2 = s, \end{matrix}$$

where $m \leq (s^2 + s + 1)$, always exists.

Proof. Consider the geometry PG(3, s) and choose a point 0, say, with the coordinates $(0, 0, 0, 1)$. Through 0 there pass $s^2 + s + 1$ lines, and on each line there are $s + 1$ points. Choose any $m \leq s^2 + s + 1$ lines through 0 and let the points other than 0 on these lines correspond to symbols. We then have ms symbols divided into m groups of s symbols. Let the planes not passing through 0 correspond to sets. Then these are s^3 sets, and each set contains exactly one symbol from each group. Also each symbol is contained in s^2 sets. We can easily verify that symbols from the same group do not occur together, whereas symbols from different groups occur together s times. Thus the proof is complete.

Another interesting geometric method of constructing group-divisible designs, due to Sprott (1959), is given in the following theorem:

Theorem 8.6.6. If s is a prime or a prime power, a regular, symmetrical, group-divisible design with the parameters

$$(8.6.13) \quad \begin{aligned} v = b = s^4 - s = (s^2 - s)(s^2 + s + 1), \quad & m = s^2 + s + 1, \\ n = s^2 - s, \quad r = k = s^2, \quad & \lambda_1 = 0, \quad \lambda_2 = 1 \end{aligned}$$

always exists.

Proof. By identifying the points as symbols and the lines as sets, we can construct from $PG(2, s^2)$ and $PG(2, s)$ BIB designs with the parameters

$$(8.6.14) \quad v = b = s^4 + s^2 + 1, \quad r = k = s^2 + 1, \quad \lambda = 1$$

and

$$(8.6.15) \quad v = b = s^2 + s + 1, \quad r = k = s + 1, \quad \lambda = 1.$$

It has been shown by Carmichael (1956) that the geometry $PG(m, s^n)$ contains $PG(m, s^k)$ if k is a factor of n. Hence the BIB design with the parameters given in (8.6.14) contains the design with the parameters given in (8.6.15).

Divide the sets of (8.6.14) into four components as follows:

1. Sets and symbols of (8.6.15).
2. Sets of (8.6.15) but the other $s^4 - s$ symbols.
3. Symbols of (8.6.15) but the other $s^4 - s$ sets.
4. Symbols and sets other than those of (8.6.15).

It can easily be seen that plan 4 is the required plan and the sets in plan 2 provide the groups of the association scheme.

It is still an open problem whether the group-divisible design with the parameters given in (8.6.13) and the BIB design with the parameters given in (8.6.15) can be suitably embedded to get the solution of (8.6.14).

Dembowski (1968) showed that finite, proper uniform projective Hjelmslev planes are precisely the regular group-divisible designs with the parameters

$$(8.6.16) \quad \begin{aligned} v = b = s^2(s^2 + s + 1), \quad & r = k = s(s + 1), \\ m = s^2 + s + 1, \quad n = s^2, \quad & \lambda_1 = s, \quad \lambda_2 = 1, \end{aligned}$$

and their duals are likewise regular group-divisible designs with the same parameters.

The series of group-divisible designs with the parameters given in (8.6.16) can also be constructed by a new combinatorial method. For this purpose we need R-form BIB designs as defined below.

Let the sets of a BIB design with the parameters

$$(8.6.17) \quad v^* = s^2, \qquad b^* = s^2(s+1), \qquad r^* = s(s+1), \qquad k^* = s, \qquad \lambda^* = s$$

be divided into s^2 classes of $s+1$ sets each, and let T_{ij} denote the jth set of the ith class ($j = 1, 2, \cdots, s+1; i = 1, 2, \cdots, s^2$). We call the s^2 sets for a constant j the jth segment of sets. We arrange the symbols in such a way that each symbol occurs in s sets of each segment. Choose any symbol θ and any segment j. Let θ occur in $T_{i_1 j}, T_{i_2 j}, \cdots, T_{i_s j}$ sets of the jth segment. Then the solution of the BIB design with the parameters given in (8.6.17) is said to be in the R-form if the symbols of the sets $T_{i_1 j'}, T_{i_2 j'}, \cdots, T_{i_s j'}$ form a complete replication of the s^2 symbols when $j' \neq j$ and $j' = 1, 2, \cdots, s+1$.

Consider the 0S2 series BIB design where each symbol occurs once in each column. Let $(a_1, a_2, \cdots, a_{s+1})$ be a set of the 0S2 BIB design. We introduce a new method of combining these sets with the sets of the R-form BIB design with the parameters given in (8.6.17). From each set $(a_1, a_2, \cdots, a_{s+1})$ we generate s^2 sets, each of size $s(s+1)$, where the ith set is

$$(a_1 T_{i1}, a_2 T_{i2}, \cdots, a_{s+1} T_{i(s+1)}) \qquad \text{for } i = 1, 2, \cdots, s^2,$$

the notation $a_j T_{ij}$ standing for s ordered pairs of symbols having a_j as the first symbol and each of the symbols of T_{ij} as the second. By generating s^2 sets from each set of an 0S2 BIB design in the above manner, we get a group-divisible design with the parameters given in (8.6.16).

We illustrate this method by constructing the group-divisible design with the parameters given in (8.6.16) for $s = 2$. The solution for 0S2 series of BIB designs for $s = 2$ is

$$(8.6.18) \qquad \begin{array}{llll} (0, 1, 3); & (1, 2, 4); & (2, 3, 5); & (3, 4, 6); \\ (4, 5, 0); & (5, 6, 1); & (6, 0, 2). \end{array}$$

The R-form solution of the BIB design with the parameters given in (8.6.17) for $s = 2$ in the symbols a, b, c, d is

$$(8.6.19) \qquad \begin{array}{lllll} (a, b); & (a, b); & (c, d); & (c, d) & \text{first segment} \\ (a, c); & (b, d); & (a, c); & (b, d) & \text{second segment} \\ (b, c); & (a, d); & (a, d); & (b, c) & \text{third segment} \\ \text{first} & \text{second} & \text{third} & \text{fourth} \\ \text{class} & \text{class} & \text{class} & \text{class} \end{array}$$

Combining the solutions (8.6.18) and (8.6.19) in the above described method, we get

(8.6.20)

$(0a, 0b, 1a, 1c, 3b, 3c)$;	$(0a, 0b, 1b, 1d, 3a, 3d)$;
$(0c, 0d, 1a, 1c, 3a, 3d)$;	$(0c, 0d, 1b, 1d, 3b, 3c)$;
$(1a, 1b, 2a, 2c, 4b, 4c)$;	$(1a, 1b, 2b, 2d, 4a, 4d)$;
$(1c, 1d, 2a, 2c, 4a, 4d)$;	$(1c, 1d, 2b, 2d, 4b, 4c)$;
$(2a, 2b, 3a, 3c, 5b, 5c)$;	$(2a, 2b, 3b, 3d, 5a, 5d)$;
$(2c, 2d, 3a, 3c, 5a, 5d)$;	$(2c, 2d, 3b, 3d, 5b, 5c)$;
$(3a, 3b, 4a, 4c, 6b, 6c)$;	$(3a, 3b, 4b, 4d, 6a, 6d)$;
$(3c, 3d, 4a, 4c, 6a, 6d)$;	$(3c, 3d, 4b, 4d, 6b, 6c)$;
$(4a, 4b, 5a, 5c, 0b, 0c)$;	$(4a, 4b, 5b, 5d, 0a, 0d)$;
$(4c, 4d, 5a, 5c, 0a, 0d)$;	$(4c, 4d, 5b, 5d, 0b, 0c)$;
$(5a, 5b, 6a, 6c, 1b, 1c)$;	$(5a, 5b, 6b, 6d, 1a, 1d)$;
$(5c, 5d, 6a, 6c, 1a, 1d)$;	$(5c, 5d, 6b, 6d, 1b, 1c)$;
$(6a, 6b, 0a, 0c, 2b, 2c)$;	$(6a, 6b, 0b, 0d, 2a, 2d)$;
$(6c, 6d, 0a, 0c, 2a, 2d)$;	$(6c, 6d, 0b, 0d, 2b, 2c)$.

This can be seen to be the solution of a regular group-divisible design with the parameters

$$(8.6.21)\quad v = 28 = b,\quad r = 6 = k,\quad m = 7,\quad n = 4,\quad \lambda_1 = 2,\quad \lambda_2 = 1,$$

where the seven groups of the association scheme are

(8.6.22)

$$0a, 0b, 0c, 0d;$$
$$1a, 1b, 1c, 1d;$$
$$2a, 2b, 2c, 2d;$$
$$3a, 3b, 3c, 3d;$$
$$4a, 4b, 4c, 4d;$$
$$5a, 5b, 5c, 5d;$$
$$6a, 6b, 6c, 6d.$$

The combination of parameters given in (8.6.21) and its solution are not listed in the 1954 tables of two-associate-class PBIB designs.

8.7. COMBINATORIAL PROBLEMS OF TRIANGULAR DESIGNS

8.7.1. Alternative Definition

We have defined the triangular association scheme in Definition 8.4.3. An alternative definition, provided by P. W. M. John (1966), can be easily extended to higher associate classes and is given below:

Let the $n(n-1)/2$ symbols be represented by ordered pairs (θ, ϕ), where θ and ϕ satisfy $0 < \theta < \phi \leq n$. Then two symbols are said to be first associates if their representations have an integer in common. Otherwise they are said to be second associates.

We can easily verify that this characterization is the same as Definition 8.4.3.

8.7.2. Uniqueness of the Association Scheme When $n \neq 8$

The parameters of the triangular association scheme uniquely determine the association scheme when $n \neq 8$, and this result was proved by Connor (1958), Shrikhande (1959a), Hoffman (1960), and L. C. Chang (1960). If $n = 8$, there are three other possible association schemes with the parameters of the triangular association scheme; we discuss them in subsection 8.7.3.

Letting $(x_1, x_2) = i$ denote that the symbols x_1 and x_2 are ith associates and $p_{jk}{}^i(x_1, x_2)$ denote the number of common symbols that are jth associates of x_1 and kth associates of x_2, where $(x_1, x_2) = i$, we give the following characterization of the triangular association scheme due to Shrikhande (1959a):

Theorem 8.7.1. A necessary and sufficient condition for a two-associate-class association scheme for $n(n-1)/2$ symbols with $p_{11}{}^1(x_1, x_2) = n - 2$ to be a triangular association scheme is that all the first associates of any symbol x whatsoever be divisible into two sets $(y_1, y_2, \cdots, y_{n-2})$ and $(z_1, z_2, \cdots, z_{n-2})$ such that $(y_i, y_j) = (z_i, z_j) = 1$ for $i, j = 1, 2, \cdots, n-2; i \neq j$

Proof. The necessity part is obvious.

Since y_i has $n - 3$ first associates y_j and since $p_{11}{}^1(x, y_i) = n - 2$, y_i has one z as first associate. Without loss of generality, let us assume that $(y_i, z_i) = 1$ for $i = 1, 2, \cdots, n-2$. Thus we can write the symbols as

(8.7.1)
$$\begin{array}{ccccccc} * & x & y_1 & y_2 & \cdots & y_{n-2} \\ x & * & z_1 & z_2 & \cdots & z_{n-2} \end{array}$$

where symbols in the same row or column are first associates and others are second associates.

Now the first associates of y_1 being $x, y_2, \cdots, y_{n-2}, z_1$, let $u_1, u_2, \cdots, u_{n-3}$ be the remaining first associates of y_1. We assume without loss of generality that

(8.7.2) $\quad (y_2, u_1) = 1, \quad (y_3, u_2) = 1, \cdots, (y_{n-2}, u_{n-3}) = 1.$

Thus the $3n - 6$ symbols can be arranged in three rows as follows:

(8.7.3)
$$\begin{array}{ccccccc} * & x & y_1 & y_2 & \cdots & y_{n-2} \\ x & * & z_1 & z_2 & \cdots & z_{n-2} \\ y_1 & z_1 & * & u_1 & \cdots & u_{n-3} \end{array}$$

The first associates of y_2 so far enumerated are $x, y_1, y_3, \cdots, y_{n-2}, z_2$, and u_1. Let $v_1, v_2, \cdots, v_{n-4}$ be the remaining $n - 4$ first associates of y_2.

The two sets of first associates of y_2 are thus

(8.7.4)

$$
\begin{array}{cccccc}
x & y_1 & y_3 & \cdots & y_{n-2} \\
z_2 & u_1 & v_1 & \cdots & v_{n-4}
\end{array}
$$

where we assume that symbols in the same column are first associates. We thus write down the fourth row, obtaining

(8.7.5)

$$
\begin{array}{cccccccc}
* & x & y_1 & y_2 & y_3 & \cdots & y_{n-2} \\
x & * & z_1 & z_2 & z_3 & \cdots & z_{n-2} \\
y_1 & z_1 & * & u_1 & u_2 & \cdots & u_{n-3} \\
y_2 & z_2 & u_1 & * & v_1 & \cdots & v_{n-4}
\end{array}
$$

Continuing similarly, we can build up the triangular association scheme and thereby complete the proof for the sufficiency of the stated condition.

We now prove the following theorem:

Theorem 8.7.2. [Hoffman (1960)]. When $n \neq 8$, the $2(n-2)$ first associates of any symbol can be divided into two sets such that the $n-2$ symbols of one class are mutually first associates of each other; the $n-2$ symbols of the other set are mutually first associates.†

Proof. Let B_1 be the association matrix of the first associates as defined in Section 8.3. It can be easily verified that the characteristic roots of B_1 are as follows:

1. $2n-4$ with multiplicity 1.
2. $n-4$ with multiplicity $n-1$.
3. -2 with multiplicity $v-n$.

We can thus see that -2 is the least characteristic root of B_1. Thus the least characteristic root of any principal submatrix K of B_1 must be greater than or equal to -2, and if equality holds, the characteristic vector corresponding to that root of K should be orthogonal to the projection of the characteristic vectors corresponding to the least characteristic root -2 of B_1 on the subspace corresponding to K. By using this result and the parameters of the triangular association scheme, we can show that B_1 cannot contain

(8.7.6)

$$
\begin{bmatrix}
0 & 0 & 0 & 1 \\
0 & 0 & 0 & 1 \\
0 & 0 & 0 & 1 \\
1 & 1 & 1 & 0
\end{bmatrix},
$$

† Read with Theorem 8.7.1, this result implies that the triangular association scheme is uniquely determined by its parameters when $n \neq 8$.

$$(8.7.7) \quad \begin{bmatrix} 0 & 0 & 1 & 1 & 1 & 1 & 1 & 1 \\ 0 & 0 & 1 & 1 & 1 & 1 & 0 & 0 \\ 1 & 1 & 0 & 0 & 1 & 1 & 1 & 1 \\ 1 & 1 & 0 & 0 & 1 & 1 & 0 & 0 \\ 1 & 1 & 1 & 1 & 0 & 0 & 1 & 0 \\ 1 & 1 & 1 & 1 & 0 & 0 & 0 & 1 \\ 1 & 0 & 1 & 0 & 1 & 0 & 0 & a \\ 1 & 0 & 1 & 0 & 0 & 1 & a & 0 \end{bmatrix},$$

or

$$(8.7.8) \quad \begin{bmatrix} 0 & 0 & 1 & 1 & 1 & 1 & 1 & 1 \\ 0 & 0 & 1 & 1 & 1 & 1 & 0 & 0 \\ 1 & 1 & 0 & 0 & 1 & 1 & 1 & 1 \\ 1 & 1 & 0 & 0 & 1 & 1 & 0 & 0 \\ 1 & 1 & 1 & 1 & 0 & 0 & 1 & 1 \\ 1 & 1 & 1 & 1 & 0 & 0 & 0 & 0 \\ 1 & 0 & 1 & 0 & 1 & 0 & 0 & 0 \\ 1 & 0 & 1 & 0 & 1 & 0 & 0 & 0 \end{bmatrix},$$

where $a = 0$ or 1, as principal submatrices. When $n \neq 8$, we can also show that if the symbols x_1 and x_2 are second associates and the symbols x_3, x_4, x_5, x_6 are first associates of each of x_1 and x_2, then the principal submatrix of B_1 corresponding to these six symbols will necessarily be of the form[†]

$$(8.7.9) \quad \begin{bmatrix} 0 & 0 & 1 & 1 & 1 & 1 \\ 0 & 0 & 1 & 1 & 1 & 1 \\ 1 & 1 & 0 & 0 & 1 & 1 \\ 1 & 1 & 0 & 0 & 1 & 1 \\ 1 & 1 & 1 & 1 & 0 & 0 \\ 1 & 1 & 1 & 1 & 0 & 0 \end{bmatrix}.$$

Let x be a given symbol and let $(x, y) = 2$. Choose four more symbols z_1, z_2, z_3, z_4 so that the association among $x, y, z_1, z_2, z_3,$ and z_4 is as given

[†] In case $n = 8$, other associationship between x_3, x_4, x_5, and x_6 is possible, when they are first associates of x_1 and x_2, with $(x_1, x_2) = 2$.

in (8.7.9). Now z_3 and z_4 are first associates of both x and z_1, and, since $p_{11}{}^1 = n - 2$, there are $n - 4$ other first associates of both x and z_1. Each of these must be a first associate of at least one of the symbols z_3 or z_4. Otherwise it, z_3, and z_4 would be mutual second associates and x would be a first associate of each of the three, violating structure (8.7.6). Furthermore, from (8.7.7) each of these symbols is a first associate of z_3 or z_4. Without loss of generality, say it is z_3. From (8.7.8) these $n - 4$ symbols are mutual first associates; furthermore, each is a first associate of z_1 and z_3, which are themselves first associates, and thus z_1, z_3, and these $n - 4$ symbols are altogether $n - 2$ first associates of x and are mutual first associates.

Of the $n - 2$ first associates of x and z_2, the symbol z_3 is in the class already described, z_4 is not, and there are $n - 4$ others. These $n - 4$ are mutual first associates by the same reasoning as above; they are entirely different from the previous $n - 4$ symbols of the first class, since each of those was a second associate of z_2. Each is obviously a first associate of z_2 and z_4. Thus z_2, z_4, and these $n - 4$ symbols constitute our second class, and the proof is complete.

8.7.3. Pseudotriangular Association Scheme

When $n = 8$, there are three more possible association schemes other than the triangular association scheme, with the parameters of the triangular association scheme. This result was proved by L.-C. Chang (1960). These three association schemes were also reproduced by Seiden (1966). For completeness we reproduce these three association schemes in Tables 8.7.1, 8.7.2, and 8.7.3.

Table 8.7.1. First pseudotriangular association scheme when $n = 8$

Symbol	First Associates												
1	2	3	4	5	6	7	8	9	10	11	12	13
2	1	3	4	5	6	7	8	14	15	16	17	18
3	1	2	4	5	9	10	11	14	15	16	19	20
4	1	2	3	5	6	9	12	14	15	21	22	23
5	1	2	3	4	7	10	13	14	15	21	24	25
6	1	2	4	7	8	9	12	17	18	22	23	26
7	1	2	5	6	8	10	13	17	18	24	25	26
8	1	2	6	7	11	12	13	16	17	18	27	28
9	1	3	4	6	10	11	12	19	20	22	23	26
10	1	3	5	7	9	11	13	19	20	24	25	26
11	1	3	8	9	10	12	13	16	19	20	27	28
12	1	4	6	8	9	11	13	21	22	23	27	28
13	1	5	7	8	10	11	12	21	24	25	27	28
14	2	3	4	5	15	16	17	19	21	22	24	27
15	2	3	4	5	14	16	18	20	21	23	25	28
16	2	3	8	11	14	15	17	18	19	20	27	28

Table 8.7.1—(Continued)

Symbol	First Associates											
17	2	6	7	8	14	16	18	19	22	24	26	27
18	2	6	7	8	15	16	17	20	23	25	26	28
19	3	9	10	11	14	16	17	20	22	24	26	27
20	3	9	10	11	15	16	18	19	23	25	26	28
21	4	5	12	13	14	15	22	23	24	25	27	28
22	4	6	9	12	14	17	19	21	23	24	26	27
23	4	6	9	12	15	18	20	21	22	25	26	28
24	5	7	10	13	14	17	19	21	22	25	26	27
25	5	7	10	13	15	18	20	21	23	24	26	28
26	6	7	9	10	17	18	19	20	22	23	24	25
27	8	11	12	13	14	16	17	19	21	22	24	28
28	8	11	12	13	15	16	18	20	21	23	25	27

Table 8.7.2. Second pseudotriangular association scheme when $n = 8$

Symbol	First Associates											
1	2	3	4	5	6	7	8	9	10	11	12	13
2	1	3	4	5	6	7	8	14	15	16	17	18
3	1	2	4	5	9	10	11	14	15	16	19	20
4	1	2	3	6	9	12	13	14	15	16	21	22
5	1	2	3	7	8	10	11	14	17	19	23	24
6	1	2	4	7	8	12	13	15	18	21	25	26
7	1	2	5	6	8	10	12	17	18	23	25	27
8	1	2	5	6	7	11	13	17	18	24	26	28
9	1	3	4	10	11	12	13	16	20	22	27	28
10	1	3	5	7	9	11	12	19	20	23	25	27
11	1	3	5	8	9	10	13	19	20	24	26	28
12	1	4	6	7	9	10	13	21	22	23	25	27
13	1	4	6	8	9	11	12	21	22	24	26	28
14	2	3	4	5	15	16	17	19	21	22	23	24
15	2	3	4	6	14	16	18	19	20	21	25	26
16	2	3	4	9	14	15	17	18	20	22	27	28
17	2	5	7	8	14	16	18	22	23	24	27	28
18	2	6	7	8	15	16	17	20	25	26	27	28
19	3	5	10	11	14	15	20	21	23	24	25	26
20	3	9	10	11	15	16	18	19	25	26	27	28
21	4	6	12	13	14	15	19	22	23	24	25	26
22	4	9	12	13	14	16	17	21	23	24	27	28
23	5	7	10	12	14	17	19	21	22	24	25	27
24	5	8	11	13	14	17	19	21	22	23	26	28
25	6	7	10	12	15	18	19	20	21	23	26	27
26	6	8	11	13	15	18	19	20	21	24	25	28
27	7	9	10	12	16	17	18	20	22	23	25	28
28	8	9	11	13	16	17	18	20	22	24	26	27

Table 8.7.3. Third pseudotriangular association scheme when $n = 8$

Symbol	First Associates												
1	2	3	4	5	6	7	8	9	10	11	12	13
2	1	3	4	5	6	7	8	14	15	16	17	18
3	1	2	4	5	6	9	10	14	15	19	20	21
4	1	2	3	5	6	9	11	14	16	22	23	24
5	1	2	3	4	7	9	11	15	17	19	22	25
6	1	2	3	4	8	10	12	14	16	20	23	26
7	1	2	5	8	11	12	13	15	17	18	25	27
8	1	2	6	7	10	12	13	16	17	18	26	28
9	1	3	4	5	10	11	13	19	21	22	24	28
10	1	3	6	8	9	12	13	19	20	21	26	28
11	1	4	5	7	9	12	13	22	23	24	25	27
12	1	6	7	8	10	11	13	20	23	25	26	27
13	1	7	8	9	10	11	12	18	21	24	27	28
14	2	3	4	6	15	16	18	20	21	23	24	27
15	2	3	5	7	14	17	18	19	20	21	25	27
16	2	4	6	8	14	17	18	22	23	24	26	28
17	2	5	7	8	15	16	18	19	22	25	26	28
18	2	7	8	13	14	15	16	17	21	24	27	28
19	3	5	9	10	15	17	20	21	22	25	26	28
20	3	6	10	12	14	15	19	21	23	25	26	27
21	3	9	10	13	14	15	18	19	20	24	27	28
22	4	5	9	11	16	17	19	23	24	25	26	28
23	4	6	11	12	14	16	20	22	24	25	26	27
24	4	9	11	13	14	16	18	21	22	23	27	28
25	5	7	11	12	15	17	19	20	22	23	26	27
26	6	8	10	12	16	17	19	20	22	23	25	28
27	7	11	12	13	14	15	18	20	21	23	24	25
28	8	9	10	13	16	17	18	19	21	22	24	26

8.7.4. Set Structure of Certain Triangular Designs

The following result on the set structure of triangular designs was obtained by Raghavarao (1960a):

Theorem 8.7.3. If in a triangular design

$$(8.7.10) \qquad r + (n - 4)\lambda_1 - (n - 3)\lambda_2 = 0,$$

then $2k$ is divisible by n. Furthermore, every set of the design contains $2k/n$ symbols from each of the n rows of the association scheme.

The proof follows along lines similar to those used in Theorem 8.5.5 and is left as an exercise to the reader. It is to be noted that the condition stated in Raghavarao's paper and (8.7.10) are the same.

The following result due to Agrawal (1964) can be proved along lines similar to those used in Theorem 8.5.9:

Theorem 8.7.4. If $\rho_1 = r + (n - 4)\lambda_1 - (n - 3)\lambda_2$ and $\rho_2 = r - 2\lambda_1 + \lambda_2$ are the characteristic roots of NN' other than rk, where N is the incidence matrix of the triangular design, and if x is the number of symbols common to any two sets, then

$$(8.7.11) \quad \max\left[0, 2k - v, k - \rho_i\right] \leqq x \leqq \min\left[k, \rho_i + \frac{2(rk - \rho_i)}{b} - k\right],$$

where $i = 1$ if $\lambda_1 > \lambda_2$ and $i = 2$ if $\lambda_1 < \lambda_2$.

The following theorem, due to S. M. Shah (1964), can be easily proved:

Theorem 8.7.5. A given set of a triangular design with $r + (n - 4)\lambda_1 - (n - 3)\lambda_2 = 0$ cannot have more than

$$(8.7.12) \qquad b - 1 - \frac{n(v - n)(r - 1)^2}{n(b + 1 - 2r) - (v - rk)(n - 2)}$$

disjoint sets with it. If some set has that many disjoint sets, then

$$k\left[n(b + 1 - 2r) - (v - rk)(n - 2)\right]/n(v - n)(r - 1)$$

is an integer and each nondisjoint set has

$$(8.7.13) \qquad \frac{k\{n(b + 1 - 2r) - (v - rk)(n - 2)\}}{n(v - n)(r - 1)}$$

symbols in common with that given set.

Corollary 8.7.5.1. The necessary and sufficient conditions for a set of a triangular design with $r + (n - 4)\lambda_1 - (n - 3)\lambda_2 = 0$ to have the same number of symbols in common with each of the remaining sets is that (a) $b = v - n + 1$ and (b) $k(r - 1)/(v - n)$ be an integer.

8.8. METHODS OF CONSTRUCTING TRIANGULAR DESIGNS

Chang, Liu, and Liu (1965) listed 225 possible parameter combinations for triangular designs in the range $r, k \leq 10$ and showed that 97 of them have solutions, 110 have no solutions, and 18 remain unsolved. Their list of unsolved parameter combinations is presented in Table 8.8.1.

Table 8.8.1. Values for the unsolved parametric combinations for triangular designs

Series	v	b	r	k	λ_1	λ_2
1	15	27	9	5	3	2
2	15	30	10	5	2	4
3	21	42	10	5	1	3
4	21	42	10	5	3	1
5	21	21	6	6	2	1
6	21	28	8	6	3	1
7	21	35	10	6	2	3
8	21	35	10	6	3	2
9	21	30	10	7	2	4
10	21	30	10	7	4	2
11	21	30	10	7	5	1
12	21	21	10	10	4	5
13	36	63	7	4	0	1
14	45	63	7	5	0	1
15	45	45	9	9	1	2
16	55	99	9	5	0	1
17	55	55	10	10	3	1
18	66	99	9	6	0	1

We now give some methods of constructing triangular designs.

By writing the rows of the association scheme as sets of a design we get a triangular design. Hence

Theorem 8.8.1. A triangular design with the parameters

$$(8.8.1) \quad v = \frac{n(n-1)}{2}, \quad b = n, \quad r = 2, \quad k = n - 1, \quad \lambda_1 = 1, \quad \lambda_2 = 0$$

can always be constructed.

It is to be noted that the triangular design with the parameters given in (8.8.1) can be obtained by dualizing an irreducible BIB design with the parameters

$$(8.8.2) \quad v^* = n, \quad b^* = \binom{n}{2}, \quad r^* = n - 1, \quad k^* = 2, \quad \lambda^* = 1.$$

The following theorem gives another method of constructing triangular designs, the proof of which will be given in Chapter 10:

Theorem 8.8.2. [Shrikhande (1960)]. If there exists a BIB design with the parameters

$$(8.8.3) \quad v^* = \frac{(n-1)(n-2)}{2}, \quad b^* = \frac{n(n-1)}{2},$$
$$r^* = n, \quad k^* = n-2, \quad \lambda^* = 2,$$

then, by dualizing it, a triangular design with the parameters

$$(8.8.4) \quad v = \frac{n(n-1)}{2}, \quad b = \frac{(n-1)(n-2)}{2}, \quad r = n-2,$$
$$k = n, \quad \lambda_1 = 1, \quad \lambda_2 = 2$$

can be constructed.

Another method of constructing a triangular design with the parameters (8.8.4) is given by the following:

Theorem 8.8.3. [Shrikhande (1960)]. By omitting all the sets containing a particular symbol from a BIB design with the parameters

$$(8.8.5) \quad v^* = b^* = \frac{n(n-1)}{2} + 1, \quad r^* = k^* = n, \quad \lambda^* = 2$$

a triangular design with the parameters given in (8.8.4) can be constructed.

Proof. Since a BIB design with the parameters given in (8.8.5) is symmetrical, every pair of sets of the design has two symbols in common. Let θ be the particular symbol. Considering the k sets in which θ occurs and omitting θ, we observe that every symbol occurs twice and the association scheme is triangular. We can now verify that the remaining sets form the triangular design with the parameters given in (8.8.4).

We can easily show that the existence of a triangular design with the parameters given in (8.8.4) implies the existence of the two series of BIB designs with the parameters given in (8.8.3) and (8.8.5). It is to be remarked here that triangular designs with the parameters given in (8.8.4) do not exist for all possible n.

The following theorems provide other methods of constructing triangular designs:

Theorem 8.8.4. [Chang, Liu, and Liu (1965)]. If there exists a BIB design with the parameters

$$(8.8.6) \quad v^* = n-1, \quad b^*, \quad r^*, \quad k^*, \quad \lambda^*,$$

then a triangular design with the parameters

(8.8.7)
$$v = \frac{n(n-1)}{2}, \qquad b = nb^*, \qquad r = 2r^*,$$
$$k = k^*, \qquad \lambda_1 = \lambda^*, \qquad \lambda_2 = 0$$

can always be constructed.

Proof. By writing down the BIB design with the symbols of the 1st, 2nd, $\cdots$, and nth rows of the triangular association scheme, we can show that a triangular design with the parameters given in (8.8.7) can be obtained.

Theorem 8.8.5 [Shrikhande (1965)]. If there exists a triangular design with the parameters

(8.8.8)
$$v = (2n-1)n, \qquad b = (2n-1)(2n-3), \qquad r = 2n-3,$$
$$k = n, \qquad \lambda_1 = 0, \qquad \lambda_2 = 1,$$

then a triangular design with the parameters

(8.8.9)
$$v' = (2n-1)(n-1), \qquad b' = (3n-1)(2n-3), \qquad r' = 2n-3,$$
$$k' = n-1, \qquad \lambda_1 = 0, \qquad \lambda_2 = 1$$

can always be constructed.

Conversely, if there exists a triangular design with the parameters given in (8.8.9), a triangular design with the parameters given in (8.8.8) can always be constructed.

Proof. Let

(8.8.10)
$$T_{2n} = \begin{matrix} * & 1 & 2 & \cdots & 2n-1 \\ 1 & * & 2n & \cdots & 4n-3 \\ 2 & 2n & * & \cdots & 6n-6 \\ \cdot & \cdot & \cdot & \cdots & \cdot \\ 2n-1 & 4n-3 & 6n-6 & \cdots & * \end{matrix}$$

be the association scheme of the design whose parameters are given in (8.8.8). In view of Theorem 8.7.3, each set of this design contains one symbol from each row of T_{2n}. We can easily see that, by omitting the symbols $1, 2, \cdots, 2n-1$ from this design, we can get the triangular design with the parameters given in (8.8.9), whose association scheme T_{2n-1} is obtained by omitting the first row and column of T_{2n}.

To prove the second part, let T_{2n-1} and T_{2n-2} be the association schemes obtained from T_{2n} by omitting the first row and column, and the first two rows and columns, respectively. Let the triangular design with the parameters given in (8.8.9) have the T_{2n-1} association scheme. Let θ be any symbol of T_{2n-2}. Then θ has two first associates and $2n-4$ second associates from the $2n-2$ treatments in the first row of T_{2n-1}. Since $r' = 2n-3$, $\lambda_1 = 0$, and $\lambda_2 = 1$, it is clear that through each of the symbols $4n-2$, $4n-1$, $\cdots$, $6n-6$ there is a different set that contains no symbol from the first row of T_{2n-1}. Denote the class of sets thus formed by S_1'. We can easily verify that S_1' provides a complete replication of the $(2n-3)(n-1)$ symbols of T_{2n-2}. Since the scheme T_{2n-1} is unchanged by interchanging the first and ith rows and columns, it is obvious that the sets of the design whose parameters are given in (8.8.9) can be partitioned into mutually disjoint and exhaustive classes S_i', each of $2n-3$ sets, with the property that all the symbols of T_{2n-1}, excepting those in the ith row, occur exactly once in S_i' $(i = 1, 2, \cdots, 2n-1)$. To each set of class S_i' adjoin a new symbol i to form the class S_i of sets $(i = 1, 2, \cdots, 2n-1)$. Consider the design obtained by the class of sets $S_1, S_2, \cdots, S_{2n-1}$. This design can now be verified to be a triangular one with the parameters given in (8.8.8).

8.9. COMBINATORIAL PROBLEMS OF LATIN-SQUARE-TYPE DESIGNS

8.9.1. Alternative Definition

We have defined the L_i association scheme in Definition 8.4.4. Shrikhande and Jain (1962) provided the following alternative definition of this association scheme: Suppose an orthogonal array $(s^2, i, s, 2)$ exists. Identify the columns with s^2 symbols in any arbitrary manner. Define the first associates of any symbol θ as those symbols for which the corresponding columns coincide with the column corresponding to θ in exactly one position. Define the remaining symbols as the second associates of θ. The reader can easily verify that this definition is equivalent to Definition 8.4.4.

8.9.2. Uniqueness of the Association Scheme

The uniqueness of the L_i association scheme was studied by Mesner (1956), Bruck (1963) for $i \geq 2$, and Shrikhande (1959b) for $i = 2$. In this subsection we restrict ourselves to show the uniqueness of the L_2 association scheme.

Combinatorially enumerating the possibilities, Shrikhande (1959b) established the following result, for whose proof we refer to the original paper:

Theorem 8.9.1. Given the parameters of the L_2 association scheme, if $s = 2$, 3, or $s > 4$ and if the first associates of any symbol θ are $\phi_1, \phi_2, \cdots, \phi_{s-1}, \psi_1, \psi_2 \cdots \psi_{s-1}$ where the set $(\phi_2, \cdots, \phi_{s-1})$ is the set of first associates common to both θ and ϕ_1 and the set $(\psi_1, \psi_2, \cdots, \psi_{s-1})$ is the set of first associates of θ, which are the second associates of ϕ_1, then any two symbols from the set $(\phi_1, \cdots, \phi_{s-1})$ are first associates. Similarly any two symbols from the set $(\psi_1, \psi_2, \cdots, \psi_{s-1})$ are first associates, and any symbol ϕ_i is a second associate of any symbol ψ_j $(i, j = 1, 2, \cdots, s - 1)$. Then the following obtains:

Theorem 8.9.2. The parameters of the L_2 association scheme uniquely determine the L_2 association scheme if $s \neq 4$.

Proof. There is nothing to prove when $s = 2$. For $s = 3$, or $s > 4$, from Theorem 8.9.1, we can write down the first associates of a symbol θ as follows:

$$\begin{array}{ccccc} & \theta & \phi_1 & \phi_2 & \cdots & \phi_{s-1} \\ & \psi_1 & & & & \\ (8.9.1) & & & & & \\ & \psi_2 & & & & \\ & \cdot & & & & \\ & \psi_{s-1} & & & & \end{array}$$

where symbols in the first row and the first column are mutual first associates and any symbol ϕ is a second associate of any symbol ψ. Let δ be any second associate of θ. Then, since $p_{11}{}^2(\theta, \delta) = 2$, δ has two first associates from the symbols ϕ and ψ. We can easily show that two ϕ symbols or two ψ symbols cannot be first associates of δ. Thus δ has one ϕ symbol and one ψ symbol as first associates, say ϕ_i and ψ_j, respectively. Hence any δ, where $(\theta, \delta) = 2$, determines uniquely a pair (ϕ_i, ψ_j) such that $(\phi_i, \delta) = 1$, $(\psi_j, \delta) = 1$. We can easily show that conversely every pair (ϕ_i, ψ_j) uniquely determines δ such that $(\theta, \delta) = 2$, $(\phi_i, \delta) = 1 = (\psi_j, \delta)$. Thus the correspondence between δ and the pair (ϕ_i, ψ_j) is a bijection. We can therefore arrange δ in the cell determined by the intersection of the ψ_j row and the ϕ_i column. Thus the $(s - 1)^2$ blank positions of (8.9.1) can be uniquely filled by utilizing the $(s - 1)^2$ second associates of θ, to form the scheme

$$\begin{array}{ccccc} & \theta & \phi_1 & \phi_2 & \cdots & \phi_{s-1} \\ & \psi_1 & \delta_1 & \delta_2 & \cdots & \delta_{s-1} \\ (8.9.2) & \psi_2 & \delta_s & \delta_{s+1} & \cdots & \delta_{2(s-1)} \\ & \cdot & \cdot & \cdot & \cdots & \cdot \\ & \psi_{s-1} & \delta_{s^2-3s+3} & \delta_{s^2-3s+4} & \cdots & \delta_{(s-1)^2} \end{array}$$

Then the first associates of ϕ_i and ψ_j are the symbols occurring in the row and column corresponding to them. Now consider ϕ_1. The symbols θ and ϕ_1, which are first associates, have $\phi_2, \cdots, \phi_{s-1}$ as common first associates, whereas $\delta_1, \delta_s, \cdots, \delta_{s^2-3s+3}$ are first associates of ϕ_1 and second associates of θ. Then from Theorem 8.9.1 the symbols $\delta_1, \delta_s, \cdots, \delta_{s^2-3s+3}$ are mutual first associates, and each of these symbols is a second associate of $\phi_2, \phi_3, \cdots, \phi_{s-1}$, and θ. Similarly we can show that the first associates of any symbol are the symbols occurring in the same row or column and other symbols are second associates. Hence we get the L_2 association scheme, and the proof is complete.

8.9.3. Pseudo L_2 Association Scheme

Shrikhande (1959b) has shown that when $s = 4$, there is one more possible association scheme other than the L_2 association scheme; we reproduce this in Table 8.9.1.

Table 8.9.1. Pseudo L_2 association scheme when $s = 4$

Symbol	First Associates	Second Associates
1	2, 3, 4, 5, 6, 7	8, 9, 10, 11, 12, 13, 14, 15, 16
2	1, 3, 4, 8, 9, 10	5, 6, 7, 11, 12, 13, 14, 15, 16
3	1, 2, 6, 9, 11, 12	4, 5, 7, 8, 10, 13, 14, 15, 16
4	1, 2, 7, 10, 13, 14	3, 5, 6, 8, 9, 11, 12, 15, 16
5	1, 6, 7, 8, 15, 16	2, 3, 4, 9, 10, 11, 12, 13, 14
6	1, 3, 5, 12, 13, 15	2, 4, 7, 8, 9, 10, 11, 14, 16
7	1, 4, 5, 11, 14, 16	2, 3, 6, 8, 9, 10, 12, 13, 15
8	2, 5, 9, 10, 15, 16	1, 3, 4, 6, 7, 11, 12, 13, 14
9	2, 3, 8, 11, 14, 15	1, 4, 5, 6, 7, 10, 12, 13, 16
10	2, 4, 8, 12, 13, 16	1, 3, 5, 6, 7, 9, 11, 14, 15
11	3, 7, 9, 12, 14, 16	1, 2, 4, 5, 6, 8, 10, 13, 15
12	3, 6, 10, 11, 13, 16	1, 2, 4, 5, 7, 8, 9, 14, 15
13	4, 6, 10, 12, 14, 15	1, 2, 3, 5, 7, 8, 9, 11, 16
14	4, 7, 9, 11, 13, 15	1, 2, 3, 5, 6, 8, 10, 12, 16
15	5, 6, 8, 9, 13, 14	1, 2, 3, 4, 7, 10, 11, 12, 16
16	5, 7, 8, 10, 11, 12	1, 2, 3, 4, 6, 9, 13, 14, 15

8.9.4. Set Structure of $L_i(s)$ Designs

Along lines similar to those used in Theorem 8.5.5, the following theorem of Raghavarao (1960a) can be proved:

Theorem 8.9.3. If in an L_2 design

$$(8.9.3) \qquad\qquad r + (s - 2)\lambda_1 - (s - 1)\lambda_2 = 0,$$

then k is divisible by s. Furthermore, every set of the design contains k/s symbols from each of the s rows (or columns) of the association scheme.

It may be noted that condition (8.9.3) is identical with the condition in Raghavarao's paper.

The following result due to Agrawal (1964) can also be proved along lines similar to those used in Theorem 8.5.9:

Theorem 8.9.4. Let $\rho_1 = r + (s - i)\lambda_1 - (s - i + 1)\lambda_2$ and $\rho_2 = r - i\lambda_1 + (i - 1)\lambda_2$. Then x, the number of symbols common to any two sets of an L_i design satisfies

$$(8.9.4) \quad \max\,[0, 2k - v, k - \rho_i] \leqq x \leqq \min\left[k, \rho_i + \frac{2(rk - \rho_i)}{b} - k\right],$$

where $i = 1$ if $\lambda_1 > \lambda_2$ and $i = 2$ if $\lambda_1 < \lambda_2$.

We close this section by giving an upper bound to the number of disjoint sets in L_2 designs, a result due to S. M. Shah (1964).

Theorem 8.9.5. A given set of an L_2 design with

$$r + (s - 2)\lambda_1 - (s - 1)\lambda_2 = 0$$

cannot have more than

$$(8.9.5) \qquad b - 1 - \frac{v(r - 1)^2(s - 1)^2}{(b - r)(v - k) - (s - 1)^2(v - rk)}$$

disjoint sets with it. If some set has that many, then

$$k[(b - r)(v - k) - (s - 1)^2(v - rk)]/v(r - 1)(s - 1)^2$$

is an integer and each nondisjoint set has

$$(8.9.6) \qquad \frac{k[(b - r)(v - k) - (s - 1)^2(v - rk)]}{v(r - 1)(s - 1)^2}$$

symbols in common with that given set.

As a consequence of the above theorem, we have

Corollary 8.9.5.1. The necessary and sufficient conditions for a set of an L_2 design with $r + (s - 2)\lambda_1 - (s - 1)\lambda_2 = 0$ to have the same number of symbols in common with each of the remaining sets is that (a) $b = v - 2s + 2$ and (b) $k(r - 1)/(s - 1)^2$ be an integer.

8.10. METHODS OF CONSTRUCTING L_2 DESIGNS

The possible parametric combinations of L_2 designs for $r, k \leqq 10$ are 155, of which 93 have solutions, 57 are nonexistent, and 5 remain unsolved. The parametric combinations of the five unsolved L_2 designs, as listed by Chang and Liu (1964), are presented in Table 8.10.1.

Table 8.10.1. Values for the Unsolved parametric combinations for L_2 designs

Series	v	b	r	k	λ_1	λ_2
1	36	54	9	6	2	1
2	36	60	10	6	0	2
3	36	40	10	9	3	2
4	49	49	9	9	3	1
5	100	90	9	10	0	1

We now study some methods of constructing L_2 designs.

Theorem 8.10.1. An L_2 design with the parameters

$$(8.10.1) \quad v = s^2, \quad b = s(s-1), \quad r = s-1, \quad k = s, \quad \lambda_1 = 0, \quad \lambda_2 = 1$$

exists if s is a prime or a prime power.

Proof. Let us write the s^2 symbols in a square array of s rows and s columns. If s is a prime or a prime power, a complete set of MOLS of order s exists. Superimposing each of these latin squares on the square array of symbols and forming sets with symbols corresponding to the different letters of the latin squares, we can generate $s(s-1)$ sets and easily verify that they form an L_2 design with the parameters given in (8.10.1).

Let there be a BIB design with the parameters

$$(8.10.2) \quad v^* = s, \quad b^*, \quad r^*, \quad k^*, \quad \lambda^*.$$

By replacing the ith symbol of this BIB design by the symbols of the ith row of the association scheme and adjoining to this the design obtained by replacing the ith symbol by the symbols of the ith column of the association scheme, we get L_2 design with the parameters

$$(8.10.3) \quad \begin{aligned} &v = s^2, \quad b = 2b^*, \quad r = 2r^*, \\ &k = sk^*, \quad \lambda_1 = r^* + \lambda^*, \quad \lambda_2 = 2\lambda^*. \end{aligned}$$

Thus we have the following theorem:

Theorem 8.10.2. An L_2 design with the parameters given in (8.10.3) can always be constructed so long as there exists a BIB design with the parameters given in (8.10.2).

Theorem 8.10.3. If there exists a BIB design with the parameters given in (8.10.2), then an L_2 design with the parameters

$$(8.10.4) \quad v = s^2, \qquad b = 2sb^*, \qquad r = 2r^*, \qquad k = k^*, \qquad \lambda_1 = \lambda^*, \qquad \lambda_2 = 0$$

can always be constructed.

Proof. By writing down the BIB design for the symbols occurring in the ith row ($i = 1, 2, \cdots, s$) and also the jth column ($j = 1, 2, \cdots, s$), we can generate $2sb^*$ sets that can be verified to form an L_2 design with the parameters given in (8.10.4).

Theorem 8.10.4. [Clatworthy (1967)]. An L_2 design with the parameters

$$(8.10.5) \quad v = b = s^2, \qquad r = k = 2s - 1, \qquad \lambda_1 = s, \qquad \lambda_2 = 2$$

can always be constructed for every s.

Proof. The design obtained by writing every symbol of the association scheme along with its first associates in sets can easily be seen to be the required design.

Theorem 8.10.5. [Clatworthy (1967)]. An L_2 design with the parameters

$$(8.10.6) \quad \begin{aligned} v = s^2, \qquad b = s(s-1), \qquad r = 2(s-1), \\ k = 2s, \qquad \lambda_1 = s, \qquad \lambda_2 = 2 \end{aligned}$$

can always be constructed for every s.

Proof. The sets formed by combining all possible pairs of rows and all possible pairs of columns can be verified to constitute an L_2 design with the parameters given in (8.10.6).

8.11. PSEUDOCYCLIC ASSOCIATION SCHEME

We have seen in Section 8.4 that association schemes with the parameters given in (8.4.4), whatever their combinatorial structure, are of the pseudocyclic type. We prove the following theorem, which is a consequence of Theorems 5.3 and 5.5 of Connor and Clatworthy (1954):

Theorem 8.11.1. An association scheme in which v is odd and Δ as given by (8.3.24) is not a perfect square must be of the pseudocyclic type.

Proof. We have seen in Section 8.3 that the multiplicities

$$\alpha_i = \frac{n_1 + n_2}{2} + (-1)^i \left[\frac{(n_1 - n_2) + \gamma(n_1 + n_2)}{2\sqrt{\Delta}} \right] \qquad i = 1, 2$$

of the roots

$$\theta_i = r - \tfrac{1}{2}[(\lambda_1 - \lambda_2)(-\gamma + (-1)^i\sqrt{\Delta} + (\lambda_1 + \lambda_2)] \qquad i = 1, 2$$

of NN' must be integral. When Δ is not an integral square, it is necessary that

(8.11.1) $$n_1 - n_2 + \gamma(n_1 + n_2) = 0,$$

which will hold if and only if $n_1 = n_2$ and $\gamma = 0$, that is, $p_{12}{}^1 = p_{12}{}^2$. If we assume that $t = p_{12}{}^1 = p_{12}{}^2$, then $p_{22}{}^1 = t$ and $n_2 = 2t$; also $p_{11}{}^2 = t$ and hence $n_1 = 2t$ and $v = n_1 + n_2 + 1 = 4t + 1$. Thus the association scheme is pseudocyclic.

Theorem 8.11.2. [Mesner (1956, 1965)]. A two-associate-class association scheme with v equal to a prime must be of the pseudocyclic type.

Proof. For α_1 and α_2 given by (8.3.27) we can easily verify that

(8.11.2) $$vn_1 n_2 = \Delta\alpha_1 \alpha_2 .$$

Here n_1, n_2, α_1, and α_2 are all less than v. If v is a prime, say p, then the left-hand side of (8.11.2) is divisible by p but not by p^2, and so is the right-hand side and hence Δ. Thus Δ is not a perfect square, and from Theorem 8.11.1 the association scheme is pseudocyclic, completing the proof of the theorem.

Theorem 8.11.3. The existence of a pseudocyclic association scheme implies and is implied by the existence of a $(4t + 1)$th order square matrix M with each row having the elements 1, and -1 exactly $2t$ times such that

(8.11.3) $$MM' = (4t + 1)I_{4t+1} - E_{4t+1, 4t+1}.$$

Proof. Let B_1 be the association matrix of first associates. Then we can easily verify that

(8.11.4) $$M = B_1 - (E_{4t+1, 4t+1} - B_1 - I_{4t+1})$$

is our required M.

Any matrix M satisfying (8.11.3) can be shown to be a symmetrical matrix with zero in the diagonal position. We can then see that if the (i, j)th element $(i \neq j)$ of M is 1, the ith and jth rows contain the ordered pairs $(1, 1), (1, -1)$, $(-1, 1)$, and $(-1, -1)$ exactly $t - 1, t, t$, and t times, respectively. Similarly, if the (i, j)th element $(i \neq j)$ of M is -1, these pairs occur t, t, t, and $t - 1$

times, respectively. Identifying the rows and columns of M with a set of $4t + 1$ symbols and taking $1(-1)$ in position (i, j) to imply that $(i, j) = 1(2)$, we obviously have a pseudocyclic association scheme.

When augmented by

$$(8.11.5) \qquad \begin{bmatrix} 0 & E_{1, 4t+1} \\ E_{4t+1, 1} & M \end{bmatrix},$$

the matrix M will form a useful weighing design. This is demonstrated in Chapter 17, where we further show that a necessary condition for its existence is that $(4t + 1, -1)_p = 1$ for all primes p. This result in the present context can be stated as follows:

Theorem 8.11.4. A necessary condition for the existence of a pseudocyclic association scheme is that $(4t + 1, -1)_p = 1$ for all primes p. Hence the association scheme is nonexistent if the square free part of $4t + 1$ contains a prime congruent to $3 \pmod 4$.

Pseudocyclic association schemes, in view of the above theorem, do not exist for $v = 21, 33, 57$, etc.

Theorems 8.11.3 and 8.11.4 were proved by Shrikhande and Singh (1962). It may be noted that the result of Theorem 8.11.4 was implicitly contained in a paper by Raghavarao (1960c).

In Section 17.4 we shall give a method of constructing the M-matrices of Theorem 8.11.3 whenever $4t + 1$ is a prime or a prime power. Since the existence of a pseudocyclic association scheme is equivalent to the existence of the M-matrix, pseudocyclic association schemes exist when $4t + 1$ is a prime or a prime power. The pseudocyclic association scheme as determined in this manner may be identical with some other known association schemes. For $4t + 1 = 9$ the pseudocyclic association scheme can be shown to be equivalent to the L_2 association scheme. The existence of a pseudocyclic association scheme for $v = 45$ is not ruled out, and the construction of this scheme remains an open problem.

8.12. HIGHER-ASSOCIATE-CLASS ASSOCIATION SCHEMES

So far we restricted ourselves to studying only two-associate-class schemes. The schemes of higher associate classes have not been extensively and exhaustively studied in the literature. We give some general results about some well-known higher-associate-class association schemes.

8.12.1. Rectangular Association Scheme

The rectangular association scheme is a three-associate-class association scheme introduced by Vartak (1955). Let there be $v = mn$ symbols arranged in a rectangle of m rows and n columns. With respect to each symbol, the first associates are the other $n - 1$ symbols of the same row, the second associates are the other $m - 1$ symbols of the same column, and the remaining $(m - 1)(n - 1)$ symbols are third associates. For this association scheme we have

$$n_1 = n - 1, \qquad n_2 = m - 1, \qquad n_3 = (m - 1)(n - 1),$$

$$P_1 = \begin{bmatrix} n - 2 & 0 & 0 \\ 0 & 0 & m - 1 \\ 0 & m - 1 & (m - 1)(n - 2) \end{bmatrix},$$

(8.12.1)
$$P_2 = \begin{bmatrix} 0 & 0 & n - 1 \\ 0 & m - 2 & 0 \\ n - 1 & 0 & (m - 2)(n - 1) \end{bmatrix},$$

$$P_3 = \begin{bmatrix} 0 & 1 & n - 2 \\ 1 & 0 & m - 2 \\ n - 2 & m - 2 & (n - 2)(m - 2) \end{bmatrix}.$$

We can easily verify that the characteristic roots of NN' are $\theta_0 = rk$, $\theta_1 = r - \lambda_1 + (m - 1)(\lambda_2 - \lambda_3)$, $\theta_2 = r - \lambda_2 + (n - 1)(\lambda_1 - \lambda_3)$, and $\theta_3 = r - \lambda_1 - \lambda_2 + \lambda_3$, with the respective multiplicities $\alpha_0 = 1$, $\alpha_1 = n - 1$, $\alpha_2 = m - 1$, and $\alpha_3 = (n - 1)(m - 1)$. For further properties of the rectangular association scheme we refer to S. M. Shah (1964) and Vartak (1955, 1959).

8.12.2. Cubic Association Scheme

Another useful three-associate-class association scheme is the cubic association scheme introduced by Raghavarao and Chandrasekhararao (1964).

Let there be $v = s^3$ symbols denoted by (α, β, γ) $(\alpha, \beta, \gamma = 1, 2, \cdots, s)$. We define the distance δ between the symbols (α, β, γ) and $(\alpha', \beta' \gamma')$ to be the number of nonnull elements in $(\alpha - \alpha', \beta - \beta', \gamma - \gamma')$. Two symbols are called first, second or third associates, depending on whether $\delta = 1, 2,$ or 3, respectively. We verify

$$n_1 = 3(s-1) \qquad n_2 = 3(s-1)^2, \quad n_3 = (s-1)^3;$$

$$P_1 = \begin{bmatrix} s-2 & 2(s-1) & 0 \\ 2(s-1) & 2(s-1)(s-2) & (s-1)^2 \\ 0 & (s-1)^2 & (s-1)^2(s-2) \end{bmatrix};$$

$$(8.12.2) \quad P_2 = \begin{bmatrix} 2 & 2(s-2) & (s-1) \\ 2(s-2) & 2(s-1)+(s-2)^2 & 2(s-1)(s-2) \\ (s-1) & 2(s-1)(s-2) & (s-1)(s-2)^2 \end{bmatrix};$$

$$P_3 = \begin{bmatrix} 0 & 3 & 3(s-2) \\ 3 & 6(s-2) & 3(s-2)^2 \\ 3(s-2) & 3(s-2)^2 & (s-2)^3 \end{bmatrix}.$$

The characteristic roots of NN' can be obtained as

$$\theta_0 = rk, \quad \theta_1 = r + (2s-3)\lambda_1 + (s-1)(s-3)\lambda_2 - (s-1)^2\lambda_3,$$
$$\theta_2 = r + (s-3)\lambda_1 - (2s-3)\lambda_2 + (s-1)\lambda_3,$$
$$\theta_3 = r - 3\lambda_1 + 3\lambda_2 - \lambda_3,$$

with the respective multiplicities

$$\alpha_0 = 1, \quad \alpha_1 = 3(s-1), \quad \alpha_2 = 3(s-1)^2, \quad \text{and} \quad \alpha_3 = (s-1)^3.$$

Raghavarao and Chandrasekhararao (1964) showed that the design with cubic association scheme is more efficient than the usual three-dimensional lattice design for a design with the parameters

$$(8.12.3) \qquad v = 27 = b, \qquad r = 8 = k, \qquad \lambda_1 = 4, \qquad \lambda_2 = 2, \qquad \lambda_3 = 1.$$

For the plan of this design we refer to the original paper, which also presents the construction and combinatorial problems of cubic designs [see also Chandrasekhararao (1964)].

8.12.3. Extended Triangular Association Scheme

The two-associate-class triangular association scheme was generalized by P. W. M. John (1966) to three associate classes. Bose and Laskar (1967) independently gave a construction equivalent to that of P. W. M. John by considering it as a generalization of the triangular strongly regular graph. Let there be $v = (s+2)(s+3)(s+4)/6$ symbols represented by (α, β, γ), where $0 < \alpha < \beta < \gamma \leqq s + 4$. Two symbols are first associates if they have two integers in common, second associates if they have one integer in common, and third associates otherwise. Clearly

$$n_1 = 3(s + 1), \qquad n_2 = \frac{3s(s + 1)}{2}, \qquad n_3 = \frac{s(s - 1)(s + 1)}{6};$$

$$P_1 = \begin{bmatrix} s + 2 & 2s & 0 \\ 2s & s^2 & \dfrac{s(s - 1)}{2} \\ 0 & \dfrac{s(s - 1)}{2} & \dfrac{s(s - 1)(s - 2)}{6} \end{bmatrix};$$

$$(8.12.4) \quad P_2 = \begin{bmatrix} 4 & 2s & s - 1 \\ 2s & \dfrac{(s - 1)(s + 6)}{2} & (s - 1)(s - 2) \\ s - 1 & (s - 1)(s - 2) & \dfrac{(s - 1)(s - 2)(s - 3)}{6} \end{bmatrix};$$

$$P_3 = \begin{bmatrix} 0 & 9 & 3(s - 2) \\ 9 & 9(s - 2) & \dfrac{3(s - 2)(s - 3)}{2} \\ 3(s - 2) & \dfrac{3(s - 2)(s - 3)}{2} & \dfrac{(s - 2)(s - 3)(s - 4)}{6} \end{bmatrix}.$$

The characteristic roots of NN' are

$$\theta_0 = rk,$$

$$\theta_1 = r + (2s - 1)\lambda_1 + \frac{s(s - 5)\lambda_2}{2} - \frac{s(s - 1)\lambda_3}{2},$$

$$\theta_2 = r + (s - 3)\lambda_1 - (2s - 3)\lambda_2 + (s - 1)\lambda_3,$$

$$\theta_3 = r - 3\lambda_1 + 3\lambda_2 - \lambda_3,$$

with the respective multiplicities

$$\alpha_0 = 1, \qquad \alpha_1 = s + 3, \qquad \alpha_2 = \frac{(s + 1)(s + 4)}{2},$$

and

$$\alpha_3 = \frac{(s - 1)(s + 3)(s + 4)}{6}.$$

8.12.4. Right-Angular Association Scheme

This is a four-associate-class association scheme introduced by Tharthare (1963). There are $v = 2sl$ symbols arranged in l right angles of equal arms of length s. The angular positions of these right angles are kept blank, and symbols are written along the arms. For every symbol θ the following hold:

1. Symbols other than θ occurring in the same arm are first associates.

2. Symbols occurring in an arm different from θ but in the same right angle with θ are second associates.

3. Symbols occurring in the arm parallel to θ but in other right angles are third associates.

4. The remaining symbols are fourth associates. We can easily verify

$$n_1 = s - 1, \qquad n_2 = s, \qquad n_3 = s(l - 1) = n_4;$$

(8.12.5)

$$P_1 = \begin{bmatrix} s-2 & 0 & 0 & 0 \\ 0 & s & 0 & 0 \\ 0 & 0 & s(l-1) & 0 \\ 0 & 0 & 0 & s(l-1) \end{bmatrix};$$

$$P_2 = \begin{bmatrix} 0 & s-1 & 0 & 0 \\ s-1 & 0 & 0 & 0 \\ 0 & 0 & 0 & s(l-1) \\ 0 & 0 & s(l-1) & 0 \end{bmatrix};$$

$$P_3 = \begin{bmatrix} 0 & 0 & s-1 & 0 \\ 0 & 0 & 0 & s \\ s-1 & 0 & s(l-2) & 0 \\ 0 & s & 0 & s(l-1) \end{bmatrix};$$

$$P_4 = \begin{bmatrix} 0 & 0 & 0 & s-1 \\ 0 & 0 & s & 0 \\ 0 & s & 0 & s(l-2) \\ s-1 & 0 & s(l-2) & 0 \end{bmatrix}.$$

The characteristic roots of NN' are

$$\begin{aligned}
\theta_0 &= rk, \\
\theta_1 &= r - \lambda_1 + s(\lambda_1 - \lambda_2) + s(l-1)(\lambda_3 - \lambda_4), \\
\theta_2 &= r - \lambda_1, \\
\theta_3 &= r - \lambda_1 + s(\lambda_1 + \lambda_2 - \lambda_3 - \lambda_4), \\
\theta_4 &= r - \lambda_1 + s(\lambda_1 - \lambda_2 - \lambda_3 + \lambda_4),
\end{aligned}$$

with the respective multiplicities

$$\alpha_0 = 1, \quad \alpha_1 = 1, \quad \alpha_2 = 2l(s-1), \quad \alpha_3 = l-1, \quad \text{and} \quad \alpha_4 = l-1.$$

8.12.5. Generalized Right-Angular Association Scheme

This is also a four-associate-class association scheme due to Tharthare (1965). Let there be $v = pls$ symbols denoted by (α, β, γ) ($\alpha = 1, 2, \cdots, l$; $\beta = 1, 2, \cdots, p$; $\gamma = 1, 2, \cdots, s$). For the symbol (α, β, γ) first associates are those that differ in the third position; second associates are those that differ in the second position while being the same or different in the third position; third associates are those that have the same second position, a different first position, and the same or different third position; the others are fourth associates. Clearly

$$n_1 = s-1, \qquad n_2 = s(p-1), \qquad n_3 = s(l-1), \qquad n_4 = s(l-1)(p-1);$$

$$P_1 = \begin{bmatrix} s-2 & 0 & 0 & 0 \\ 0 & s(p-1) & 0 & 0 \\ 0 & 0 & s(l-1) & 0 \\ 0 & 0 & 0 & s(l-1)(p-1) \end{bmatrix};$$

(8.12.6)

$$P_2 = \begin{bmatrix} 0 & s-1 & 0 & 0 \\ s-1 & s(p-2) & 0 & 0 \\ 0 & 0 & 0 & s(l-1) \\ 0 & 0 & s(l-1) & s(l-1)(p-2) \end{bmatrix};$$

$$P_3 = \begin{bmatrix} 0 & 0 & s-1 & 0 \\ 0 & 0 & 0 & s(p-1) \\ s-1 & 0 & s(l-2) & 0 \\ 0 & s(p-1) & 0 & s(p-1)(l-2) \end{bmatrix};$$

$$P_4 = \begin{bmatrix} 0 & 0 & 0 & s-1 \\ 0 & 0 & s & s(p-2) \\ 0 & s & 0 & s(l-2) \\ s-1 & s(p-2) & s(l-2) & s(l-2)(p-2) \end{bmatrix}.$$

The characteristic roots of NN' for a generalized right-angular design are

$$\theta_0 = rk,$$
$$\theta_1 = r - \lambda_1 + s(\lambda_1 - \lambda_2) + s(l - 1)(\lambda_3 - \lambda_4),$$
$$\theta_2 = r - \lambda_1,$$
$$\theta_3 = r - \lambda_1 + s[\lambda_1 - \lambda_3 + (p - 1)(\lambda_2 - \lambda_4)],$$
$$\theta_4 = r - \lambda_1 + s(\lambda_1 - \lambda_2 - \lambda_3 + \lambda_4),$$

with the respective multiplicities

$$\alpha_0 = 1, \quad \alpha_1 = p - 1, \quad \alpha_2 = pl(s - 1), \quad \alpha_3 = l - 1,$$

and

$$\alpha_4 = (p - 1)(l - 1).$$

8.12.6. Group-Divisible m-Associate Association Scheme

This is a generalization of the group-divisible association scheme of two associate classes to m associate classes. It was introduced by Roy (1953–1954). We give the association scheme as defined by Raghavarao (1960b).

In this association scheme there are $v = N_1 N_2 \cdots N_m$ symbols denoted by $v_{i_1 i_2 \cdots i_m}$ $(i_1 = 1, 2, \cdots, N_1; i_2 = 1, 2, \cdots, N_2; \ldots; i_m = 1, 2, \cdots, N_m)$. Two symbols having only the first $(m - i)$ suffixes of $v_{i_1 i_2 \cdots i_m}$ the same are the ith associates $(i = 1, 2, \cdots, m)$.

We can easily verify that

$$n_i = N_m N_{m-1} \cdots N_{m-i+2}(N_{m-i+1} - 1),$$

$$(8.12.7) \qquad P_i = \left[\begin{array}{c|c} 0_{(i-1), (i-1)} & \mathbf{x_{i-1}} \,|\, 0_{(i-1), (m-i)} \\ \hline \mathbf{x'_{i-1}} & D_{(m-i+1), (m-i+1)} \\ \hline 0_{(m-i), (i-1)} & \end{array} \right],$$

where $i = 1, 2, \cdots, m$, $\mathbf{x_{i-1}}$ is the $(i - 1)$th order column vector with the elements $n_1, n_2, \cdots, n_{i-1}$, and $D_{(m-i+1), (m-i+1)}$ is the diagonal matrix with the diagonal elements

$$N_m N_{m-1} \cdots N_{m-i+2}(N_{m-i+1} - 2), n_{i+1}, n_{i+2}, \cdots, n_m.$$

The characteristic roots of NN' can be shown to be

$$\theta_0 = rk,$$
$$\theta_i = (r - \lambda_{m-i+1}) + (\lambda_1 - \lambda_{m-i+1})n_1 + \cdots + (\lambda_{m-i} - \lambda_{m-i+1})n_{m-i}$$
$$i = 1, 2, \cdots, m,$$

with the respective multiplicities $\alpha_0 = 1$, $\alpha_i = N_1 N_2 \cdots N_{i-1}(N_i - 1)$. For detailed study of these designs we refer to Hinkelmann and Kempthorne (1963), Raghavarao (1960b, 1962), and Roy (1953–1954, 1962).

8.12.7. Other Association Schemes

The rectangular association scheme was generalized to the extended group-divisible (EGD) association scheme by Hinkelmann and Kempthorne (1963), and some detailed study of this association scheme was made by Hinkelmann (1964). The cubic association scheme has been generalized to the hypercubic association scheme [the reader is referred to Kusumoto (1965)]. The triangular association scheme was generalized to the T_m association scheme with m associate classes by Ogasawara (1965). Some other types of higher-associate-class schemes are presented by Adhikary (1966, 1967) and Yamamoto, Fuji, and Hamada (1965).

8.13. KRONECKER-PRODUCT ASSOCIATION SCHEMES

Let $B_0 = I_{v_1}, B_1, B_2, \cdots, B_m$ be the association matrices defining an m-associate-class association scheme on v_1 symbols and let $A_0 = I_{v_2}, A_1, A_2, \cdots, A_n$ be another set of association matrices defining an n-associate-class association scheme on v_2 symbols. Then we have the following:

Theorem 8.13.1. The matrices

$$(8.13.1) \qquad A_0 \times B_0 = I_{v_1 v_2}, \qquad A_0 \times B_j, \qquad A_i \times B_0, \qquad A_i \times B_j$$
$$i = 1, 2, \cdots, n; j = 1, 2, \cdots, m,$$

where $\times$ is the Kronecker product of matrices, form the association matrices of an $(m + n + mn)$-associate-class scheme on $v_1 v_2$ symbols.

The proof is left as an exercise to the reader. The association scheme obtained by the method of Theorem 8.13.1 is called the Kronecker-product association scheme and was given by Kusumoto (1967) and Surendran (1968).

In particular taking $m = 1$, $n = 1$, we get $B_0 = I_{v_1}$, $B_1 = E_{v_1, v_1} - I_{v_1}$, $A_0 = I_{v_2}$, $A_1 = E_{v_2, v_2} - I_{v_2}$, and the Kronecker-product association scheme thereby obtained is the well-known rectangular association scheme.

8.14. TWO-ASSOCIATE-CLASS PBIB DESIGNS WITH TWO REPLICATIONS AND $k > r$

8.14.1. Nair's Condition

When $k > r$, we have $b < v$, in which case NN' will be singular. Hence NN' will have zero as a characteristic root. Since the distinct characteristic roots of NN' other than rk are the roots of the matrix $A = (a_{\alpha\beta})$, where

$$(8.14.1) \qquad a_{\alpha\beta} = r\delta_{\alpha\beta} + \sum_{\gamma=1}^{m} \lambda_\gamma p_{\beta\gamma}{}^\alpha - \lambda_\beta n_\beta,$$

$|NN'| = 0$ implies that $|A| = 0$. Thus we have the following theorem:

Theorem 8.14.1. For a PBIB design with m associate classes and $k > r$

$$(8.14.2) \qquad\qquad\qquad |A| = 0,$$

where $A = (a_{\alpha\beta})$ and $a_{\alpha\beta}$ are given by (8.14.1).

The condition of the above theorem is known as Nair's condition [K. R. Nair (1943)].

A particular case of Theorem 8.14.1 when $m = 2$ is given in the following corollary:

Corollary 8.14.1.1 For a two-associate-class PBIB design with $k > r$

$$(8.14.3) \quad (r - \lambda_1)(r - \lambda_2) + (\lambda_1 - \lambda_2)[p_{12}{}^2(r - \lambda_1) - p_{12}{}^1(r - \lambda_2)] = 0$$

8.14.2. Two-Associate-Class PBIB Designs with $r = 2$ and $k > r$

Without loss of generality, we assume that $\lambda_1 > \lambda_2$. Since $r = 2$, the various possibilities for λ_1 and λ_2 are (a) $\lambda_1 = 2$, $\lambda_2 = 0$; (b) $\lambda_1 = 2$, $\lambda_2 = 1$; and (c) $\lambda_1 = 1$, $\lambda_2 = 0$. We omit the study of case a since in this case we can easily verify that the design obtained is a disconnected one. In case b Nair's condition (8.14.3) reduces to

$$(8.14.4) \qquad\qquad\qquad p_{12}{}^1 = 0,$$

and thus the corresponding PBIB design must be a group-divisible design with the parameters

$$(8.14.5) \quad v = mn, \qquad b, \qquad r = 2, \qquad k, \qquad \lambda_1 = 2, \qquad \lambda_2 = 1.$$

Since $r = \lambda_1$, this group-divisible design is a singular group-divisible design whose existence is equivalent to the existence of a BIB design with the parameters

$$(8.14.6) \quad v' = m, \qquad b' = b, \qquad r' = 2, \qquad k' = \frac{k}{n}, \qquad \lambda' = 1.$$

Since in a BIB design $k' \leq r'$, we have $k' = 2$ and again, since $\lambda'(v' - 1) = r'(k' - 1)$, we get $m = 3$ and $v'r' = b'k'$ gives $b' = 3$. Thus the series of PBIB designs with the parameters given in (8.14.5) reduces to group-divisible designs with the parameters

$$(8.14.7) \quad \begin{matrix} v = 3n, & b = 3, & r = 2, & k = 2n, & \lambda_1 = 2, \\ \lambda_2 = 1, & n = n, & m = 3, & p_{12}{}^1 = 0. \end{matrix}$$

In case c Nair's condition (8.14.3) reduces to

$$(8.14.8) \qquad\qquad\qquad 2p_{12}{}^1 - p_{12}{}^2 = 2.$$

By working the relations on the parameters of the design and association scheme and noting that all the parameters must be integral, we can show that the following two series are possible in this case:

$$(8.14.9) \quad \begin{array}{llllll} v = s^2, & b = 2s, & r = 2, & k = s, & n_1 = 2(s-1), \\ n_2 = (s-1)^2, & \lambda_1 = 1, & \lambda_2 = 0, & p_{12}{}^1 = s - 1 \end{array}$$

and

$$(8.14.10) \quad \begin{array}{llllll} v = \dfrac{n(n-1)}{2}, & b = n, & r = 2, & k = n-1, & n_1 = 2(n-2), \\ n_2 = \dfrac{(n-2)(n-3)}{2}, & \lambda_1 = 1, & \lambda_2 = 0, & p_{12}{}^1 = n - 3. \end{array}$$

Thus we have the following theorem:

Theorem 8.14.2. There are three series of connected two-associate-class PBIB designs with $k > r = 2$ given by (8.14.7), (8.14.9), and (8.14.10).

These three series were obtained by Bose (1950–1951) [see also K. R. Nair (1950, 1951a)].

8.15. LATTICE DESIGNS AS PBIB DESIGNS

8.15.1. Square-Lattice Designs

For the square-lattice class of designs $v = s^2$ and each set contains s symbols. The symbols are arranged in the form of an $s \times s$ square. Then, if we assign the symbols occurring in the rows and columns to different sets, we have $2s$ sets each of size s. Again, if $i - 2(i - 2 \leq s - 1)$ MOLS exist, we can superimpose them on the s^2 symbols arranged in the form of a square and assign to sets symbols corresponding to the letters of these squares. Such an arrangement is called an i-ple lattice design. A 2-ple lattice design is known as a simple lattice, and a 3-ple lattice design is called a triple lattice. Let the basic pattern be repeated n times. Then we get an incomplete block design with the parameters

$$(8.15.1) \quad v = s^2, \quad b = nsi, \quad r = ni, \quad k = s, \quad \lambda_1 = n, \quad \lambda_2 = 0,$$

which can be easily shown to be a PBIB design of the L_i association scheme, with the usual association-scheme parameters,

$$(8.15.2) \quad n_1 = i(s-1), \quad n_2 = (s-i+1)(s-1), \quad p_{11}{}^1 = i^2 - 3i + s.$$

8.15.2. Cubic-Lattice Designs

A three-dimensional, or cubic, lattice has $v = s^3$ symbols arranged on the points of an $s \times s \times s$ cube. A replicate of s^2 sets of size s is obtained by taking the lines parallel to each of the three possible directions. We can easily verify that such an arrangement is a PBIB design with cubic association scheme having the parameters

$$\text{(8.15.3)} \quad \begin{aligned} v &= s^3, & b &= 3s^2, & r &= 3, & k &= s, & n_1 &= 3(s-1), \\ n_2 &= 3(s-1)^2, & n_3 &= (s-1)^3, & \lambda_1 &= 1, & \lambda_2 &= 0 = \lambda_3. \end{aligned}$$

8.15.3. Simple Rectangular-Lattice Designs

For this class of designs $v = p(p-1)$ and the $p(p-1)$ symbols will be identified by a pair of digits x and y ($x \neq y$; $x, y = 1, 2, \cdots, p$). All symbols having the same value for x will form the xth set of the first replication, and all symbols having the same value for y will form the yth set of the second replication. K. R. Nair (1951b) has shown that such simple rectangular-lattice designs form four-associate-class PBIB designs with the parameters

$$\begin{aligned} v &= p(p-1), & k &= p-1, & r &= 2, & b &= 2p, \\ \lambda_1 &= 1, & \lambda_2 &= 0 = \lambda_3 = \lambda_4, \\ n_1 &= 2(p-2), & n_2 &= (p-2)(p-3), & n_3 &= 2(p-2), & n_4 &= 1; \end{aligned}$$

$$P_1 = \begin{bmatrix} p-3 & p-3 & 1 & 0 \\ p-3 & (p-3)(p-4) & p-3 & 0 \\ 1 & p-3 & p-3 & 1 \\ 0 & 0 & 1 & 0 \end{bmatrix};$$

$$P_2 = \begin{bmatrix} 2 & 2(p-4) & 2 & 0 \\ 2(p-4) & (p-4)(p-5) & 2(p-4) & 1 \\ 2 & 2(p-4) & 2 & 0 \\ 0 & 1 & 0 & 0 \end{bmatrix};$$

$$\text{(8.15.4)}$$

$$P_3 = \begin{bmatrix} 1 & p-3 & p-3 & 1 \\ p-3 & (p-3)(p-4) & p-3 & 1 \\ p-3 & p-3 & 1 & 0 \\ 1 & 0 & 0 & 0 \end{bmatrix};$$

$$P_4 = \begin{bmatrix} 0 & 0 & 2(p-2) & 0 \\ 0 & (p-2)(p-3) & 0 & 0 \\ 2(p-2) & 0 & 0 & 0 \\ 0 & 0 & 0 & 0 \end{bmatrix}.$$

8.15.4. Triple Rectangular-Lattice Designs

In a triple rectangular lattice design there are $v = p(p - 1)$ symbols, and each symbol may be identified by ordered triplets (x, y, z) $(x \neq y, x \neq z, y \neq z;$ $x, y, z, = 1, 2, \cdots, p)$. All symbols having the same x value will form the xth set of the first replicate, all symbols having the same value for y will form the yth set of the second replicate, and all symbols having the same value for z will form the zth set of the third replicate. K. R. Nair (1951b) showed that triple rectangular lattice designs are PBIB designs with the parameters

$$(8.15.5) \quad \begin{matrix} v = 6, & b = 9, & r = 3, & k = 2, & \lambda_1 = 2, \\ \lambda_2 = 0, & n_1 = 3, & n_2 = 2, & p_{11}{}^1 = 0 \end{matrix}$$

when $p = 3$, and with the parameters

$$v = 12 = b, \quad r = 3 = k, \quad \lambda_1 = 1, \quad \lambda_2 = 0 = \lambda_3,$$

$$n_1 = 6, \quad n_2 = 3, \quad n_3 = 2;$$

$$(8.15.6) \quad P_1 = \begin{bmatrix} 2 & 2 & 1 \\ 2 & 0 & 1 \\ 1 & 1 & 0 \end{bmatrix}; \quad P_2 = \begin{bmatrix} 4 & 0 & 2 \\ 0 & 2 & 0 \\ 2 & 0 & 0 \end{bmatrix}; \quad P_3 = \begin{bmatrix} 3 & 3 & 0 \\ 3 & 0 & 0 \\ 0 & 0 & 1 \end{bmatrix},$$

when $p = 4$, but they are not PBIB designs for $p \geq 5$. Thus it appears that there is a more general class of partially balanced designs with less stringent conditions than those imposed on PBIB designs, and we shall study some generalizations of PBIB designs in Section 8.16.

8.16. GENERALIZATIONS OF PBIB DESIGNS

B. V. Shah (1959) generalized PBIB designs by relaxing the condition $p_{jk}{}^i = p_{kj}{}^i$. He assumed that if two symbols are ith associates, then the number of symbols common to the jth associates of the first symbol and the kth associates of the second symbol, plus the number of symbols common to the kth associates of the first symbol and the jth associates of the second symbol, is equal to $2q_{jk}{}^i$ and is the same for all pairs of symbols that are ith associates. He further showed that his condition is equivalent to a PBIB design association scheme, for a two-associate class. The analysis of these designs was also shown to be similar to the usual PBIB design analysis. To illustrate the notion, let us consider a four-associate-class association scheme given by B. V. Shah (1959) for six symbols, reproduced in Table 8.16.1.

*Table 8.16.1. Four-associate-class association scheme
for six symbols*

Symbol	Associates			
	First	Second	Third	Fourth
1	2	6	4	3, 5
2	1	3	5	4, 6
3	4	2	6	1, 5
4	3	5	1	2, 6
5	6	4	2	1, 3
6	5	1	3	2, 4

Here we observe that $p_{12}^4(1, 3) = 1$ and $p_{21}^4(1, 3) = 0$, thereby showing that it is not the association scheme of PBIB designs. However, we can easily verify that it is the association scheme of generalized PBIB designs as given by B. V. Shah.

C. R. Nair (1964) generalized PBIB designs by relaxing the condition of symmetry in the relation of association. For nine symbols the association scheme defined in Table 8.16.2 is of the type of association scheme defined by C. R. Nair.

*Table 8.16.2. Three-associate-class association scheme
for nine symbols*

Symbol	Associates		
	First	Second	Third
0	8	4	1, 2, 3, 5, 6, 7
1	7	6	0, 2, 3, 4, 5, 8
2	3	5	0, 1, 4, 6, 7, 8
3	5	2	0, 1, 4, 6, 7, 8,
4	0	8	1, 2, 3, 5, 6, 7
5	2	3	0, 1, 4, 6, 7, 8
6	1	7	0, 2, 3, 4, 5, 8
7	6	1	0, 2, 3, 4, 5, 8
8	4	0	1, 2, 3, 5, 6, 7

Further types of generalizations were obtained by M. B. Rao (1966a,b) by varying the number of replications.

8.17. CONCLUDING REMARKS

The work on PBIB designs has been so extensive that a complete book can be written on this subject. Due to the limitation of space, we covered the two-associate-class designs at greater length and briefly surveyed other results. The end of this chapter contains a list of references and a bibliography that covers not only the material of this chapter but also other results on PBIB designs. No claim for the completeness of the references is made.

Finite geometries are very useful for the construction of PBIB designs. In this connection we refer to Bose and Chakrarti (1966); Clatworthy (1964); Dai and Feng (1964a,b); Raychaudhuri (1962, 1965); Seiden (1966); Wan (1964a,b; 1965); Wan and Yang (1964); Yamamoto, Fukuda, and Hamada (1966); and Yang (1965a,b). Other construction methods of interest are presented by Addelman and Bush (1964), Archbold and Johnson (1956), Bose and Clatworthy (1955), Clatworthy (1967), Masuyama (1964, 1965, 1967), Okuno and Okuno (1961), B. V. Shah (1960), Shrikhande and Raghavarao (1963b), and Sprott (1955, 1959). Methods of constructing PBIB designs by dualizing known designs have been given by P. V. Rao (1960), Roy (1954a,b), and Shrikhande (1952, 1960, 1965). Cyclic designs studied by David and Wolock (1965) and J. A. John (1966) are practically useful PBIB designs with two or more associate classes. A complete enumeration of the series of three-replicate PBIB designs with two-associate classes has been made by Laha and Roy (1956) and some series of three-replicate designs have been given by K. R. Nair (1952). Affine α-resolvability of PBIB designs has been studied by Shrikhande and Raghavarao (1963a). The papers of Bose (1963), and B. Chang (1964) are mainly expository in nature, but they also contain many new results. For the analysis of PBIB designs and detailed treatment of lattice designs we refer to Kempthorne (1952).

In spite of the vast development that has taken place in this area over the last three decades, many research problems remain open. We have indicated some such open problems in this chapter. We close this chapter by remarking that the classification of higher-associate-class schemes, deserves the attention of future research workers.

REFERENCES

1. Addelman, S., and Bush, S. (1964). A procedure for constructing incomplete block designs. *Technometrics*, **6**, 389–404.
2. Adhikary, B. (1966). Some types of *m*-associate P.B.I.B. association schemes. *Calcutta Stat. Assn. Bull.*, **15**, 47–74.
3. Adhikary, B. (1967). A new type of higher associate cyclical association schemes. *Calcutta Stat. Assn. Bull.*, **16**, 40–44.

4. Agrawal, H. L. (1964). On the bounds of the number of common treatments between blocks of certain two associate designs. *Calcutta Stat. Assn. Bull.*, **13**, 76–79.

5. Archbold, J. W., and Johnson, N. L. (1956). A method of constructing partially balanced incomplete block designs. *Ann. Math. Stat.*, **27**, 624–632.

6. Bose, R. C. (1950–1951). Partially balanced incomplete block designs with two associate classes involving only two replications. *Calcutta Stat. Assn. Bull.*, **3**, 120–125.

7. Bose, R. C. (1963). Combinatorial properties of partially balanced designs and association schemes. *Contributions to Statistics.* Presented to Prof. P. C. Mahalanobis on his 70th Birthday. Pergamon Press, New York, pp. 21–48.

8. Bose, R. C., and Chakravarti, I. M. (1966). Hermitian varieties in a finite projective space PG(N, q^2). *Can. J. Math.*, **18**, 1161–1182.

9. Bose, R. C., and Clatworthy, W. H. (1955). Some classes of partially balanced designs. *Ann. Math. Stat.*, **26**, 212–232.

10. Bose, R. C., Clatworthy, W. H., and Shrikhande, S. S. (1954). Tables of partially balanced incomplete block designs with two associate classes. *North Carolina Agricultural Experimental Bulletin*, No. 107.

11. Bose, R. C., and Connor, W. S. (1952). Combinatorial properties of group divisible incomplete block designs. *Ann. Math. Stat.*, **23**, 367–383.

12. Bose, R. C., and Laskar, R. (1967). A characterization of tetrahedral graphs. *J. Comb. Theory*, **2**, 366–385.

13. Bose, R. C., and Mesner, D. M. (1959). On linear associative algebras corresponding to association schemes of partially balanced designs. *Ann. Math. Stat.*, **30**, 21–38.

14. Bose, R. C., and Nair, K. R. (1939). Partially balanced incomplete block designs. *Sankhyā*, **4**, 337–372.

15. Bose, R. C., and Shimamoto, T. (1952). Classification and analysis of partially balanced incomplete block designs with two associate classes. *J. Am. Stat. Assn.*, **47**, 151–184.

16. Bose, R. C., Shrikhande, S. S., and Bhattacharya, K. N. (1953). On the construction of group divisible incomplete block designs. *Ann. Math. Stat.*, **24**, 167–195.

17. Bruck, R. H. (1963). Finite nets. II. Uniqueness and imbedding. *Pacific J. Math.*, **13**, 421–457.

18. Carmichael, R. D. (1956). *Introduction to the Theory of Groups of Finite Order.* Dover, New York.

19. Chandrasekhararao, K. (1964). On further constructions of cubic designs. *Calcutta Stat. Assn. Bull.*, **13**, 71–75.

20. Chang, B. (1964). Partially balanced incomplete block designs. *Shuxue Jinzhan*, **7**, 240–281.

21. Chang, L.-C. (1960). Association schemes of partially balanced designs with parameters $v = 28$, $n_1 = 12$, $n_2 = 15$ and $p_{11}^2 = 4$. *Science Record,* new series, **4**, 12–18.

22. Chang, L.-C., and Liu, W.-R. (1964). Incomplete block designs with square parameters for $k \leqq 10$ and $r \leqq 10$. *Scientia Sinica*, **13**, 1493–1495.

23. Chang, L.-C., Liu, C.-W., and Liu, W.-R. (1965). Incomplete block designs with triangular parameters for which $k \leq 10$ and $r \leq 10$. *Scientia Sinica*, **14**, 329–338.

24. Clatworthy, W. H. (1954). A geometrical configuration which is a partially balanced design. *Proc. Am. Math. Soc.*, **5**, 47.

25. Clatworthy, W. H. (1955). Partially balanced incomplete block designs with two associate classes and two treatments per block. *J. Res. Natl. Bur. Std.*, **54**, 177–190.

26. Clatworthy, W. H. (1956). *Contributions on Partially balanced Incomplete Block Designs with Two Associate Classes.* National Bureau of Standards Applied Mathematics Series No. 47, Washington, D.C.

27. Clatworthy, W. H. (1967). Some new families of partially balanced designs of the latin square type and related designs. *Technometrics*, **9**, 229–243.

28. Connor, W. S. (1952). Some relations among the blocks of symmetrical group divisible designs. *Ann. Math. Stat.*, **23**, 602–609.

29. Connor, W. S. (1958). The uniqueness of the triangular association scheme. *Ann. Math. Stat.*, **29**, 262–266.

30. Connor, W. S., and Clatworthy, W. H. (1954). Some theorems for partially balanced designs. *Ann. Math. Stat.*, **25**, 100–112.

31. Dai, Z.-D., and Feng, X.-N. (1964a). Notes on finite geometries and the construction of PBIB designs. IV. Some "Anzahl" theorems in orthogonal geometry over finite fields of characteristic not 2. *Scientia Sinica*, **13**, 2001–2004.

32. Dai, Z.-D., and Feng, X.-N. (1964b). Notes on finite geometries and the construction of PBIB designs. V. Some "Anzahl" theorems in orthogonal geometry over finite fields of characteristic 2. *Scientia Sinica*, **13**, 2005–2008.

33. David, H. A., and Wolock, F. W. (1965). Cyclic designs. *Ann. Math. Stat.*, **36**, 1526–1534.

34. Dembowski, P. (1968). *Finite Geometries.* Springer Verlag.

35. Hinkelmann, K. (1964). Extended group divisible partially balanced incomplete block designs. *Ann. Math. Stat.*, **35**, 681–695.

36. Hinkelmann, K., and Kempthorne, O. (1963). Two classes of group divisible partial diallel crosses. *Biometrika*, **50**, 281–291.

37. Hoffman, A. J. (1960). On the uniqueness of the triangular association scheme. *Ann. Math. Stat.*, **31**, 492–497.

38. Hoffman, A. J., and Singleton, R. R. (1960). On Moore graphs with diameters 2 and 3. *IBM J. Res. Develop.*, **4**, 497–504.

39. John, J. A. (1966). Cyclic incomplete block designs. *J. Roy. Stat. Soc.*, **B28**, 345–360.

40. John, P. W. M. (1966). An extension of the triangular association scheme to three associate classes. *J. Roy. Stat. Soc.*, **B28**, 361–365.

41. Kapadia, C. H. (1966). On the block structure of singular group divisible designs. *Ann. Math. Stat.*, **37**, 1398–1400.

42. Kempthorne, O. (1952). *The Design and Analysis of Experiments.* Wiley, New York.

43. Kusumoto, K. (1965). Hyper cubic designs. *Wakayama Medical Reports*, **9**, 123–132.

44. Kusumoto, K. (1967). Association schemes of new types and necessary conditions for existence of regular and symmetrical PBIB designs with those association schemes. *Ann. Inst. Stat. Math.*, **19**, 73–100.

45. Laha, R. G., and Roy, J. (1956). Two associate partially balanced designs involving three replications. *Sankhyā*, **17**, 175–184.

46. Masuyama, M. (1964). Construction of PBIB designs by fractional development. *Rep. Stat. Appl. Res. Un. Japan Sci. Engrs.*, **11**, 47–54.

47. Masuyama, M. (1965). Cyclic generation of triangular PBIB designs. *Rep. Stat. Appl. Res. Un. Japan, Sci. Engrs.*, **12**, 73–81.

48. Masuyama, M. (1967). Construction cyclique de blocs incomplete partielle-ment equilibres. *Rev. Intern. Stat. Inst.*, **35**, 107–124.

49. Mesner, D. M. (1956). An investigation of certain combinatorial properties of partially balanced incomplete block designs and association schemes with a detailed study of designs of latin square and related types. Unpublished Ph.D. thesis, Michigan State University.

50. Mesner, D. M. (1965). A note on the parameters of PBIB association schemes. *Ann. Math. Stat.*, **36**, 331–336.

51. Mesner, D. M. (1967). A new family of partially balanced incomplete block designs with some latin square design properties. *Ann. Math. Stat.*, **38**, 571–581.

52. Nair, C. R. (1964). A new class of designs. *J. Am. Stat. Assn.*, **59**, 817–833.

53. Nair, K. R. (1943). Certain inequality relations among the combinatorial parameters of incomplete block designs. *Sankhyā*, **6**, 255–259.

54. Nair, K. R. (1950). Partially balanced incomplete block designs involving only two replications. *Calcutta Stat. Assn. Bull.*, **3**, 83–86.

55. Nair, K. R. (1951a). Some two replicate partially balanced designs. *Calcutta Stat. Assn. Bull.*, **3**, 174–176.

56. Nair, K. R. (1951b). Rectangular lattices and partially balanced incomplete block designs. *Biometrics*, **7**, 145–154.

57. Nair, K. R. (1952). Some three replicate partially balanced incomplete block designs. *Calcutta Stat. Assn. Bull.*, **4**, 39–42.

58. Nair, K. R., and Rao, C. R. (1942). A note on partially balanced incomplete block designs. *Science and Culture*, **7**, 568–569.

59. Nandi, H. K., and Adhikari, B. (1966). On the definition of Bose–Shimamoto cyclical association scheme. *Calcutta Stat. Assn. Bull.*, **15**, 165–168.

60. Ogasawara, M. (1965). A necessary condition for the existence of regular and symmetrical PBIB designs of T_m type. Inst. Stat. Mimeo. Series 418, Chapel Hill, N.C.

61. Okuno, C., and Okuno, T. (1961). On the construction of a class of partially balanced incomplete block designs by calculus of blocks. *Rep. Stat. Appl. Res. Un. Japan Sci. Engrs.*, **8**, 113–139.

62. Raghavarao, D. (1959). A note on the construction of group divisible designs from hyper graeco latin cubes of the first order. *Calcutta Stat. Assn. Bull.*, **9**, 67–70.

63. Raghavarao, D. (1960a). On the block structure of certain PBIB designs with triangular and L_2 association schemes. *Ann. Math. Stat.*, **31**, 787–791.

64. Raghavarao, D. (1960b). A generalization of group divisible designs. *Ann. Math. Stat.*, **31**, 756–771.

65. Raghavarao, D. (1960c). Some aspects of weighing designs. *Ann. Math. Stat.*, **31**, 878–884.

66. Raghavarao, D. (1962). Some results for GD *m*-associate designs. *Calcutta Stat. Assn. Bull.*, **11**, 150–154.

67. Raghavarao, D., and Chandrasekhararao, K. (1964). Cubic designs. *Ann. Math. Stat.*, **35**, 389–397.

68. Rao, C. R. (1946). Difference sets and combinatorial arrangements derivable from finite geometries. *Proc. Natl. Inst. Sci. India*, **12**, 123–135.

69. Rao, M. B. (1966a). Partially balanced block designs with two different number of replications. *J. Indian Stat. Assn.*, **4**, 1–9.

70. Rao, M. B. (1966b). Group divisible family of PBIB designs. *J. Indian Stat. Assn.*, **4**, 14–28.

71. Rao, P. V. (1960). On the construction of some partially balanced incomplete block designs with more than three associate classes. *Calcutta Stat. Assn. Bull.*, **9**, 87–92.

72. Raychaudhuri, D. K. (1962). Application of the geometry of quadrics for constructing PBIB designs. *Ann. Math. Stat.*, **33**, 1175–1186.

73. Raychaudhuri, D. K. (1965). Some configurations in finite projective spaces and partially balanced incomplete block designs. *Can. J. Math.*, **17**, 114–123.

74. Roy, P. M. (1953–1954). Hierarchical group divisible incomplete block designs with *m*-associate classes. *Science and Culture*, **19**, 210–211.

75. Roy, P. M. (1954a). On the relation between BIB and PBIB designs. *J. Indian Soc. Agric. Stat.*, **6**, 30–47.

76. Roy, P. M. (1954b). On a method of inversion in the construction of partially balanced incomplete block designs from the corresponding BIB designs. *Sankhyā*, **14**, 39–52.

77. Roy, P. M. (1962). On the properties and construction of HGD designs with *m*-associate classes. *Calcutta Stat. Assn. Bull.*, **11**, 10–38.

78. Seiden, E. (1966). A note on the construction of partially balanced incomplete block designs with parameters $v = 28$, $n_1 = 12$, $n_2 = 15$ and $p_{11}^2 = 4$. *Ann. Math. Stat.*, **37**, 1783–1789.

79. Shah, B. V. (1959). A generalization of partially balanced incomplete block designs. *Ann. Math. Stat.*, **30**, 1041–1050.

80. Shah, B. V. (1960). A matrix substitution method of constructing partially balanced designs. *Ann. Math. Stat.*, **31**, 34–42.

81. Shah, S. M. (1964). An upper bound for the number of disjoint blocks in certain PBIB designs. *Ann. Math. Stat.*, **35**, 398–407.

82. Shrikhande, S. S. (1952). On the dual of some balanced incomplete block designs. *Biometrics*, **8**, 66–72.

83. Shrikhande, S. S. (1959a). On a characterization of the triangular association scheme. *Ann. Math. Stat.*, **30**, 39–47.

84. Shrikhande, S. S. (1959b). The uniqueness of the L_2 association scheme. *Ann. Math. Stat.*, **30**, 781–798.

85. Shrikhande, S. S. (1960). Relations between certain incomplete block designs. *Contributions to Probability and Statistics*, Stanford University Press, Stanford, pp. 388–395.
86. Shrikhande, S. S. (1965). On a class of partially balanced incomplete block designs. *Ann. Math. Stat.*, **36**, 1807–1814.
87. Shrikhande, S. S., and Jain, N. C. (1962). The nonexistence of some partially balanced incomplete block designs with latin square type association scheme. *Sankhyā*, **B24**, 259–268.
88. Shrikhande, S. S., and Raghavarao, D. (1963a). Affine α-resolvable incomplete block designs. *Contributions to Statistics*. Presented to Prof. P. C. Mahalanobis on his 70th Birthday. Pergamon Press, New York, pp. 471–480.
89. Shrikhande, S. S., and Raghavarao, D. (1963b). A method of construction of incomplete block designs. *Sankhyā*, **A25**, 399–402.
90. Shrikhande, S. S., and Singh, N. K. (1962). On a method of constructing symmetrical balanced incomplete block designs. *Sankhyā*, **A24**, 25–32.
91. Sprott, D. A. (1955). Some series of partially balanced incomplete block designs. *Can. J. Math.*, **7**, 369–380.
92. Sprott, D. A. (1959). A series of symmetric group divisible designs. *Ann. Math. Stat.*, **30**, 249–251.
93. Surendran, P. U. (1968). Association matrices and the Kronecker product of designs. *Ann. Math. Stat.*, **39**, 676–680.
94. Tharthare, S. K. (1963). Right angular designs. *Ann. Math. Stat.*, **34**, 1057–1067.
95. Tharthare, S. K. (1965). Generalized right angular designs. *Ann. Math. Stat.*, **36**, 1535–1553.
96. Vartak, M. N. (1955). On an application of Kronecker product of matrices to statistical designs. *Ann. Math. Stat.*, **26**, 420–438.
97. Vartak, M. N. (1959). The non-existence of certain PBIB designs. *Ann. Math. Stat.*, **30**, 1051–1062.
98. Wan, C.-H. (1964a). Notes on finite geometries and the constructions of PBIB designs. I. Some "Anzahl" theorems in symplectic geometry over finite fields. *Scientia Sinica*, **13**, 515–516.
99. Wan, C.-H. (1964b). Notes on finite geometries and the constructions of PBIB designs. II. Some PBIB designs with two associate classes based on the symplectic geometry over finite fields. *Scientia Sinica*, **13**, 516–517.
100. Wan, C.-H. (1965). Notes on finite geometries and the constructions of PBIB designs. VI. Some association schemes and PBIB designs based on finite geometries. *Scientia Sinica*, **14**, 1872–1876.
101. Wan, C.-H., and Yang, B.-F. (1964). Notes on finite geometries and the construction of PBIB designs. III. Some "Anzahl" theorems in unitary geometry over finite fields and their applications. *Scientia Sinica*, **13**, 1006–1007.
102. Yamamoto, S., Fuji, Y., and Hamada, N. (1965). Composition of some series of association algebras. *J. Sci. Hiroshima Univ.*, **A29**, 181–215.
103. Yamamoto, S., Fukuda, T., and Hamada, N. (1966). On finite geometries and cyclically generated incomplete block designs. *J. Sci. Hiroshima Univ.*, Series A **30**, 137–149.

104. Yang, B.-F. (1965a). Studies in finite geometries and construction of incomplete block designs. VII. An association scheme with many associate classes constructed from maximal completely isotropic subspaces in symplectic geometry over finite fields. *Acta Math. Sinica*, **15**, 812–825.

105. Yang, B.-F. (1965b). Studies in finite geometries and the construction of incomplete block designs. VIII. An association scheme with many associate classes constructed from maximal completely isotropic subspaces in unitary geometry over finite fields. *Acta Math. Sinica*, **15**, 826–841.

BIBLIOGRAPHY

Agrawal, H. L. On the bounds of the number of common treatments between blocks of semi-regular group divisible designs. *J. Amer. Stat. Assn.*, **59**, 867–871 (1964).

Agrawal, H. L. Comparison of the bounds of the number of common treatments between blocks of certain partially balanced incomplete block designs. *Ann. Math. Stat.*, **37**, 739–740 (1966).

Berman, G. A three-parameter family of partially balanced incomplete block designs with two associate classes. *Proc. Am. Math. Soc.*, **6**, 490–493 (1955).

Bhattacharya, K. N. Problems in partially balanced incomplete block designs. *Calcutta Stat. Assn. Bull.*, **2**, 177–182 (1950).

Bose, R. C. A note on Nair's condition for partially balanced incomplete block designs with $k > r$. *Calcutta Stat. Assn. Bull.*, **4**, 123–126 (1952).

Bose, R. C., and Bush, K. A. Orthogonal arrays of strength two and three. *Ann. Math. Stat.*, **23**, 508–524 (1952).

Chang, L.-C. The uniqueness and the non-uniqueness of the triangular association schemes. *Science Record*, new series, **3**, 604–613 (1959).

Freeman, G. H. Some further methods of constructing regular group divisible incomplete block designs. *Ann. Math. Stat.*, **28**, 479–487 (1957).

Guerin, R. Vue d'ensemble sur les plans en blocs incomplets equilibres et partiellement equilibres. *Rev. Intern. Stat. Inst.*, **33**, 24–58 (1965).

John, J. A. Reduced group divisible paired comparison designs. *Ann. Math. Stat.*, **38**, 1887–1893 (1967).

Heuze, G. Contribution a l'etude des schemas d'association. Publ. Inst. Stat. Univ. Paris, **15**, 1–60 (1966).

Lafon, M. Construction de blocs incomplets partiellement equilibres à $s + 1$ classes associees. *Compt. Rend.*, **244**, 1714–1717 (1957).

Lafon, M. Blocs incomplets partiellement equilibres à deux classes associees avec quatre repetitions. *Compt. Rend.*, **244**, 1875–1877 (1957).

Liu, C.-W. A method of constructing certain symmetrical partially balanced designs. *Scientia Sinica*, **12**, 1935–1936 (1963).

Liu, C.-W., and Chang, L.-C. Some PBIB(2)-designs induced by association schemes. *Scientia Sinica*, **13**, 840–841 (1964).

Liu, W.-R., and Chang, L.-C. Group divisible incomplete block designs with parameters $v \leqq 100$ and $r \leqq 10$. *Scientia Sinica*, **13**, 839–840 (1964).

Masuyama, M. Calculus of blocks and a class of partially balanced incomplete block designs. *Rep. Stat. Appl. Res. Un. Japan Sci. Engrs.*, **8**, 59–69 (1961).

Murty, V. N. On the block structure of PBIB designs with two associate classes. *Sankhyā*, **A26**, 381–382 (1964).

Nair, C. R. On the method of block section and block intersection applied to certain PBIB designs. *Calcutta Stat. Assn. Bull.*, **11**, 49–54 (1962).

Nair, K. R. A note on group divisible incomplete block designs. *Calcutta Stat. Assn. Bull.*, **5**, 30–35 (1953).

Ramanujacharyulu, C. Non-linear spaces in the construction of symmetric PBIB designs. *Sankhyā*, **A27**, 409–414 (1965).

Rohatgi, V. K. A comparison of bounds for the number of common treatments between any two blocks of certain two associate PBIB designs. *Calcutta Stat. Assn. Bull.*, **15**, 39–42. (1966).

Saraf, W. S. On the structure and combinatorial properties of certain semi-regular group divisible designs. *Sankhyā*, **A23**, 287–296 (1961).

Seiden, E. On a geometrical method of construction of partially balanced designs with two associate classes. *Ann. Math. Stat.*, **32**, 1177–1180 (1961).

Shah, B. V. On a generalization of the Kronecker product designs. *Ann. Math. Stat.*, **30**, 48–54 (1959).

Shah, S. M. Bounds for the number of common treatments between any two blocks of certain PBIB designs. *Ann. Math. Stat.*, **36**, 337–342 (1965).

Shah, S. M. On the block structures of certain partially balanced incomplete block designs. *Ann. Math. Stat.*, **37**, 1016–1020 (1966).

Shrikhande, S. S. A.R.B.I.B. designs and nonsingular GD designs. *Calcutta Stat. Assn. Bull.*, **5**, 139–141 (1953).

Shrikhande, S. S. Cyclic solutions of symmetrical group divisible designs. *Calcutta Stat. Assn. Bull.*, **5**, 36–39 (1953).

Surendran, P. U. Common treatments between blocks of certain partially balanced incomplete block designs. *Ann. Math. Stat.*, **39**, 999–1006 (1968).

Surendran, P. U. Partially balanced designs and the sum of Kronecker products of designs. *Calcutta Stat. Assn. Bull.*, **17**, 161–166 (1968).

Thompson, H. R. On a new class of partially balanced incomplete block designs. *Calcutta Stat. Assn. Bull.*, **6**, 193–195 (1956).

Thompson, W. A., Jr. A note on PBIB design matrices. *Ann. Math. Stat.*, **29**, 919–922 (1958).

Zelen, M. A note on partially balanced designs. *Ann. Math. Stat.*, **25**, 599–602 (1954).

CHAPTER 9

Graph Theory and Partial Geometries

9.1. INTRODUCTION

During the last decade research on the combinatorial problems of BIB and PBIB designs has been oriented toward graph theory. Pioneering work in this direction has been done by Bose, Bruck, and Hoffman. In the next section we introduce some graph-theory results, necessary for understanding the combinatorial problems of designs, and devote the rest of the chapter to studying the application of graph theory to partial geometries, BIB designs, and PBIB designs and their association schemes.

9.2. SOME RESULTS ON GRAPH THEORY

Let X be a set of points called *vertices*. If $x, y \in X$, then a directed segment $u = (x, y)$ connecting x to y is called an *arc*. A system $G = (X, U)$, where U is the set of all arcs, is called a *graph*. In a graph an arc $u = (x, y)$ may exist, whereas the arc $u' = (y, x)$ need not necessarily exist. If the direction of the arc is ignored and if x and y are joined, then $u = (x, y)$ is called an *edge*. Two vertices are said to be adjacent (or nonadjacent) if there is an edge (or no edge) between them. We can say that a vertex is *incident* on an arc (or edge) according as it lies on the arc (or edge). A vertex on which more than two arcs are incident is called a *node*. In a graph there will be at most one edge connecting any two given vertices x and y. However, if there are multiple edges connecting any two vertices x and y, and if s is the maximal number of edges joining any two vertices, then G is called an *s-graph*. A *path* is a sequence $(u_1, u_2, \cdots)$ of arcs of a graph (X, U) such that the terminal vertex of each arc coincides with the initial vertex of the succeeding arc. The number of arcs constituting a path is called its *length*. A *circuit* is a finite path in which the initial vertex coincides with the terminal vertex. A *loop* is defined as a circuit of length 1. A graph G is said to be *connected* if, given two vertices x and y, there is a sequence of vertices $x_0 = x, x_1, \cdots, x_n = y$ such that there

183

is an edge connecting x_a to x_{a+1} for $a = 0, 1, \cdots, n - 1$. In a connected graph G the *distance* between two vertices x and y, denoted by $d(x, y)$, is the length of the shortest path from x to y. A graph $G = (X, U)$ is said to be *complete* if every pair of vertices is connected in at least one of the two possible directions. The term $\Delta(x, y)$, called the *edge degree* of the edge (x, y), is the number of vertices adjacent to both vertices x and y.

In this chapter by a graph we mean a graph with no multiple edges and loops unless otherwise stated. If $G = (X, U)$ is a graph, then the *complementary* graph $\bar{G}$ is defined on the same set of vertices X, where there are no loops, and two vertices are joined if and only if they are not joined in G. A subset of vertices of a graph G, any two of which are joined, is called a *clique* of G. A *claw* (P, S) of a graph G is an ordered pair consisting of a vertex P, the *vertex of the claw*, and a nonempty set S of vertices, distinct from P, that are not joined among themselves, but each of which is joined to P. The order of the claw (P, S) is $s = |S|$.

The number of edges passing through a vertex is called its *valence* (or *degree*). Clearly the sum of the degrees of the vertices is twice the number of edges, since every edge is accounted for twice in enumerating the sum of the degrees. Consequently the number of vertices of odd degree is even. A graph G is said to be *regular* if the valence of each of its vertices is constant, say d. If any two adjacent (or nonadjacent) vertices of a regular graph are joined to exactly $p_{11}{}^1$ (or $p_{11}{}^2$) other vertices, the graph is said to be *strongly regular*. The concept of strongly regular graphs is due to Bose (1963a).

The *adjacency matrix* $A = (a_{ij})$ of a graph G is a $v \times v$ matrix, where

(9.2.1)
$$a_{ij} = 1 \quad \text{if the vertices } i \text{ and } j \text{ are joined by an edge;}$$
$$= 0 \quad \text{otherwise.}$$

The adjacency matrix plays a prominent role in the theory of graphs, as does the incidence matrix in the combinatorial problems of incomplete block designs. It may be noted that although it is customary to define the adjacency matrix of a graph as a $(0, 1)$ matrix, as above, Seidel (1967) defined it as a $(0, 1, -1)$ matrix with zero in the diagonal positions and $a_{ij} = -1$ or $+1$, depending on whether the vertices are joined by an edge or not. To distinguish the matrices Bhagwandas and Shrikhande (1968) called the latter the *Seidel matrix*. The adjacency matrix of a graph G is a symmetrical matrix and has constant row and column sums if G is a regular graph.

The following theorem can be easily proved:

Theorem 9.2.1. The adjacency matrix A of a regular graph G of valence d is the adjacency matrix of a strongly regular graph if there exist nonnegative integers e and f such that

(9.2.2)
$$A^2 = (e - f)A + fE_{v, v} + (d - f)I_v.$$

We note that the $p_{jk}{}^i$ parameters of the strongly regular graph given by Theorem 9.2.1 are $p_{11}{}^1 = e$ and $p_{11}{}^2 = f$.

Between any two vertices of G there may be more than one path (or none at all). Berge (1962) gave the following theorem, proving it by the method of induction on n:

Theorem 9.2.2. Let G be a graph and A be its adjacency matrix. Then the (i, j)th element $p_{ij(n)}$ of $P = A^n$ is equal to the number of distinct paths of length n that go from the ith vertex to the jth vertex.

An immediate corollary to this theorem is as follows:

Corollary 9.2.1 [Adhikary (1968)]. The total number of paths from the ith vertex to the jth vertex is $\sum_{n=1}^{l\max} p_{ij(n)}$, where l_{max} is the length of the longest path from the ith to the jth vertices.

Adhikary (1968) generalized the definition of a strongly regular graph to a strongly regular graph of order m as follows:

Definition 9.2.1. A regular graph G is said to be strongly regular of order m if the following conditions are satisfied:

1. Between any two vertices there exist m distinct relationships. Two vertices are said to have a relationship of the ith type between them if they are connected by exactly λ_i paths $(i = 1, 2, \cdots, m)$.
2. For any two vertices connected by λ_i paths the number of vertices connected by λ_j paths to one vertex and by λ_k paths to the other vertex is a constant $p_{jk}{}^i$, independently of the particular vertices chosen.

9.2.1. Operations on Graphs.

Operation 9.2.1. Given n graphs $G_i = (X_i, U_i)$ for $i = 1, 2, \cdots, n$, their product is the graph $G = G_1 \otimes G_2 \otimes \cdots \otimes G_n = (X, U)$, where $X = X_1 \times X_2 \times \cdots \times X_n$ and the edges of U are $u = \{(x_1, x_2, \cdots, x_n), (y_1, y_2, \cdots, y_n)\}$ whenever $u_i = (x_i, y_i)$ is an edge of U_i for every $i = 1, 2, \cdots, n$.

Operation 9.2.2. Given two graphs $G_i = (X_i, U_i)$ for $i = 1, 2$, their product is the graph $G = G_1 \times G_2 = (X, U)$, where $X = X_1 \times X_2$ and the edges of U are $u = \{(x_1, x_2)\} (y_1, y_2)\}$ whenever (a) $x_1 = y_1$, $(x_2, y_2) \in U_2$; (b) $x_2 = y_2$, $(x_1, y_1) \in U_1$; (c) $(x_1, y_1) \in U_1$, $(x_2, y_2) \in U_2$.

We observe that the above two definitions of graph products are slightly different. The second represents the first after the addition of loops.

Operation 9.2.3. Given n graphs $G_i = (X_i, U_i)$ for $i = 1, 2, \cdots, n$, their sum of the first kind is the graph $G = G_1 + G_2 + \cdots + G_n = (X, U)$, where $X = X_1 \times X_2 \times \cdots \times X_n$, and the edges connect the vertex $(x_1, x_2, \cdots, x_n)$

to each of the vertices $(y_1, x_2, \cdots, x_n), (x_1, y_2, \cdots, x_n), \cdots, (x_1, x_2, \cdots, y_n)$, where $u_i = (x_i, y_i)$ are edges in G_i for $i = 1, 2, \cdots, n$.

Operation 9.2.4. Given n graphs $G_i = (X, U_i)$ for $i = 1, 2, \cdots, n$, their sum of the second kind is the graph $G = G_1 + G_2 + \cdots + G_n = (X, U)$, where the edge $u = (x, y) \in U$ if $u \in U_1 \cup U_2 \cup \cdots \cup U_n$.

Operation 9.2.5. Given n graphs $G_i = (X, U_i)$ for $i = 1, 2$, their sum of the third kind is the graph $G = G_1 \triangle G_2 = (X, U)$, where the edge $u = (x, y) \in U$ if $u \in U_1 \ominus U_2$, where $\ominus$ denotes the symmetric difference of the sets.

Operation 9.2.6. Given n graphs $G_i = (X, U_i)$ for $i = 1, 2, \cdots, n$, their sum of the fourth kind is the graph $G = G_1 | G_2 | \cdots | G_n | = (X, U)$, where the edge $u = (x, y)$ if $u \in \overline{U_1 \cup U_2 \cup \cdots \cup U_n}$, where $\bar{S}$ denotes the Complement of the set S.

Operations 9.2.1, 9.2.2, and 9.2.3 are given by Berge (1962); operations 9.2.4, and 9.2.6 are generalizations of the results of Adhikary (1968) for $n = 2$. We also refer in this connection to Harary and Trauth (1966), and Harary and Wilcox (1967). For general results on graph theory we refer to Berge (1962) and Ore (1962).

9.3. LINE GRAPH OF BIB DESIGNS

Let D be a BIB design with the parameters v, b, r, k, and λ. The graph $H(D)$ of the design D is the bipartite graph with $v + b$ vertices, which are the symbols and sets of D, with two vertices adjacent if and only if one is a symbol, the other a set, and the symbol belongs to that set in D.

The *line graph* $L(D)$ of D is the graph whose vertices are the edges of $H(D)$, two vertices of $L(D)$ being joined if and only if the corresponding edges of $H(D)$ have a common vertex.

There are various characterizations of the line graph of BIB designs. We state the following result, due to S. B. Rao and A. R. Rao (1969), without proof:

Theorem 9.3.1. Given positive integers v, b, r, k satisying $vr = bk$, $v - 1 = r(k - 1)$, and $r > k \geqq 2$, a graph G is the line graph of a BIB design with the parameters v, b, r, k, and $\lambda = 1$ if and only if the following four conditions hold provided $r - 2k + 1 < 0$:

1. The number of vertices in G is vr.
2. G is regular, of valence $r + k - 2$.
3. For any edge (x, y), $\Delta(x, y)$ is either $r - 2$ or $k - 2$.
4. If $d(x, y) = 2$, then the number of vertices adjacent to both x and y is 1.

A similar characterization of the line graph of a BIB design was obtained by Dowling and Laskar (1967) for $r = k \geqq 2$. When $r = k + 1$, we get a BIB design with the parameters $v = k^2$, $b = k(k + 1)$, $r = k + 1$, k, $\lambda = 1$, which is in bijection with an affine plane. Hence, putting $r = k + 1$ in the above theorem, we get a characterization of a finite affine plane.

Other characterizations of projective planes, affine planes, and symmetrical BIB designs have been given by Hoffman (1965) and Hoffman and Ray-Chaudhuri (1965a,b), who showed that the characteristic roots and their multiplicities of the adjacency matrices of their line graphs determine the corresponding configurations up to isomorphisms, excepting the case of the symmetrical BIB design with the parameters $v = b = 4$, $r = k = 3$, $\lambda = 2$.

9.4. CONNECTION OF PBIB DESIGNS AND THEIR ASSOCIATION SCHEMES WITH GRAPH THEORY

Let G be a graph with v vertices having the adjacency matrix A. Let us identify the vertices of the graph with v symbols of a PBIB design association scheme and define the symbols x and y to be first associates if the vertices x and y are adjacent in G; otherwise they are second associates.

Theorem 9.4.1. A strongly regular graph determines a two-associate-class BPIB design association scheme; the converse is also true.

Proof. By the very definition of strongly regular graphs and the identification described in the first paragraph of this section the first part follows. Conversely let us consider a two-associate-class association scheme on v symbols, letting A be the association matrix of the first associates, as defined in Section 8.3. Then A can be shown to be the adjacency matrix of a strongly regular graph on v vertices.

The graph of a group-divisible association scheme with $v = mn$ symbols arranged in m groups of n symbols each consists of m cliques of n vertices each and does not contain claws of any order. The graph of an L_2 association scheme contains claws of order 2, but not of order 3.

The following theorem is obvious:

Theorem 9.4.2. Let N be a matrix of order $v \times b$, with elements 0 and 1, and constant row and column totals. Then N is the incidence matrix of a PBIB design with two associate classes if and only if

$$(9.4.1) \qquad NN' = rI_v + \lambda_1 A + \lambda_2(E_{v,v} - I_v - A),$$

where $\lambda_1 \neq \lambda_2$, $r \geqq \max(\lambda_1, \lambda_2)$, and A is the adjacency matrix of a strongly regular graph.

We use this theorem in studying the duals of incomplete block designs in the next chapter.

In analogy to Theorem 9.4.1, after defining two symbols x and y to be the ith associates if the relation between the vertices x and y is of the ith type in a strongly regular graph of order m, we have the following:

Theorem 9.4.3. A strongly regular graph of order m determines an m-associate-class PBIB design association scheme.

Consider the cubic association scheme of Raghavarao and Chandrasekhararao given in Subsection 8.12.2. Let the symbols of the association scheme be identified with the s^3 vertices of a graph G, whose vertices are joined if and only if the corresponding symbols are first associates. Such a graph has the following properties:

1. It has s^3 nodes.
2. It is connected and regular, with valence $3(s - 1)$.
3. For every pair of vertices, the number of vertices adjacent to both x and y is $s - 2$.
4. If $d(x, y) = 2$, then the number of vertices adjacent to both x and y is 2.
5. If $d(x, y) = 2$, then there are exactly $s - 1$ nodes z, adjacent to x, for which $d(y, z) = 3$.

Such a graph was called a cubic-lattice graph by Laskar (1967). Conversely every cubic-lattice graph can be easily verified to be a strongly regular graph of order 3 and hence defines a three-associate-class association scheme, namely, the cubic association scheme. Thus we have the following theorem:

Theorem 9.4.4. The existence of a cubic-lattice graph implies and is implied by the corresponding cubic association scheme.

Consider the extended triangular association scheme of Subsection 8.12.3 and let its symbols correspond to $(s + 2)(s + 3)(s + 4)/6$ vertices of a graph G, whose vertices are joined if and only if they are first associates. Such a graph has the following properties:

1. It has $(s + 2)(s + 3)(s + 4)/6$ nodes.
2. It is connected and regular, of valence $3(s - 1)$.
3. $\Delta(x, y) = s + 2$.
4. If $d(x, y) = 2$, then the number of vertices adjacent to both x and y is 4.

Such a graph was called a tetrahedral graph by Bose and Laskar (1967). Conversely every tetrahedral graph can be easily verified to be a strongly regular graph of order 3 and hence defines a three-associate-class association scheme, namely, the extended triangular association scheme. Thus we have the following theorem:

Theorem 9.4.5. The existence of a tetrahedral graph implies and is implied by the corresponding extended triangular association scheme.

Seidel (1967) introduced the concept of equivalence in graphs. On any graph $G = (X, U)$ the operation of *complementation with respect to a vertex* $x \in X$ results in the graph with the same set of vertices X, but all existing edges with x will be canceled and all nonexisting edges with x will be introduced. The effect of complementation with respect to a vertex x on the adjacency matrix A is that in its row and column corresponding to the vertex x the elements 0 and 1 will be interchanged in the nondiagonal positions.

Complementation with respect to the vertices generates an equivalence relation on the class of all graphs with the same set of vertices. We introduce the following definition:

Definition 9.4.1. Two graphs $G_1 = (X, U_1)$ and $G_2 = (X, U_2)$ are said to be Seidel-equivalent (or S-equivalent) if one is obtained from the other by complementation with respect to each vertex x of a nonempty subset S of X.

Seidel proved the following theorem:

Theorem 9.4.6. The graphs of pseudotriangular association schemes for $v = 28$ are S-equivalent to the graph of the triangular association scheme for $v = 28$, and the graph of the pseudo L_2 association scheme for $v = 16$ is S-equivalent to the graph of the L_2 association scheme for $v = 16$.

Proof. Complementation with respect to the vertices 1, 16, 21, 26 of the pseudotriangular association scheme of Table 8.7.1; the vertices 2, 7, 8, 9, 19, 20, 21, 22 of the pseudotriangular association scheme of Table 8.7.2; and the vertices 1, 2, 4, 10, 12, 15, 17, 21, 22, 23, 27, 28 of the pseudotriangular association scheme of Table 8.7.3 gives the usual triangular association scheme. Complementation with respect to the vertices 1, 8, 11, 13 of the pseudo L_2 association scheme of Table 8.9.1 gives the usual L_2 association scheme.

The gramian of the characteristic vectors associated with the adjacency matrices of S-equivalent graphs has been given by Bhagwandas and Shrikhande (1968).

The six operations studied in Subsection 9.2.1 can be used to generate new association schemes from known association schemes; we illustrate this through an example.

Illustration 9.4.1. Let there be mn symbols a_{ij} arranged in the following array:

(9.4.2)

$$
\begin{array}{cccc}
a_{11} & a_{12} & \cdots & a_{1n} \\
a_{21} & a_{22} & \cdots & a_{2n} \\
\cdot & \cdot & \cdots & \cdot \\
a_{m1} & a_{m2} & \cdots & a_{mn}
\end{array}
$$

Let G_1 (or G_2) be the graph associated with the group-divisible-design association scheme on the above symbols where the m groups (or n groups) are taken as rows (or columns) of (9.4.2). If $m = n$, clearly $G = G_1 + G_2$ is the graph of an L_2 association scheme with $v = m^2$.

For further results in this direction we refer to Adhikary (1968).

We close this section by giving the characterization of the Kronecker product of designs in terms of graph products. Let D_i be a design (not necessarily a PBIB design) on v_i symbols and let G_i be a graph of D_i, with v_i vertices corresponding to the symbols, in which two vertices x and y are adjacent if there is a set in D_i containing the symbols x and y ($i = 1, 2$). Clearly G_i is an s-graph with multiple edges and with loops. The graph G of the Kronecker product $D = D_1 \times D_2$ of designs as defined in Section 4.4 is the graph $G = G_1 \times G_2$.

9.5. FINITE NETS

Bruck (1951, 1963) defined a finite i-net as follows:

Definition 9.5.1. An i-net N^* is a system of undefined points and lines, with their incidence relation satisfying the following conditions:

1. N^* has at least one point.
2. The lines of N^* are partitioned into i disjoint, nonempty, "parallel classes" such that (a) each point of N^* is incident with exactly one line of each class and (b) to two lines belonging to distinct classes there corresponds exactly one point of N^* that is incident with both lines.

If some line has $s \geq 1$ points, we can then easily verify that (a) each line of N^* contains exactly s distinct points; (b) each point of N^* is incident on i distinct lines; (c) there are si distinct lines in N^* and these form i parallel classes of s lines each, such that distinct lines of the same parallel class have no common points while two lines of different classes have a common point; and (d) there are s^2 points in N^*.

Here s is called the order, i the degree of the finite net N^*. Furthermore, $d = s + 1 - i$ is called the deficiency of the net N^*. The i-net will be called degenerate if $i = 1$ or 2 and trivial if $s = 1$. Hence i-nets will be considered only for $i \geq 3$.

By treating the points as symbols and the lines as sets, we can easily see that an i-net implies the existence of a two-associate-class PBIB design of the L_i association scheme with the parameters

$$v = s^2, \quad b = si, \quad r = i, \quad k = s, \quad \lambda_1 = 1, \quad \lambda_2 = 0,$$
$$(9.5.1) \quad n_1 = i(s - 1), \quad n_2 = (s - i + 1)(s - 1),$$
$$p_{11}{}^1 = (s - 2) + (i - 1)(i - 2), \quad p_{11}{}^2 = i(i - 1).$$

Given two points P and Q belonging to the i-net N^*, we say that they are "joined" in N^* if there exists a line of N^*, containing both P and Q, that is necessarily unique. Otherwise P and Q are said to be "not joined" in N^*.

Definition 9.5.2. A subset S of points of N^* is said to be a partial transversal if every two distinct points of S are not joined in N^*. A partial transversal with exactly s distinct points is called a transversal.

Definition 9.5.3. The complementary net, $\overline{N^*}$ of N^* is a d-net whose points are identical with those of N^* and whose ds lines are a suitably selected set of transversals of N^*.

We observe that a complementary d-net is again a PBIB design with the parameters similar to (9.5.1) where the relationship of association is interchanged. Clearly we can obtain an affine plane and hence an $0S1$ BIB design if a complementary net exists for any given i-net. It may be further added that a given i-net may or may not have a complementary net, and if it has one, it need not necessarily be unique. The following two theorems regarding i-nets of order s and deficiency $d = s + 1 - i$, proved by Bruck (1963), are in order here:

Theorem 9.5.1. If $s > (d - 1)^2$ and if the i-net can be completed at all into an affine plane, then it can be completed uniquely, aside from trivialities.

Theorem 9.5.2. If

$$(9.5.2) \qquad s > p(d - 1) = \tfrac{1}{2}(d - 1)^4 + (d - 1)^3 + (d - 1)^2 + \tfrac{3}{2}(d - 1),$$

then the i-net can always be completed into an affine plane.†

For the proof of the above theorems we refer to Bruck (1963). We comment that the condition of Theorem 9.5.1 does not ensure the completion of the i-net into an affine plane. Since the existence of an i-net of order s is equivalent to the existence of $(i - 2)$ MOLS of order s, the result of Theorems 9.5.1 and 9.5.2, read in the language of orthogonal latin squares, give the embedding results referred to at the end of Chapter 1. This result for $d = 2$ was obtained by Shrikhande (1961).

Bruck's conjecture that every pseudonet graph‡ of order s, degree i, deficiency d subject to $i > 1$, $s > i + 1$, $(i - 1)^2 < s \leq p(i - 1) = \tfrac{1}{2}(i - 1)^4 + (i - 1)^3 + (i - 1)^2 + \tfrac{3}{2}(i - 1)$ is either the graph of a net of order s, degree i, or the complementary graph of a net of order s, degree d, or both, was disproved by Shrikhande (1968) by exhibiting a counter example.

† In (9.5.2) $p(d - 1)$ is a function of argument $d - 1$.

‡ A pseudonet graph is a graph whose parameters are those of a net graph, whatever its combinatorial structure.

9.6. PARTIAL GEOMETRY (r, k, t)

Bose (1963a) defined a partial geometry (r, k, t) that contains the nets of Section 9.5 as special cases. He defined the following:

Definition 9.6.1. A partial geometry (r, k, t) is a system of points and lines, and their incidence relation satisfies the following axioms:

Axiom 1. Any two distinct points are incident with not more than one line.
Axiom 2. Each point is incident with r lines.
Axiom 3. Each line is incident with k points.
Axiom 4. If the point P is not incident on l, then there are exactly t lines $(t \geq 1)$ that are incident with P and also incident with some point incident with l. [Clearly $1 \leq t \leq \min (r, k)$.]

As a by-product of axiom 1, we have the following:

Axiom 1′. Any two distinct lines are incident with not more than one point.

Theorem 9.6.1. By writing the points as symbols and the lines as sets, from a partial geometry (r, k, t) we can construct a two-associate-class PBIB design with the parameters

$$v = \frac{k\{(r-1)(k-1) + t\}}{t}, \qquad b = \frac{r\{(r-1)(k-1) + t\}}{t},$$

(9.6.1)

$$r, k, n_1 = r(k-1), \qquad n_2 = \frac{(r-1)(k-1)(k-t)}{t},$$

$$\lambda_1 = 1, \qquad \lambda_2 = 0, \qquad p_{11}{}^1 = (t-1)(r-1) + k - 2, \qquad p_{11}{}^2 = rt.$$

Proof. Let us define two symbols to be first associates if they occur together in a set; otherwise they are second associates. We use the terminology of sets and symbols in talking of PBIB designs and their association schemes, and the terms "lines" and "points" are used with reference to the geometrical concepts. Since there pass r lines through a given point P, clearly

(9.6.2) $n_1 = r(k-1).$

Each of the $(b - r)$ lines not passing through P contains t points lying on lines containing P. Obviously any such point lies on $r - 1$ lines. Thus we get

(9.6.3) $n_1 = \frac{t(b - r).}{(r - 1)}$

Equating (9.6.2) and (9.6.3), we get

(9.6.4) $b = \frac{r\{(r-1)(k-1) + t\}}{t}.$

Using the trivial relation $vr = bk$, we have

(9.6.5)
$$v = \frac{k\{(r-1)(k-1)+t\}}{t}.$$

Again, each of the $b - r$ sets not containing P has exactly $(k - t)$ second associates of P, and each such second associate is contained in r sets. Hence

(9.6.6)
$$n_2 = \frac{(k-t)(b-r)}{r},$$
$$= \frac{(r-1)(k-1)(k-t)}{t}.$$

Let the symbols P and Q be first associates and occur together in a set l. We count their common first associates. The $k - 2$ other points on l along with the $(r-1)(t-1)$ first associates of Q lying on the $r - 1$ lines through P other than l are the only first associates that are common to both P and Q. Hence

(9.6.7)
$$p_{11}{}^1 = (r-1)(t-1) + k - 2.$$

Lastly, let P and R be two second-associate symbols. No line passes through P and R. Each of the r lines passing through P has t first associates of R, as ensured by axiom 4, and hence

(9.6.8)
$$p_{11}{}^2 = rt,$$

which completes the proof of the theorem.

From the above proof it is clear that a partial geometry can be interpreted as a two-associate-class association scheme with the parameters

(9.6.9)
$$v = \frac{k\{(r-1)(k-1)+t\}}{t}, \qquad n_1 = r(k-1),$$
$$n_2 = \frac{(r-1)(k-1)(k-t)}{t},$$
$$p_{11}{}^1 = (t-1)(r-1) + k - 2, \qquad p_{11}{}^2 = rt$$

if two symbols are first associates (or second associates) if their corresponding points in the geometry are joined (or not joined) by a line of the geometry.

Let us consider two-associate-class PBIB designs with the parameters $\lambda_1 = 1$, $\lambda_2 = 0$, and $k > r$. Then from (8.14.3) we have

(9.6.10)
$$(r-1)r + \{p_{12}{}^2(r-1) - rp_{12}{}^1\} = 0.$$

Bose and Clatworthy (1955) showed that such a PBIB design necessarily has the parameters given in (9.6.1). We now prove the converse of Theorem 9.6.1, established by Bose (1963a).

Theorem 9.6.2. A two-associate-class PBIB design with the parameters $\lambda_1 = 1$, $\lambda_2 = 0$, and $k > r$ is necessarily a partial geometry.

Proof. Without loss of generality, let us assume that the PBIB design of the hypothesis has the parameters given in (9.6.1). Then let us identify the symbols of the PBIB design as points and the sets as lines. From the very definition of a PBIB design axioms 1, 2, and 3 are satisfied. We now establish axiom 4.

Let us take a particular set K of k symbols of the PBIB design and let $\overline{K}$ be the complement of K, consisting of the remaining $v - k$ symbols. Let $g(x)$ denote the number of symbols in $\overline{K}$ that have exactly x first associates in K. Then

$$(9.6.11) \qquad \sum_{x=0}^{k} g(x) = v - k = \frac{k(k-1)(r-1)}{t}.$$

By enumerating the pairs of first associates P, Q, where $P \in K$, $Q \in \overline{K}$ in two ways, and equating, we get

$$(9.6.12) \qquad \sum_{x=0}^{k} x g(x) = k(n_1 - k + 1) = k(r-1)(k-1).$$

Again, counting the number of triplets (P_1, P_2, Q), where P_1, P_2 is an ordered pair of distinct symbols in K and Q is a symbol in $\overline{K}$ that is a first associate of both P_1 and P_2 in two ways, we get

$$(9.6.13) \quad \sum_{x=0}^{k} x(x-1) g(x) = k(k-1)(p_{11}^{1} - k + 2) = k(k-1)(t-1)(r-1).$$

Put

$$(9.6.14) \qquad \overline{x} = \frac{\sum_{x=0}^{k} x g(x)}{\sum_{x=0}^{k} g(x)} = t.$$

Then

$$(9.6.15) \qquad \sum_{x=0}^{k} (x-t)^2 g(x) = 0,$$

which implies that every symbol in $\overline{K}$ has exactly t first associates in K, which is our axiom 4.

In a partial geometry (r, k, t) by interchanging the role of lines and points we obtain the dual geometry, which is again a partial geometry (k, r, t).

Let us call the PBIB design with the parameters given in (9.6.1) a geometric design. If M is the usual incidence matrix of a geometric design, the characteristic roots of MM' can be obtained from Section 8.3 and are $\theta_0 = rk$, $\theta_1 = r + k - t - 1$, and $\theta_2 = 0$, with the respective multiplicities $\alpha_0 = 1$, $\alpha_1 = [rk(r-1)(k-1)]/[t(k+r-t-1)]$, and $\alpha_2 = v - 1 - \alpha_1$. Since the multiplicities of the roots need be integral, we have the following theorem:

Theorem 9.6.3. A necessary condition for the existence of a partial geometry (r, k, t) is that

$$(9.6.16) \qquad \frac{rk(r-1)(k-1)}{t(k+r-t-1)}$$

be a positive integer.

By applying this theorem, we can rule out the existence of partial geometries $(3, 4, 1)$, $(4, 5, 1)$, etc.

The partial geometries $(r, k, r-1)$ and $(2, n-1, 2)$ can be interpreted respectively as latin-square-type and triangular association schemes.

The following theorem due to Bose (1963a) can be considered as a generalized uniqueness theorem, for whose proof we refer to the original paper:

Theorem 9.6.4. A psuedogeometric design with the parameters given in (9.6.1) is a geometric design if

$$(9.6.17) \qquad k > \tfrac{1}{2}[r(r-1) + t(r+1)(r^2 - 2r + 2)].$$

We can easily verify that this theorem covers the uniqueness of triangular and L_2 designs, discussed in Chapter 8, and of L_i designs, discussed in Section 9.5.

For further results on the use of partial geometries in the construction of PBIB designs we refer to Benson (1966), Bhagwandas (1968), and Bose (1963b).

9.7. EXTENDED PARTIAL GEOMETRY

Raghavarao (1966) extended the scope of partial geometry to include three-associate geometric designs. Consider a system of v points and b lines, with the relation of incidence satisfying the following axioms:

Axiom 1. Any two points are incident with not more than one line.
Axiom 2. Each point is incident with r lines.
Axiom 3. Each line is incident with k points.
Axiom 4. Given a point P, the lines not passing through P can be divided into two disjoint sets S_1 and S_2 with cardinalities c_1 and c_2 such that every line of the set S_1 can be intersected by exactly one line passing through P and no line of set S_2 can be intersected by a line passing through P.

Let P be a point of the geometry, and let us number the points lying on the r lines passing through P by (i, j) $(i = 1, 2, \cdots, r; j = 1, 2, \cdots, k - 1)$. Through the point (i, j) there pass $r - 1$ lines other than the line connecting it with P; we call these $r - 1$ lines the pencil $\{i, j\}$, the tth line of the pencil $\{i, j\}$ being designated by $\{i, j; t\}$. A moment's consideration leads us to the following assumption:

Axiom 5. Exactly one line of the pencil $\{i, j\}$ intersects exactly one line of the pencil $\{i', j'\}$, where $i \neq i'$; furthermore, if $\{i, j; t\}$ intersects a line of $\{i', j'\}$, where $i \neq i'$, then it intersects one line from each of the pencils $\{i', j''\}$, where $i' \neq i$ and $j'' = 1, 2, \cdots, k - 1$.

The geometry satisfying the above axioms 1 through 5 was called by Raghavarao extended partial geometry and was denoted by $[r, k; 0, 1]$. He proved the following theorem:

Theorem 9.7.1. The geometric design derivable from the extended partial geometry $[r, k; 0, 1]$ by identifying the points as symbols and the lines as sets is a three-associate-class PBIB design with the parameters

$$v = 1 + n_1 + n_2 + n_3, \qquad n_1 = r(k - 1), \qquad n_2 = \frac{r(r - 1)(k - 1)^2}{2},$$

$$n_3 = \frac{(r - 1)(r - 2)(k - 1)^3}{2}, \qquad b = \frac{vr}{k}, \qquad r, \qquad k, \qquad \lambda_1 = 1, \qquad \lambda_2 = 0 = \lambda_3;$$

$$P_1 = \begin{bmatrix} k - 2 & (r - 1)(k - 1) & 0 \\ (r - 1)(k - 1) & (r - 1)(k - 1)(k - 2) & \dfrac{(r - 1)(r - 2)(k - 1)^2}{2} \\ 0 & \dfrac{(r - 1)(r - 2)(k - 1)^2}{2} & \dfrac{(r - 1)(r - 2)(k - 1)^2(k - 2)}{2} \end{bmatrix};$$

$$P_2 = \begin{bmatrix} 2 & 2(k - 2) & (r - 2)(k - 1) \\ 2(k - 2) & \dfrac{(k - 2)^2 + (r + 1)}{2}(r - 2)(k - 1) & \dfrac{(r + 1)(r - 2)(k - 1)}{2}(k - 2) \\ (r - 2)(k - 1) & \dfrac{(r + 1)(r - 2)}{2}(k - 1)(k - 2) & \dfrac{(r - 1)(r - 2)(k - 1)}{2}(k - 2)^2 + \dfrac{(r - 2)(r - 3)(k - 1)^2}{2} \end{bmatrix};$$

$$P_3 = \begin{bmatrix} 0 & r & r(k-2) \\[2ex] r & \dfrac{r(r+1)(k-2)}{2} & \dfrac{r[(r-1)(k-2)^2 + (r-3)(k-1)]}{2} \\[3ex] r(k-2) & \dfrac{r[(r-1)(k-2)^2 + (r-3)(k-1)]}{2} & \begin{matrix} \left[\dfrac{(r-1)(r-2)(k-1)^3}{2}\right] \\ -1 - r(k-2) - r \\ \times \dfrac{(r-1)(k-2)^2 + (r-3)(k-1)}{2} \end{matrix} \end{bmatrix}.$$

We observe that the geometric design $[3, s; 0, 1]$ is the cubic design defined by Raghavarao and Chandrasekhararao (1964).

REFERENCES

1. Adhikary, B. (1968). Graph product and new P.B.I.B. association schemes. *Calcutta Stat. Assn. Bull.*, **17**, 49–56.
2. Benson, C. T. (1966). A partial geometry $(q^3 + 1, q^2 + 1, 1)$ and corresponding PBIB designs. *Proc. Am. Math. Soc.*, **17**, 747–749.
3. Berge, C. (1962). *Theory of Graphs and Its Applications*. Wiley, New York.
4. Bhagwandas (1968). Group divisible designs and partial geometries. *Calcutta Stat. Assn. Bull.*, **17**, 115–122.
5. Bhagwandas and Shrikhande, S. S. (1968). Seidel equivalence of strongly regular graphs. *Sankhyā*, A**30**, 359–368.
6. Bose, R. C. (1963a). Strongly regular graphs, partial geometries and partially balanced designs. *Pacific J. Math.*, **13**, 389–419.
7. Bose, R. C. (1963b). Combinatorial properties of partially balanced incomplete block designs and association schemes. *Contributions to Statistics*. Presented to Prof. P. C. Mahalanobis, on his 70th Birthday. Pergamon Press, New York, pp. 21–48.
8. Bose, R. C., and Clatworthy, W. H. (1955). Some classes of partially balanced designs. *Ann. Math. Stat.*, **26**, 212–232.
9. Bose, R. C., and Laskar, R. (1967). A characterization of tetrahedral graphs. *J. Comb. Theory*, **2**, 366–385.
10. Bruck, R. H. (1951). Finite nets. I. Numerical invariants. *Can. J. Math.*, **3**, 94–107.
11. Bruck, R. H. (1963). Finite nets. II. Uniqueness and imbedding. *Pacific J. Math.*, **13**, 421–457.
12. Dowling, T. A., and Laskar, R. (1967). A geometric characterization of the line graph of a finite projective plane. *J. Comb. Theory*, **2**, 402–410.

13. Harary, F., and Trauth, C. A. (1966). Connectedness of products of two directed graphs. *SIAM J.*, **14**, 250–254.
14. Harary, F., and Wilcox, G. W. (1967). Boolean operations on graphs. *Math. Scand.*, **20**, 41–51.
15. Hoffman, A. J. (1965). On the line graph of a projective plane. *Proc. Am. Math. Soc.*, **16**, 297–302.
16. Hoffman, A. J., and Ray-Chaudhuri, D. K. (1965a). On the line graph of a finite affine plane. *Can. J. Math.*, **17**, 687–694.
17. Hoffman, A. J., and Ray-Chaudhuri, D. K. (1965b). On the line graph of a symmetric balanced incomplete block design. *Trans. Am. Math. Soc.*, **116**, 238–252.
18. Laskar, R. (1967). A characterization of cubic lattice graphs. *J. Comb. Theory*, **3**, 386–401.
19. Ore, O. (1962). *Theory of Graphs*. American Mathematical Society, Providence, R.I.
20. Raghavarao, D. (1966). An extended partial geometry. *J. Indian Soc. Agric. Stat.*, **18**, 99–107.
21. Raghavarao, D., and Chandrasekhararao, K. (1964). Cubic designs. *Ann. Math. Stat.*, **35**, 389–397.
22. Rao, S. B., and Rao, A. R. (1969). A characterization on the line graph of a BIBD with $\lambda = 1$. *Sankhyā*, **A31**, 369–370.
23. Seidel, J. J. (1967). Strongly regular graphs of L_2-type and of triangular type. *Indagationes Mathematicae*, **29**, 188–196.
24. Shrikhande, S. S. (1961). A note on mutually orthogonal latin squares. *Sankhyā*, **A23**, 115–116.
25. Shrikhande, S. S. (1968). On a conjecture of Bruck on a pseudo net-graph. *J. Algebra*, **8**, 418–422.

BIBLIOGRAPHY

Bhagwandas. Tactical configurations and graph theory. *Calcutta Stat. Assn. Bull.*, **16**, 136–138 (1967).
Di Paola, J. W. Block designs and graph theory. *J. Comb. Theory*, **1**, 132–148 (1966).
Hoffman, A. J. On the polynomial of a graph. *Am. Math. Monthly*, **70**, 30–36 (1963).
Hoffman, A. J., and Singleton, R. R. On Moore graphs with diameters 2 and 3. *I.B.M. J. Res. Develop.*, **4**, 497–504 (1960).
Shrikhande, S. S., and Bhagwandas. Duals of incomplete block designs. *J. Indian Stat. Assn.*, **3**, 30–37 (1965).

CHAPTER 10

Duals of Incomplete Block Designs

10.1. INTRODUCTION

In plane projective geometry, if the roles of lines and points are interchanged, the dual geometry is obtained. In this chapter we introduce a similar concept for incomplete block designs.

Definition 10.1.1. If N is the incidence matrix of a design D, the design D^*, which has N' as its incidence matrix, is said to be the dual design of D.

Clearly by interchanging sets and symbols in D, we obtain D^*. Dualization of known designs sometimes yields new designs and sometimes only already known designs. We have already seen in Chapter 5 that the duals of symmetrical BIB designs are again symmetrical BIB designs. In this chapter we systematically study the duality property of incomplete block designs, restricting our study to equiset-sized, equireplicated designs, except in Definition 10.5.2.

10.2. DUAL OF AN INCOMPLETE BLOCK DESIGN TO BE A SPECIFIED DESIGN

Given two designs D and E, one may be interested to know whether the design E can be the dual of D. For this purpose it is necessary that the number of symbols of D be equal to the number of sets in E, the number of sets of D be equal to the number of symbols in E, and the number of replications of D be equal to the set size in E. Furthermore, if N is the incidence matrix of D and N^* is the incidence matrix of E, then the $s + 1$ nonzero distinct characteristic roots $\rho_0, \rho_1, \rho_2, \cdots, \rho_s (s \leq m$, the number of associate classes of D), along with their multiplicities, should be the same for NN' and $N^*N^{*'}$. We now derive further additional conditions by using the Hasse–Minkowski invariant.

Let $\xi_{i1}, \xi_{i2}, \cdots, \xi_{i\alpha_i}$ be a complete set of independent, rational, characteristic vectors corresponding to the root $\rho_i (\neq 0)$ of NN'. Then the gramian of the independent, rational, characteristic vectors corresponding to the root ρ_i of NN' is

$$(10.2.1) \qquad\qquad Q_i = X_i' X_i,$$

where

$$(10.2.2) \qquad\qquad X_i = [\xi_{i1}, \xi_{i2}, \cdots, \xi_{i\alpha_i}].$$

Clearly the column vectors of $N'X_i$ are the independent, rational, characteristic vectors of $N'N$ corresponding to the same root ρ_i. Hence the gramian of the independent, rational, characteristic vectors corresponding to the root ρ_i of $N'N$ is

$$(10.2.3) \qquad\qquad Q_i^* = X_i'NN'X_i = X_i'\rho_i X_i = \rho_i Q_i.$$

We can easily verify that

$$(10.2.4) \qquad\qquad |Q_i^*| = \rho_i^{\alpha_i}|Q_i|$$

and

$$(10.2.5) \qquad C_p(Q_i^*) = (-1, \rho_i)_p^{\alpha_i(\alpha_i + 1)/2}(\rho_i, |Q_i|)_p^{\alpha_i - 1}C_p(Q_i).$$

For the given design D we know the gramian Q_i of the independent, rational, characteristic vectors corresponding to the nonzero roots ρ_i of NN' ($i = 1, 2, \cdots, s$). For the design E we can separately find the gramian P_i of the independent, rational, characteristic vectors corresponding to the roots ρ_i of $N^*N^{*'}$ ($i = 1, 2, \cdots, s$). In order for E to be the dual of D, it is clearly necessary that P_i be rationally congruent to Q_i^*, the gramian of the independent, rational, characteristic vectors corresponding to the root ρ_i of $N'N$, and hence

$$(10.2.6) \qquad\qquad |P_i| \sim |Q_i^*| \qquad (i = 1, 2, \cdots, s),$$

and

$$(10.2.7) \qquad\qquad C_p(P_i) = C_p(Q_i^*) \qquad (i = 1, 2, \cdots, s).$$

In particular we prove the following result due to Raghavarao (1966):

Theorem 10.2.1. Necessary conditions for the dual of a symmetrical PBIB design with the association scheme A to be a PBIB design with the same set of parameters and the same association scheme A are

$$(10.2.8) \qquad\qquad \rho_i^{\alpha_i} \sim 1 \qquad i = 1, 2, \cdots, s$$

and

$$(10.2.9) \qquad (-1, \rho_i)_p^{\alpha_i(\alpha_i + 1)/2}(\rho_i, |Q_i|)_p^{\alpha_i - 1} = 1 \qquad i = 1, 2, \cdots, s$$

for every prime p, where $\rho_0, \rho_1, \rho_2, \cdots, \rho_s$ are the distinct nonzero roots of NN', with the multiplicities $\alpha_0 = 1, \alpha_1, \alpha_2, \cdots, \alpha_s$; Q_i is the gramian of the independent, rational, characteristic vectors corresponding to ρ_i of NN'.

Proof. Since the dual is also specified to be of the same type as the original design,

$$(10.2.10) \qquad Q_i^* \sim Q_i \qquad (i = 1, 2, \cdots, s).$$

By using (10.2.6) and (10.2.7), where $P_i = Q_i$, and making use of the Hilbert norm residue symbol, we get our required results.

We now illustrate our theorem.

Illustration 10.2.1.1. Consider the PBIB design with the triangular association scheme T16 (see the PBIB design tables introduced in Chapter 8), with the parameters

$$(10.2.11) \qquad v = 10 = b, \qquad r = 6 = k, \qquad \lambda_1 = 4, \qquad \lambda_2 = 2.$$

For this design the characteristic roots are $\rho_0 = 36$, $\rho_1 = 6$, $\rho_2 = 0$, with the respective multiplicities $\alpha_0 = 1$, $\alpha_1 = 4$, $\alpha_2 = 5$. The gramian Q_1 of the independent, rational, characteristic vectors corresponding to ρ_1 can be calculated by the method given by Corsten (1960) and satisfies

$$(10.2.12) \qquad |Q_1| \sim 5.$$

The left-hand side of (10.2.9) for $i = 1$ becomes

$$(10.2.13) \qquad (5, 6)_p,$$

which has the value -1 for $p = 3$. Thus condition (10.2.9) is violated, and the dual of the design with the parameters given in (10.2.11) cannot be a PBIB design with the same set of parameters and a triangular association scheme. In fact its dual is the singular group-divisible design with the parameters

$$(10.2.14) \qquad \begin{array}{lll} v = 10 = b, & r = 6 = k, & m = 5, \\ n = 2, & \lambda_1 = 6, & \lambda_2 = 3. \end{array}$$

We now derive a sufficient condition for the dual of a given symmetrical PBIB design to be again a PBIB design with the same set of parameters, whatever the association scheme. This result was obtained by Hoffman (1963).

Let D be a symmetrical PBIB design with m associate classes and B_i the ith association matrix $(i = 1, 2, \cdots, m)$. Define an $(m + 2) \times m$ matrix $K = (k_{ij})$, where

$$(10.2.15) \qquad k_{-1, j} = 1 \qquad\qquad\qquad j = 1, 2, \cdots, m;$$

(10.2.16) $k_{0,j} = \lambda_j$ $j = 1, 2, \cdots, m;$

(10.2.17) $k_{i,j} = r\delta_{ij} + \sum_{s=1}^{m} \lambda_s p_{ij}^s;$ $i, j = 1, 2, \cdots, m.$

Theorem 10.2.2. [Hoffman (1963)]. The dual of a regular,† symmetrical PBIB design with m associate classes is again a PBIB design with the same set of parameters, whatever the association scheme, if the greatest common divisor of the $\binom{m+2}{m}$ determinants of order m contained in K is 1.

Proof. Since the B_i-matrices are the association matrices of the PBIB design, clearly

(10.2.18) $$NN' = rI_v + \sum_{s=1}^{m} \lambda_s B_s;$$

(10.2.19) $$E_{v,v} = I_v + \sum_{s=1}^{m} B_s;$$

(10.2.20) $B_i B_j = B_j B_i = n_i \delta_{ij} I_v + \sum_{s=1}^{m} p_{ij}^s B_s$ $i, j = 1, 2, \cdots, m.$

Define

(10.2.21) $L_s = N^{-1} B_s N$ $s = 1, 2, \cdots, m.$

This definition is permissible because the design is symmetrical and regular, and hence N is nonsingular. Now we can easily verify that if the B_s-matrices are replaced by L_s and N, N' are interchanged, (10.2.18), (10.2.19), and (10.2.20) will be satisfied. Thus, if we can show that L_s ($s = 1, 2, \cdots, m$) are association matrices, it follows that N' is the incidence matrix of a PBIB design with the same set of parameters as N, with the association scheme determined by the matrices L_s ($s = 1, 2, \cdots, m$). To show that L_s are association matrices we have to show them to be symmetrical matrices with elements 0 and 1. Now

(10.2.22) $NN' B_s = B_s NN'$ $s = 1, 2, \cdots, m,$

and hence

(10.2.23) $L_s' = L_s$ $s = 1, 2, \cdots, m,$

thereby implying that L_s ($s = 1, 2, \cdots, m$) are symmetrical matrices. Denote

---

† A regular PBIB design is one whose incidence matrix N is such that NN' is nonsingular. We shall have occasion to use this concept repeatedly in Chapter 12.

the entry in the (i, j)th position of L_s by $L_s(i, j)$. To show that the elements of L_s are 0 and 1 it is sufficient to show that L_s has integral elements such that

$$(10.2.24) \qquad \sum_{i, j} L_s(i, j) = \text{tr } (L_s^2).$$

Now

$$(10.2.25) \qquad \begin{aligned} \sum_{i, j} L_s(i, j) &= E_{1, v} N^{-1} B_s N E_{v, 1} \\ &= \text{tr } (B_s N E_{v, v} N^{-1}) \\ &= \text{tr } (B_s E_{v, v} N N^{-1}) \\ &= E_{1, v} B_s E_{v, 1} \\ &= n_s v \end{aligned}$$

and

$$(10.2.26) \qquad \text{tr } (L_s^2) = \text{tr } (N^{-1} B_s^2 N) = \text{tr } (B_s^2) = n_s v$$

from (10.2.20). Thus (10.2.24) holds, and if we can show that L_s has integral elements, we shall have the complete proof.

From (10.2.19) we have

$$(10.2.27) \qquad E_{v, v} - I_v = \sum_{s=1}^{m} L_s$$

and from (10.2.18) we have

$$(10.2.28) \qquad N'N - rI_v = \sum_{s=1}^{m} \lambda_s L_s .$$

Next, multiplying (10.2.18) on the left by N^{-1} and on the right by $B_s N$, we get

$$(10.2.29) \quad N'B_s N - n_s \lambda_s I_v = \sum_{t=1}^{m} \left\{ r\delta_{st} + \sum_{i=1}^{m} \lambda_i p_{si}^t \right\} L_t \qquad s = 1, 2, \cdots, m.$$

We observe that the coefficients of L_s on the right-hand sides of (10.2.27), (10.2.28), and (10.2.29) are the entries of the matrix K. Now, let λ_{ij} be the vector of $m + 2$ components that are the entries in position (i, j) of the left hand sides of (10.2.27), (10.2.28), and (10.2.29). Let y_{ij} be the vector with m components whose sth component is the entry in position (i, j) of L_s. Then

$$(10.2.30) \qquad \lambda_{ij} = K y_{ij} .$$

The assumption that the greatest common divisor of the $\binom{m + 2}{m}$ determinants of K is 1 makes y_{ij} an integral component vector, which in turn implies that L_s has integral elements.

It is to be noted that Theorem 10.2.2 does not ensure that the association scheme of the dual design is the same as the association scheme of the original design. However, if N commutes with each B_s, then $L_s = B_s$ $(s = 1, 2, \cdots, m)$ and the association scheme of the dual design is the same as the association scheme of the original design.

It is also to be noted that since many of the association schemes are uniquely determined by the secondary parameters and since the dual design has the same set of parameters as the original design, in all the cases where the uniqueness of the association scheme is guaranteed the association scheme of the dual design is the same as the original design.

The conditions imposed in Theorem 10.2.2 are too strong and certainly not necessary. Consider the semiregular group-divisible design SR7 with the parameters

$$
(10.2.31) \qquad \begin{aligned} v &= 8 = b, & r &= 4 = k, & m &= 4, \\ n &= 2, & \lambda_1 &= 0, & \lambda_2 &= 2, \end{aligned}
$$

which is self dual with even the same association scheme but does not satisfy any of the conditions of Theorem 10.2.2. But at the same time some sort of hypothesis is essential for the result to hold true, as was shown by Hoffman (1963) with a counterexample where neither the hypothesis nor the conclusion of the theorem holds.

10.3. DUALS OF ASYMMETRICAL BIB DESIGNS WITH $\lambda = 1$ OR $\lambda = 2$

Shrikhande (1952) proved that the duals of asymmetrical BIB designs with $\lambda = 1$ or 2 are PBIB designs with two associate classes. The same results were reestablished by Shrikhande and Bhagwandas (1965) through results on graph theory, which we now discuss.

Theorem 10.3.1. If D is an asymmetrical BIB design with the parameters $v, b, r, k, \lambda = 1$, then its dual D^* is a two-associate-class PBIB design with the parameters†

$$
(10.3.1) \qquad \begin{aligned} v^* &= b, & b^* &= v, & r^* &= k, & k^* &= r, & n_1^* &= k(r-1), \\ n_2^* &= b - 1 - n_1, & \lambda_1^* &= 1, & \lambda_2^* &= 0, \\ p_{11}{}^{1*} &= r - 2 + (k-1)^2, & p_{11}{}^{2*} &= k^2. \end{aligned}
$$

Proof. Let N be the incidence matrix of the BIB design under consideration. Then

$$
(10.3.2) \qquad NN' = (r-1)I_v + E_{v,v}.
$$

---

† We use an asterisk on the parameters of dual designs in this chapter.

Now it is easy to see that

$$(10.3.3) \qquad N'N = kI_b + B,$$

where B is a $b \times b$ symmetrical matrix with elements $0, 1$ such that its diagonal elements are zero and each row (or column) sum is $k(r - 1)$. Thus B is a regular adjacency matrix of valence $k(r - 1)$. The matrix B is strongly regular, in view of Theorem 9.2.1, because

$$(10.3.4) \qquad \begin{aligned} B^2 &= (N'N - kI_b)(N'N - kI_b) \\ &= (r - 1 - 2k)B + k^2 E_{b, b} + k(r - k - 1)I_b. \end{aligned}$$

For this strongly regular graph $p_{11}{}^{2*} = k^2$ and $p_{11}{}^{1*} = r - 2 + (k - 1)^2$. Since in (10.3.3) $N'N$ is of the form required by Theorem 9.4.2, with $\lambda_1 = 1$, $\lambda_2 = 0$, N' is the incidence matrix of a PBIB design with two associate classes; this completes the proof of the theorem.

Theorem 10.3.2. If D is a BIB design with the parameters

$$(10.3.5) \quad v = \binom{r - 1}{2}, \qquad b = \binom{r}{2}, \qquad r, \qquad k = r - 2, \qquad \lambda = 2,$$

then its dual design D^* is a PBIB design with the parameters

$$v^* = b, \qquad b^* = v, \qquad r^* = k, \qquad k^* = r, \qquad n_1^* = 2(r - 2),$$

$$(10.3.6) \quad n_2^* = \binom{r - 2}{2}, \qquad \lambda_1^* = 1, \qquad \lambda_2^* = 2, \qquad p_{11}{}^{1*} = r - 2, \qquad p_{11}{}^{2*} = 4.$$

Proof. The incidence matrix N of the BIB design here satisfies

$$(10.3.7) \qquad NN' = (r - 2)I_v + 2E_{v, v}.$$

Now consider the $b \times b$ symmetrical matrix

$$(10.3.8) \qquad B = (r - 4)I_b + 2E_{b, b} - N'N.$$

Clearly the diagonal elements of B are 0 and B has integral elements. Furthermore,

$$(10.3.9) \qquad E_{1, b} B E_{b, 1} = \text{tr}(B^2),$$

and hence B has the elements 0 and 1 only. Each row sum in B is clearly $(r - 4) + 2b - rk = 2(r - 2) = n_1$. Hence B is a regular adjacency matrix of valence $n_1 = 2(r - 2)$. It is also strongly regular. In fact

$$(10.3.10) \qquad B^2 = (r - 6)B + 2(r - 4)I_b + 4E_{b, b}.$$

For this strongly regular adjacency matrix $p_{11}{}^{1*} = r - 2$ and $p_{11}{}^{2*} = 4$.

From (10.3.8) we can easily see that

$$(10.3.11) \qquad N'N = kI_b + B + 2(E_{b,\,b} - I_b - B),$$

and hence the dual design is a PBIB design with two associate classes and with the parameters given in (10.3.6).

From the definitions of affine α-resolvable BIB designs and group-divisible designs the following theorem is obvious:

Theorem 10.3.3. The dual of an affine α-resolvable BIB design is a semi-regular group-divisible design.

In fact we can have a stronger result than that of the above theorem.

Theorem 10.3.4. [Shrikhande and Bhagwandas (1965)]. In a BIB design, if every set has n_1 sets that have q_1 symbols in common and the other sets have q_2 symbols in common, then its dual design is a PBIB design with two associate classes.

The proof for this theorem can be constructed along lines similar to those used in Theorems 10.3.1 and 10.3.2, and is left as an exercise.

10.4. DUALS OF SOME PBIB DESIGNS

In the next section we study PBIB designs whose duals are BIB designs.

Since the dual of a partial geometry (r, k, t) is again a partial geometry (k, r, t), we have the following:

Theorem 10.4.1. The dual of a PBIB design with two associate classes and the parameters

$$v = \frac{k[(r-1)(k-1)+t]}{t}, \qquad b = \frac{r[(r-1)(k-1)+t]}{t},$$

$$(10.4.1) \qquad r, \qquad k, \qquad n_1 = r(k-1), \qquad n_2 = \frac{(r-1)(k-1)(k-t)}{t},$$

$$\lambda_1 = 1, \qquad \lambda_2 = 0, \qquad p_{11}{}^1 = (t-1)(r-1)(k-2), \qquad p_{11}{}^2 = rt$$

constructed from a partial geometry is a PBIB design with two associate classes and the parameters

$$v^* = b, \qquad b^* = v, \qquad r^* = k, \qquad k^* = r, \qquad n_1^* = k(r-1),$$

$$(10.4.2) \qquad n_2^* = \frac{(r-1)(k-1)(r-t)}{t} \qquad \lambda_1^* = 1, \qquad \lambda_2^* = 0,$$

$$p_{11}{}^{1*} = (t-1)(r-2)(k-1), \qquad p_{11}{}^{2*} = kt.$$

In particular the following applies:

Corollary 10.4.1.1. [Shrikhande (1965)]. For $n > 2$ the dual of a two-associate-class PBIB design with the parameters

$$(10.4.3) \quad v = \binom{2n}{2}, \quad b = (2n-1)(2n-3), \quad r = 2n-3, \quad k = n,$$

$$\lambda_1 = 0, \quad \lambda_2 = 1, \quad n_1 = 2(2n-2), \quad p_{11}^{1} = 2n-2, \quad p_{11}^{2} = 4$$

is a two-associate-class PBIB design with the parameters

$$(10.4.4) \quad v^* = (2n-1)(2n-3), \quad b^* = \binom{2n}{2}, \quad r^* = n, \quad k^* = 2n-3,$$

$$\lambda_1^* = 0, \quad \lambda_2^* = 1, \quad n_1^* = 2(n-1)^2, \quad p_{11}^{1*} = p_{11}^{2*} = (n-1)^2.$$

Analogously to Theorems 10.3.1 and 10.3.2, Shrikhande and Bhagwandas (1965) proved the following theorems:

Theorem 10.4.2. The dual of a symmetrical two-associate-class PBIB design D, with $\lambda_1 = \lambda_2 \pm 1$ is again a PBIB design D^* with the same set of parameters.

Theorem 10.4.3. The dual of a three-associate-class PBIB design with $v > b$ and values 0 or 1 for $\lambda_1, \lambda_2, \lambda_3$ satisfying rank $(NN') = b$, where N is the incidence matrix of D, is a two-associate-class PBIB design.

Analogously to Theorem 10.3.3, we have

Theorem 10.4.4. The dual of an affine α-resolvable PBIB design D is a semiregular group-divisible design D^*.

We now give an interesting result [cf. Ramakrishnan (1956)] of finding a series of five-associate-class PBIB designs with two replications by dualizing the design D, consisting of $v = 2m$ symbols written down in all possible pairs and omitting pairs of the form $(2i-1, 2i)$ for $i = 1, 2, \cdots, m$. Thus the sets of D are

$$(1, 3); (1, 4); \cdots; (1, 2m);$$

$$(2, 3); (2, 4); \cdots; (2, 2m);$$

$$(10.4.5) \qquad (3, 5); \cdots; (3, 2m);$$

$$\cdots \quad \cdots \quad \cdots$$

$$(2m-2, 2m-1); \quad (2m-2, 2m).$$

We can easily verify that D is a group-divisible design with the parameters

$$(10.4.6) \quad \begin{aligned} &v = 2m, \quad m = m, \quad n = 2, \quad \lambda_1 = 0, \quad \lambda_2 = 1, \\ &n_1 = 1, \quad n_2 = 2(m-1), \quad k = 2, \quad r = 2(m-1), \\ &b = 2m(m-1), \end{aligned}$$

where the groups are

$$1, 2$$

(10.4.7) $$3, 4$$
$$\cdots$$
$$2m - 1, 2m$$

Clearly the first associate of the ith symbol is $i + 1$ if i is odd, and $i - 1$ if i is even $(i = 1, 2, \cdots, 2m)$.

In the dual design D^* there are $v^* = 2m(m - 1)$ symbols. We define a five-associate-class association scheme for these symbols as follows: Let p and q be the sets in which a particular symbol θ occurs in D^*, where the set size is $k^* = 2(m - 1)$. For p, q as symbols of D let p_a and q_a be the first associates, respectively. We now define the association scheme in terms of the symbols occurring in the four sets p, q, p_a, and q_a. Clearly the sets p and q intersect only in one symbol, namely, θ. Let θ_1 be the symbol common to p and q_a, and let θ_2 be the symbol common to q and p_a. Then we define θ_1 and θ_2 to be the first associates of θ, and hence $n_1^* = 2$. The symbols, other than $\theta, \theta_1, \theta_2$, occurring in p and q are defined as the second associates of θ, and hence $n_2^* = 2(k^* - 2) = 4(m - 2)$. Let ϕ be the symbol common to p_a and q_a, and let us define it to be the third associate of θ. Thus $n_3^* = 1$. The symbols, other than θ_1, θ_2, and ϕ, occurring in p_a and q_a are defined to be the fourth-associates of θ, and hence $n_4^* = 2(k^* - 2) = 4(m - 2)$. The symbols not occurring in p, q, p_a, q_a are defined as the fifth associates of θ, and hence $n_5^* = 2(m - 2)(m - 3)$. We can easily verify that the $p_{jk}{}^{i*}$ values are as follows:

$$P_1^* = \begin{bmatrix} 0 & 0 & 1 & 0 & 0 \\ 0 & 2(m-2) & 0 & 2(m-2) & 0 \\ 1 & 0 & 0 & 0 & 0 \\ 0 & 2(m-2) & 0 & 2(m-2) & 0 \\ 0 & 0 & 0 & 0 & 2(m-2)(m-3) \end{bmatrix};$$

$$P_2^* = \begin{bmatrix} 0 & 1 & 0 & 1 & 0 \\ 1 & 2(m-5) & 0 & 1 & 2(m-3) \\ 0 & 0 & 0 & 1 & 0 \\ 1 & 1 & 1 & 2(m-5) & 2(m-3) \\ 0 & 2(m-3) & 0 & 2(m-3) & 2(m-3)(m-4) \end{bmatrix};$$

$$(10.4.8) \quad P_3^* = \begin{bmatrix} 2 & 0 & 0 & 0 & 0 \\ 0 & 0 & 0 & 4(m-2) & 0 \\ 0 & 0 & 0 & 0 & 0 \\ 0 & 4(m-2) & 0 & 0 & 0 \\ 0 & 0 & 0 & 0 & 2(m-2)(m-3) \end{bmatrix};$$

$$P_4^* = \begin{bmatrix} 0 & 1 & 0 & 1 & 0 \\ 1 & 1 & 1 & 2(m-5) & 2(m-3) \\ 0 & 1 & 0 & 0 & 0 \\ 1 & 2(m-5) & 0 & 1 & 2(m-3) \\ 0 & 2(m-3) & 0 & 2(m-3) & 2(m-3)(m-4) \end{bmatrix};$$

$$P_5^* = \begin{bmatrix} 0 & 0 & 0 & 0 & 2 \\ 0 & 4 & 0 & 4 & 4(m-4) \\ 0 & 0 & 0 & 0 & 1 \\ 0 & 4 & 0 & 4 & 4(m-4) \\ 2 & 4(m-4) & 1 & 4(m-4) & 2(m-4)(m-5) \end{bmatrix}.$$

Thus we have proved the following theorem:

Theorem 10.4.5. The dual of the group-divisible design with the parameters given in (10.4.6) is a five-associate-class PBIB design with the parameters

$$(10.4.9) \quad \begin{aligned} & v^* = 2m(m-1), \quad b^* = 2m, \quad r^* = 2, \quad k^* = 2(m-1), \\ & \lambda_1^* = \lambda_2^* = 1, \quad \lambda_3^* = \lambda_4^* = \lambda_5^* = 0, \quad n_1^* = 2, \\ & n_2^* = 4(m-2), \quad n_3^* = 1, \quad n_4^* = 4(m-2), \\ & n_5^* = 2(m-2)(m-3), \end{aligned}$$

with $p_{jk}{}^{i*}$ parameters given by (10.4.8).

10.5. LINKED BLOCK DESIGNS

In this section we study another class of designs, known as linked block (LB) designs. These designs were first introduced by Youden (1951), who obtained them as duals of BIB designs.

Definition 10.5.1. A design is said to be a linked block design if every pair of sets has exactly μ symbols in common.

Thus from the above definition we see that linked block designs can be obtained by dualizing known BIB designs. Obviously the symmetrical BIB design is also a linked block design. Linked block designs were classified by Roy and Laha (1956–1957) into three categories as follows: (a) symmetrical BIB designs; (b) PBIB designs; and (c) irregular designs not belonging to any of the known types.

Symmetrical BIB designs do not present any new feature. We now derive the necessary and sufficient conditions for PBIB designs to be linked block designs.

Theorem 10.5.1. If N is the incidence matrix of a PBIB design with m associate classes and NN' has only two nonzero roots $\theta_0 = rk$ and θ_1, with the respective multiplicities $\alpha_0 = 1$ and $\alpha_1 = b - 1$, then and only then the PBIB design is a linked block design.

The proof of this can be trivially obtained by noting that the nonzero characteristic roots of NN' and $N'N$ and their multiplicities are the same and $N'N$ for a linked block design has two distinct characteristic roots $\theta_0 = rk$ and $\theta_1 = k - \mu$, with the respective multiplicities $\alpha_0 = 1$ and $\alpha_1 = b - 1$.

By noting the characteristic roots and their multiplicities of group-divisible, triangular, and L_i designs, we deduce the following corollaries of Theorem 10.5.1:

Corollary 10.5.1.1. A singular group-divisible design is of the linked block type if and only if $b = m$.

Corollary 10.5.1.2. A semiregular group-divisible design is of the linked block type if and only if $b = v - m + 1$.

Corollary 10.5.1.3. A triangular design is of the linked block type if and only if (a) $r = 2\lambda_1 - \lambda_2$ and $b = n$ or (b) $r = (n - 3)\lambda_2 - (n - 4)\lambda_1$ and $b = (n - 1)(n - 2)/2$.

Corollary 10.5.1.4. An L_i design is of the linked block type if and only if (a) $r = (i - s)(\lambda_1 - \lambda_2) + \lambda_2$ and $b = s(s - i) + i$ or (b) $r = i(\lambda_1 - \lambda_2) + \lambda_2$ and $b = i(s - 1) + 1$.

Certain irregular designs of the linked block type were also listed by Roy and Laha (1956–1957).

The idea of linked block designs has been extended to sets of different sizes by Adhikary (1965), who defined a multiply linked block (MLB) design as follows:

Definition 10.5.2. An arrangement of v symbols into b sets of which b_i are of size k_i ($i = 1, 2, \cdots, s$; $b_1 + b_2 + \cdots + b_s = b$) is said to be a multiply linked block design if

1. Every symbol occurs in r sets.

2. Every pair of sets of size k_i has μ_{ii} symbols in common and every pair of sets where one set is of size k_i and the other of size k_j has μ_{ij} symbols in common for $i \neq j$; $i, j = 1, 2, \cdots, s$.

Adhikary further considered methods of constructing MLB designs from known linked block designs and latin squares.

The concept of linked block designs was generalized by Roy and Laha (1957), and Nair (1966) to partially linked block designs, which are duals of PBIB designs.

REFERENCES

1. Adhikary, B. (1965). On the properties and construction of balanced block designs with variable replications. *Calcutta Stat. Assn. Bull.*, **14**, 36–64.
2. Corsten, L. C. A. (1960). Proper spaces related to triangular partially balanced incomplete block designs. *Ann. Math. Stat.*, **31**, 498–501.
3. Hoffman, A. J. (1963). On the duals of symmetric partially balanced incomplete block designs. *Ann. Math. Stat.*, **34**, 528–531.
4. Nair, C. R. (1966). On partially linked block designs. *Ann. Math. Stat.*, **37**, 1401–1406.
5. Raghavarao, D. (1966). Duals of partially balanced incomplete block designs and some non-existence theorems. *Ann. Math. Stat.*, **37**, 1048–1052.
6. Ramakrishnan, C. S. (1956). On the dual of a PBIB design and a new class of designs with two replications. *Sankhyā*, **17**, 133–142.
7. Roy, J., and Laha, R. G. (1956). Classification and analysis of linked block designs. *Sankhyā*, **17**, 115–132.
8. Roy, J., and Laha, R. G. (1957). On partially balanced linked block designs. *Ann. Math. Stat.*, **28**, 488–493.
9. Shrikhande, S. S. (1952). On the dual of some balanced incomplete block designs. *Biometrics*, **8**, 66–72.
10. Shrikhande, S. S. (1965). On a class of partially balanced incomplete block designs. *Ann. Math. Stat.*, **36**, 1807–1814.
11. Shrikhande, S. S., and Bhagwandas (1965). Duals of incomplete block designs. *J. Indian Stat. Assn.*, **3**, 30–37.
12. Youden, W. G. (1951). Linked blocks: a new class of incomplete block designs (abstract). *Biometrics*, **7**, 124.

BIBLIOGRAPHY

Agrawal, H. L. On the dual of BIB designs. *Calcutta Stat. Assn. Bull.*, **12**, 104–105 (1963).

Bose, R. C. Strongly regular graphs, partial geometrics, and partially balanced designs. *Pacific J. Math.*, **13**, 389–419 (1963).

Shrikhande, S. S., and Raghavarao, D. Affine α-resolvable incomplete block designs. *Contributions to Statistics*. Presented to Prof. P. C. Mahalanobis on his 70th birthday. Pergamon Press, New York, 1963, pp. 471–480.

CHAPTER 11

Symmetrical Unequal-Block Arrangements with Two Unequal Block Sizes

11.1. INTRODUCTION

We have studied various incomplete block designs of constant set size in Chapters 5 through 10. In experimenting in hilly areas or for want of experimental material it is not always possible to accommodate sets of equal size. To meet this difficulty Kishen (1940–1941) introduced designs, called symmetrical unequal-block (SUB) arrangements, that have two different set sizes. Raghavarao (1962) has thoroughly investigated their constructions and analysis, and we discuss these results in this chapter.

Let us define a two-associate-class association scheme on v symbols with the parameters $n_1, n_2, p_{jk}{}^i$ $(i, j, k = 1, 2)$ as given by Definition 8.1.1.

Definition 11.1.1. The arrangeme nt of v symbols in b sets, where b_1 sets are of size k_1 and b_2 sets are of size k_2 $(b_1 + b_2 = b)$, is said to be a SUB arrangement with two unequal block sizes if

1. Every symbol occurs in $b_i k_i/v$ sets of size k_i $(i = 1, 2)$.
2. Every pair of first-associate† symbols occurs together in u sets of size k_1 and in $\lambda - u$ sets of size k_2 and every pair of second-associate symbols occurs together in λ sets of size k_2.

From condition 1 of Definition 11.1.1 we can see that every symbol occurs in $b_1 k_1/v + b_2 k_2/v = r$, say, sets. The terms v, b, r, k_i, b_i, n_i, $p_{jk}{}^i$, u, λ

---

† Kishen and Raghavarao used the term "first block associates," but because of the analogy of SUB arrangements to PBIB designs, we use the notations of PBIB designs only while studying SUB arrangements.

$(i, j, k = 1, 2)$ may be called the parameters of a SUB arrangement. They satisfy

$$v - 1 = n_1 + n_2, \qquad \sum_{k=1}^{2} p_{jk}{}^i = n_j - \delta_{ij},$$

$$(11.1.1) \quad n_i p_{jk}{}^i = n_j p_{ki}{}^j = n_k p_{ij}{}^k, \qquad b = b_1 + b_2,$$

$$vr = b_1 k_1 + b_2 k_2, \qquad v(v - 1)\lambda = b_1 k_1 (k_1 - 1) + b_2 k_2 (k_2 = 1).$$

It can easily be seen that the SUB arrangement as defined above is equivalent to adjoining two PBIB designs with two associate classes and the same association scheme but of different set sizes, with certain restrictions on the λ-parameters.

Though SUB arrangements are combinatorially balanced like pairwise balanced designs, as indicated in Chapter 4, the following differences between them are noteworthy:

1. The SUB arrangements are equireplicated designs, whereas pairwise balanced designs may not be so.

2. The symbols of pairwise balanced designs need not possess an association scheme as the symbols of SUB arrangements.

It can be further seen that SUB arrangements form a special class of pairwise balanced designs that can be conveniently analyzed, whereas the analysis of a general pairwise balanced design is very cumbersome.

The analysis of these designs was worked out by Kishen (1940–1941) and Raghavarao (1962) on the assumption that the same intrablock error variance holds for blocks of all sizes. In fact it seems more appropriate to assume equal intrablock error variance only for the blocks of the same size, but this leads to a complicated analysis and is a problem for further research. The assumption of equal intrablock variance is of very doubtful general validity but may be reasonable in some cases—for example, if the experiment is an agronomic one and the block sizes do not vary much. Thus we consider these arrangements in only those cases where the block sizes are very close. For the analysis of these designs we refer the reader to the paper of Raghavarao (1962).

We shall have occasion to use SUB arrangements in the construction of rotatable designs in Chapter 16.

11.2. COMBINATORIAL PROBLEMS IN SUB ARRANGEMENTS

Theorem 11.2.1. In a SUB arrangement of two block sizes the inequality

$$(11.2.1) \qquad\qquad b > v$$

holds.

Proof. Let $N = (n_{ij})$ be the $v \times b$ incidence matrix of a SUB arrangement, where $n_{ij} = 1$ or 0, depending on whether the ith symbol occurs in the jth set or not. Then

$$(11.2.2) \qquad NN' = (r - \lambda)I_v + \lambda E_{v,v}$$

and hence

$$(11.2.3) \qquad v = \text{rank } (NN') = \text{rank } (N) \leqq b.$$

If $b = v$, then, since

$$(11.2.4) \qquad \begin{aligned} NE_{v,1} &= rE_{v,1}, \\ NN'E_{v,1} &= [r + (v - 1)\lambda]E_{v,1}, \end{aligned}$$

it follows that

$$(11.2.5) \qquad N'E_{v,1} = \frac{[r + (v - 1)\lambda]}{r} E_{v,1},$$

which implies that all the sets of the design are of the same size, contradicting the definition of a SUB arrangement. Thus $b \neq v$, and inequality (11.2.3) boils down to the required inequality (11.2.1).

By following an argument similar to that in Section 8.3, we can show that the characteristic roots of the C-matrix of a SUB arrangement are 0, ϕ_1, and ϕ_2, where

$$(11.2.6) \qquad \begin{aligned} \phi_i &= a_0 - \tfrac{1}{2}[(a_1 - a_2)(-\gamma + (-1)^i\sqrt{\Delta}) + (a_1 + a_2)], \\ a_0 &= r - v^{-1}b, \qquad a_1 = -k_1^{-1}k_2^{-1}(uk_2 + (\lambda - u)k_1), \\ a_2 &= -k_2^{-1}\lambda, \qquad \gamma = p_{12}^{\;2} - p_{12}^{\;1}, \\ \beta &= p_{12}^{\;2} + p_{12}^{\;1}, \qquad \Delta = \gamma^2 + 2\beta + 1. \end{aligned}$$

The SUB arrangement is not a variance-balanced design, since $\phi_1 \neq \phi_2$. In fact, if ϕ_1 and ϕ_2 are equal, then

$$(11.2.7) \qquad (a_1 - a_2)^2 \Delta = 0,$$

and since $p_{jk}^{\;i}$'s are nonnegative, Δ is always nonzero and (11.2.7) is impossible. Thus we have the following:

Theorem 11.2.2. No SUB arrangement with two unequal block sizes is a variance-balanced design.

11.3. CONSTRUCTION OF SUB ARRANGEMENTS FROM KNOWN PBIB DESIGNS

Symmetrical unequal-block arrangements with two unequal block sizes can be constructed from existing group-divisible, triangular, and L_i, PBIB designs. The following theorems can be easily proved (we leave their proofs as exercises):

Theorem 11.3.1. If there exists a group-divisible design with the parameters

$$(11.3.1) \qquad v^* = mn, \qquad b^*, \qquad r^*, \qquad k^* \neq n, \qquad \lambda_1^* = \lambda - 1, \qquad \lambda_2^* = \lambda,$$

then a SUB arrangement with the parameters

$$
\begin{aligned}
& v = mn, \qquad b = b^* + m, \qquad r = r^* + 1, \qquad k_1 = n, \qquad k_2 = k^*, \\
(11.3.2) \quad & n_1 = n - 1, \qquad n_2 = n(m - 1), \qquad b_1 = m, \qquad b_2 = b^*, \\
& p_{11}{}^1 = n - 2, \qquad u = 1, \qquad \lambda = \lambda
\end{aligned}
$$

can be constructed by adding m sets to the plan of the group-divisible design where the ith added set contains the symbols of the ith group ($i = 1, 2, \cdots, m$). Conversely, if there exists a SUB arrangement with the parameters given in (11.3.2), then, by removing the sets of size k_1, we get a group-divisible design with the parameters given in (11.3.1), where the ith group contains symbols of the ith set of size k_1 ($i = 1, 2, \cdots, m$).

Theorem 11.3.2. If there exists a triangular design with the parameters

$$(11.3.3) \qquad v^* = \frac{n(n - 1)}{2}, \qquad b^*, \qquad r^*, \qquad k^* \neq n - 1,$$

$$\lambda_1^* = \lambda - 1, \qquad \lambda_2^* = \lambda,$$

then a SUB arrangement with the parameters

$$
\begin{aligned}
& v = \frac{n(n - 1)}{2}, \qquad b = b^* + n, \qquad r = r^* + 2, \qquad k_1 = n - 1, \\
(11.3.4) \quad & k_2 = k^*, \qquad b_1 = n, \qquad b_2 = b^*, \qquad n_1 = 2n - 4, \\
& n_2 = \frac{(n - 2)(n - 3)}{2}, \qquad p_{11}^1 = n - 2, \qquad u = 1, \qquad \lambda = \lambda
\end{aligned}
$$

can be constructed by adjoining n new sets to the plan of the PBIB design, where the ith added set contains symbols of the ith row of the association scheme ($i = 1, 2, \cdots, n$). Conversely, if there exists a SUB arrangement with the parameters given in (11.3.4), by removing the sets of size k_1, we obtain a triangular design.

Theorem 11.3.3. If there exists an L_2 design with the parameters

$$(11.3.5) \qquad v^* = s^2, \qquad b^*, \qquad r^*, \qquad k^* \neq s, \qquad \lambda_1^* = \lambda - 1, \qquad \lambda_2^* = \lambda,$$

Table 11.3.1. *Parameters of SUB arrangements obtainable by the method of Theorem 11.3.1*

Series	v	b	r	k_1	k_2	b_1	b_2	n_1	n_2	$p_{11}{}^1$	λ	Type
1	6	7	3	2	3	3	4	1	4	0	1	SR1
2	6	11	7	3	4	2	9	2	3	1	4	SR4
3	6	15	9	2	4	3	12	1	4	0	5	R4
4	8	12	4	2	3	4	8	1	6	0	1	R5
5	8	26	10	4	3	2	24	3	4	2	3	R7
6	9	12	7	3	6	3	9	2	6	1	4	SR14
7	12	13	4	3	4	4	9	2	9	1	1	SR20
8	12	15	7	4	6	3	12	3	8	2	3	SR25
9	12	19	5	4	3	3	16	3	8	2	1	SR21
10	12	26	6	2	3	6	20	1	10	0	1	R16
11	14	21	5	2	4	7	14	1	12	0	1	R24
12	14	35	7	2	3	7	28	1	12	0	1	R25
13	15	20	5	3	4	5	15	2	12	1	1	R27
14	15	28	6	5	3	3	25	4	10	3	1	SR36
15	15	33	9	5	4	3	30	4	10	3	2	R31
16	16	36	7	4	3	4	32	3	12	2	1	R35
17	18	39	7	6	3	3	36	5	12	4	1	SR45
18	18	57	9	2	3	9	48	1	16	0	1	R40
19	20	21	5	4	5	5	16	3	16	2	1	SR51
20	20	29	6	5	4	4	25	4	15	3	1	SR52
21	20	70	10	2	3	10	60	1	18	0	1	R42
22	24	30	6	4	5	6	24	3	20	2	1	R45
23	24	50	8	3	4	8	42	2	21	1	1	R46
24	24	76	10	6	3	4	72	5	18	4	1	R47
25	26	65	9	2	4	13	52	1	24	0	1	R53
26	27	63	9	3	4	9	54	2	24	1	1	R54
27	28	53	8	7	4	4	49	6	21	5	1	SR68
28	30	31	6	5	6	6	25	4	25	3	1	SR70
29	35	54	8	7	5	5	49	6	28	5	1	SR76
30	40	69	9	8	5	5	64	7	32	6	1	SR79
31	40	82	10	4	5	10	72	3	36	2	1	R61
32	42	55	8	7	6	6	49	6	35	5	1	SR80
33	48	56	8	6	7	8	48	5	42	4	1	R63
34	48	70	9	8	6	6	64	7	40	6	1	SR82
35	54	87	10	9	6	6	81	8	45	7	1	SR84
36	56	57	8	7	8	8	49	6	49	5	1	SR85
37	56	71	9	8	7	7	64	7	48	6	1	SR86
38	63	72	9	7	8	9	63	6	56	5	1	R66
39	63	88	10	9	7	7	81	8	54	7	1	SR87
40	72	73	9	8	9	9	64	7	64	6	1	SR89
41	72	89	10	9	8	8	81	8	63	7	1	SR90
42	80	90	10	8	9	10	80	7	72	6	1	R68

then a SUB arrangement with the parameters

$$v = s^2, \qquad b = b^* + 2s, \qquad r = r^* + 2, \qquad k_1 = s,$$

(11.3.6) $\quad k_2 = k^*, \qquad n_1 = s(s-1), \qquad n_2 = (s-1)^2, \qquad b_1 = 2s,$

$$b_2 = b^*, \qquad p_{11}{}^1 = s - 2, \qquad u = 1, \qquad \lambda = \lambda$$

can be constructed by adding $2s$ more sets to the plan of the L_2 design, where the ith added set contains the symbols from the ith row of the association scheme $(i = 1, 2, \cdots, s)$ and the $(s+j)$th added set contains the symbols from the jth column $(j = 1, 2, \cdots, s)$ of the association scheme. Conversely, if there exists a SUB arrangement with the parameters given in (11.3.6), then by removing the sets of size k_1, we obtain an L_2 design with the parameters given in (11.3.5).

Theorem 11.3.4. If there exists an L_i $(i > 2)$ design with the parameters

(11.3.7) $\quad v^* = s^2, \qquad b^*, \qquad r^*, \qquad k^* \neq s, \qquad \lambda_1^* = \lambda - 1, \qquad \lambda_2^* = \lambda,$

then a SUB arrangement with the parameters

$$v = s^2, \qquad b = b^* + si, \qquad r = r^* + i, \qquad k_1 = s, \qquad k_2 = k^*,$$

(11.3.8) $\quad n_1 = i(s-1), \qquad n_2 = (s-1)(s-i+1), \qquad b_1 = si, \qquad b_2 = b^*,$

$$p_{11}{}^1 = i^2 - 3i + s, \qquad u = 1, \qquad \lambda = \lambda$$

can always be constructed by adding si sets to the plan of the L_i design. The newly added sets are formed by picking up the symbols occurring in the same row or the same column, or the same letter of each of the latin squares.

The parameters of SUB arrangements obtainable by the above methods are listed in Tables 11.3.1, 11.3.2, and 11.3.3. For all the designs listed in these tables the parameter $u = 1$ and the reference is to the tables of PBIB designs of Bose, Clatworthy, and Shrikhande (1954).

Table 11.3.2. Parameters of SUB arrangements obtainable by the method of Theorem 11.3.2

	Parameters											
Series	v	b	r	k_1	k_2	b_1	b_2	n_1	n_2	$p_{11}{}^1$	λ	Type
1	10	11	5	4	5	5	6	6	3	3	2	T9
2	10	15	8	4	6	5	10	6	3	3	4	T18
3	10	25	8	4	3	5	20	6	3	3	2	T14
4	15	16	6	5	6	6	10	8	6	4	2	T22
5	15	21	5	5	3	6	15	8	6	4	1	T28

Table 11.3.3. Parameters of SUB arrangements obtainable by the method of Theorems 11.3.3 and 11.3.4

						Parameters						
Series	v	b	r	k_1	k_2	b_1	b_2	n_1	n_2	$p_{11}{}^1$	λ	Type
1	9	15	6	3	4	6	9	4	4	1	2	LS1
2	9	15	7	3	5	6	9	4	4	1	3	LS10
3	16	28	6	4	3	12	16	9	6	4	1	LS14
4	36	54	8	6	5	18	36	15	20	6	1	LS17
5	49	70	9	7	6	21	49	18	30	7	1	LS18
6	64	88	10	8	7	24	64	21	42	8	1	LS19

11.4. CONSTRUCTION OF SUB ARRANGEMENTS FROM KNOWN BIB AND PBIB DESIGNS

So far we have discussed methods of constructing SUB arrangements from PBIB designs. We can also construct SUB arrangements by using both BIB and PBIB designs.

Theorem 11.4.1. Let N_2^* be the incidence matrix of a group-divisible design with the parameters

$$(11.4.1) \qquad v^* = mn, \qquad b^*, r^*, k^*, \lambda_1^*, \lambda_2^* > \lambda_1^*$$

and let N_1^* be the incidence matrix of a BIBD design with the parameters

$$(11.4.2) \qquad v' = n, \qquad b', r', k' \neq k^*, \qquad \lambda' = \lambda_2^* - \lambda_1^*.$$

Then the v^* symbols can be so arranged that

$$(11.4.3) \qquad N = [I_m \times N_1^* \mid N_2^*],$$

where $\times$ denotes the Kronecker product of matrices, is the incidence matrix of a SUB arrangement with the parameters

$$v = mn, \qquad b = b^* + mb', \qquad r = r^* + r', \qquad k_1 = k',$$
$$(11.4.4) \quad k_2 = k^*, \qquad b_1 = mb', \qquad b_2 = b^*, \qquad n_1 = n - 1,$$
$$n_2 = n(m-1), \qquad p_{11}{}^1 = n - 2, \qquad \lambda = \lambda_2^*, \qquad u = \lambda_2^* - \lambda_1^*.$$

Proof. The parameters v, b, r, k_1, k_2, n_1, n_2, and λ need no explanation. From the method of construction we see that $n - 1$ symbols occur together $\lambda_2^* - \lambda_1^*$ times with a particular symbol θ in sets of size k_1 and λ_1^* times with θ in sets of size k_2, whereas the other $n(m - 1)$ symbols occur together λ_2^* times with θ in sets of size k_2. Hence n_1 and n_2. From the association scheme of a group-divisible design we obtain $p_{11}{}^1 = n - 2$, completing the proof of the theorem.

We now numerically illustrate this theorem.

Illustration 11.4.1.1. Consider design SR12 of the tables of Bose, Clatworthy, and Shrikhande (1954), with the parameters

$$(11.4.5) \quad v^* = 9 = b^*, \quad r^* = 3 = k^*, \quad m = 3 = n, \quad \lambda_1^* = 0, \quad \lambda_2^* = 1.$$

The plan of the design is

$$(2, 3, 1); \quad (6, 8, 4); \quad (5, 9, 7);$$
$$(11.4.6) \qquad (1, 5, 6); \quad (4, 2, 9); \quad (8, 7, 3);$$
$$(9, 1, 8); \quad (7, 6, 2); \quad (3, 4, 5);$$

with the association scheme

$$1 \quad 4 \quad 7$$
$$(11.4.7) \qquad \text{groups} \quad 2 \quad 5 \quad 8$$
$$3 \quad 6 \quad 9$$

Making use of the BIB design with the parameters

$$(11.4.8) \qquad v' = 3 = b', \quad r' = 2 = k', \quad \lambda' = 1$$

and using the theorem, we get the arrangement

$$(1, 4); \quad (1, 7); \quad (4, 7); \quad (2, 5); \quad (2, 8); \quad (5, 8);$$
$$(11.4.9) \quad (3, 6); \quad (3, 9); \quad (6, 9); \quad (2, 3, 1); \quad (6, 8, 4); \quad (5, 9, 7);$$
$$(1, 5, 6); \quad (4, 2, 9); \quad (8, 7, 3); \quad (9, 1, 8); \quad (7, 6, 2); \quad (3, 4, 5);$$

which can easily be verified to be a SUB arrangement with the parameters

$$v = 9, \quad b = 18, \quad r = 5, \quad k_1 = 2, \quad k_2 = 3,$$
$$(11.4.10) \qquad b_1 = 9 = b_2, \quad n_1 = 2, \quad n_2 = 6, \quad p_{11}{}^1 = 1,$$
$$\lambda = 1, \quad u = 1.$$

The parameters of SUB arrangements that can be constructed by the method of Theorem 11.4.1 are given in Table 11.4.1.

We close this section and also the chapter with the comment that, in analogy with Theorem 11.4.1, we can construct SUB arrangements from triangular or L_i designs and BIB designs.

Table 11.4.1. Parameters of SUB arrangements obtainable by the method of Theorem 11.5.1

	Parameters											
Series	v	b	r	k_1	k_2	b_1	b_2	n_1	n_2	p_{11}^{1}	u	λ	Type†
1	6	15	8	2	4	6	9	2	3	1	1	4	SR4, a
2	8	24	9	2	4	12	12	3	4	2	1	3	SR9, b
3	9	18	5	2	3	9	9	2	6	1	1	1	SR12, a
4	9	30	9	2	3	9	21	2	6	1	1	2	R11, a
5	12	21	5	2	4	12	9	2	9	1	1	1	SR20, a
6	12	34	7	2	3	18	16	3	8	2	1	1	SR21, b
7	15	30	6	2	4	15	15	2	12	1	1	1	R27, a
8	15	45	8	2	3	15	30	2	12	1	1	1	R28, a
9	15	55	9	2	3	30	25	4	10	3	1	1	SR36, c
10	16	40	7	2	4	24	16	3	12	2	1	1	SR40, b
11	16	56	9	2	3	24	32	3	12	2	1	1	R35, b
12	20	46	7	2	5	30	16	3	16	2	1	1	SR51, b
13	20	65	9	2	4	40	25	4	15	3	1	1	SR52, c
14	24	60	8	2	5	36	24	3	20	2	1	1	R45, b
15	24	66	9	2	4	24	42	2	21	1	1	1	R46, a
16	25	75	9	2	5	50	25	4	20	3	1	1	SR64, c
17	27	81	10	2	4	27	54	2	24	1	1	1	R54, a
18	28	77	10	3	4	28	49	6	21	5	1	1	SR68, d
19	30	85	9	2	6	60	25	4	25	3	1	1	SR70, c
20	35	84	10	3	5	35	49	6	28	5	1	1	SR76, d
21	42	91	10	3	6	42	49	6	35	5	1	1	SR80, d

† Reference to group-divisible designs in the tables of PBIB designs of Bose et al. (1954). The letters a, b, c, d stand for BIB designs with the following parameters:

a. $v' = 3 = b', r' = 2 = k', \lambda' = 1.$
b. $v' = 4, b' = 6, r' = 3, k' = 2, \lambda' = 1.$
c. $v' = 5, b' = 10, r' = 4, k' = 2, \lambda' = 1.$
d. $v' = 7 = b', r' = 3 = k', \lambda' = 1.$

REFERENCES

1. Bose, R. C., Clatworthy, W. H., and Shrikhande, S. S. (1954). *Tables of Partially Balanced Incomplete Block Designs.* Institute of Statistics, University of North Carolina, Reprint Series No. 50.
2. Kishen, K. (1940–1941). Symmetrical unequal block arrangements. *Sankhyā,* **5,** 329–344.
3. Raghavarao, D. (1962). Symmetrical unequal block arrangements with two unequal block sizes. *Ann. Math. Stat.,* **33,** 620–633.

BIBLIOGRAPHY

Bhaskararao, M. A note on incomplete block designs with $b = v$. *Ann. Math. Stat.,* **36,** 1877 (1965).

CHAPTER 12

Nonexistence of Incomplete Block Designs

12.1. INTRODUCTION

If v, b, r, k, and λ are the parameters of a BIB design, we know from Chapter 5 that they satisfy the relations

$$(12.1.1) \qquad vr = bk, \qquad \lambda(v-1) = r(k-1), \qquad b \geq v.$$

Similarly, if v, b, r, k, λ_i, n_i, p_{jk}^i $(i, j, k = 1, 2, \cdots, m)$ are the parameters of a PBIB design with m associate classes, they satisfy the relations

$$(12.1.2) \qquad vr = bk, \qquad \sum_{i=1}^{m} \lambda_i n_i = r(k-1), \qquad \sum_{k=1}^{m} p_{jk}^i = n_j - \delta_{ij},$$

$$n_i p_{jk}^i = n_j p_{ki}^j = n_k p_{ij}^k.$$

Conditions (12.1.1) or (12.1.2) are only the necessary conditions for the existence of BIB or PBIB designs. Thus the corresponding configurations of BIB or PBIB designs may or may not exist even if a possible set of parameters satisfying (12.1.1) or (12.1.2) is given. Further necessary conditions for the existence of the designs can be derived through the use of the Hasse–Minkowski invariant, and we devote this chapter to the study of the nonexistence of designs. In the text we freely use the properties of the Legendre symbol (a/p), the Hilbert norm residue symbol $(a, b)_p$, and the Hasse–Minkowski invariant of a matrix A, $C_p(A)$, as given in Appendix A. The symbol $\sim$ denotes either that the square free parts of the integers are the same or that they are rationally congruent, depending on whether it is used for scalars or matrices. The author recommends the use of the Hasse-Minkowski invariant for odd primes only.

12.2. NONEXISTENCE OF SYMMETRICAL BIB DESIGNS

The nonexistence of certain symmetrical BIB designs was demonstrated by Chowla and Ryser (1950) and Shrikhande (1950) by two different methods that give essentially the same results, as remarked in Chapter 1.

Theorem 12.2.1. The necessary conditions for the existence of a symmetrical BIB design are that

(12.2.1) $$(r - \lambda)^{v-1} \sim 1$$

and, if (12.2.1) is satisfied, then

(12.2.2) $$(r - \lambda, (-1)^{(v-1)/2} v)_p = 1$$

for all primes p, where $(a, b)_p$ is the Hilbert norm residue symbol.

Proof. Let N be the incidence matrix of a symmetrical BIB design. Then

(12.2.3) $$NN' = (r - \lambda)I_v + \lambda E_{v,v}.$$

Furthermore, as $(N^{-1})(NN')(N')^{-1} = I_v$, we have

(12.2.4) $$NN' \sim I_v.$$

Now, applying Theorem A.6.4 of Appendix A and noting that

(12.2.5)
$$\begin{aligned}
C_p(NN') &= (-1, -1)_p(-1, r - \lambda)_p^{v(v-1)/2}(v, r - \lambda)_p \\
&= (-1, -1)_p(r - \lambda, (-1)^{v(v-1)/2} v)_p \\
&= (-1, -1)_p(r - \lambda, (-1)^{(v-1)/2} v)_p,
\end{aligned}$$

when v is odd, and

(12.2.6) $$C_p(I_v) = (-1, -1)_p,$$

the required result follows.

As a consequence of the above theorem, observing that

(12.2.7) $$((-1)^{(v-1)/2} v, r - \lambda)_p = ((-1)^{(v-1)/2} \lambda, r - \lambda)_p,$$

we have the following corollaries:

Corollary 12.2.1.1. The symmetrical BIB design with even v is nonexistent if $r - \lambda$ is not a perfect square.

Corollary 12.2.1.2. Let v be odd and let s and t be the square free parts of $r - \lambda$ and λ, respectively. Then the following will obtain:

1. If p is an odd prime dividing s but not t, then the design is nonexistent if $((-1)^{(v-1)/2} t/p) = -1$.

2. If p is an odd prime dividing t but not s, then the design is nonexistent if $(s/p) = -1$.

3. If p is an odd prime dividing both s and t, then the design is nonexistent if $((-1)^{(v+1)/2} s_0 t_0/p) = -1$, where $s = s_0 p$ and $t = t_0 p$.

The following four designs are nonexistent and illustrate the above corollaries:

$$v = b = 22, \qquad r = k = 7, \qquad \lambda = 2 \qquad \text{Corollary } 12.2.1.1;$$
$$v = b = 43, \qquad r = k = 7, \qquad \lambda = 1 \qquad \text{Corollary } 12.2.1.2(1);$$
$$v = b = 93, \qquad r = k = 24, \qquad \lambda = 6 \qquad \text{Corollary } 12.2.1.2(2);$$
$$v = b = 103, \qquad r = k = 18, \qquad \lambda = 3 \qquad \text{Corollary } 12.2.1.2(3).$$

12.3. NONEXISTENCE OF AFFINE α-RESOLVABLE BIB DESIGNS

Let N_1 be the incidence matrix of an affine α-resolvable BIB design with the parameters v, b, r, k, and λ satisfying

$$(12.3.1) \qquad v\alpha = k\beta, \qquad b = t\beta, \qquad r = t\alpha, \qquad b = v + t - 1,$$

where the b sets are grouped into t classes $S_1, S_2, \cdots, S_t$ each of β sets such that in each class every symbol is replicated α times. Let

$$(12.3.2) \qquad N = \begin{bmatrix} & & N_1 & & \\ E_{1,\beta} & -E_{1,\beta} & 0_{1,\beta} & \cdots & 0_{1,\beta} \\ E_{1,\beta} & 0_{1,\beta} & -E_{1,\beta} & \cdots & 0_{1,\beta} \\ \cdot & \cdot & \cdot & & \\ E_{1,\beta} & 0_{1,\beta} & 0_{1,\beta} & \cdots & -E_{1,\beta} \end{bmatrix}.$$

Obviously N is a square matrix with rational elements. We note that

$$(12.3.3) \qquad NN' = \begin{bmatrix} N_1 N_1' & 0_{v,\,t-1} \\ 0_{t-1,\,v} & \beta I_{t-1} + \beta E_{t-1,\,t-1} \end{bmatrix}.$$

Now, as

$$(12.3.4) \qquad |NN'| \sim v\beta^{t-2}(r - \lambda)^{v-1},$$

NN' and in turn N are nonsingular, and we have

$$(12.3.5) \qquad NN' \sim I_b.$$

Noting that

$$(12.3.6) \qquad \begin{aligned} C_p(NN') &= C_p(N_1 N_1') \, C_p\{\beta I_{t-1} + \beta E_{t-1,\,t-1}\}(rk(r - \lambda)^{v-1}, t\beta^{t-1})_p \\ &= (rkt, -1)_p(\beta, -1)_p^{(t-1)(t-2)/2}(t, \beta)_p^{t-2} \\ &\quad \times (r - \lambda, -1)^{v(v-1)/2}(r - \lambda, v)_p(rk, \beta)_p^{t-2} \end{aligned}$$

and applying Theorem A.6.4 of Appendix A, we get the following:

Theorem 12.3.1. The necessary conditions for the existence of an affine α-resolvable BIB design are that

$$(12.3.7) \qquad v\beta^{t-2}(r - \lambda)^{v-1} \sim 1$$

and, if (12.3.7) is satisfied, then

$$(12.3.8) \quad \begin{aligned} (rkt, \, -1)_p(\beta, \, -1)_p^{(t-1)(t-2)/2}(t, \, \beta)_p^{t-2} \\ \times (r - \lambda, \, -1)_p^{v(v-1)/2}(r - \lambda, \, v)_p(rk, \, \beta)_p^{t-2} = 1. \end{aligned}$$

This theorem was obtained by Shrikhande and Raghavarao (1963), whereas the particular case when α = 1 was obtained by Shrikhande (1953).

As an application of the above theorem, we prove

Theorem 12.3.2. An affine resolvable BIB design with the parameters

$$(12.3.9) \qquad v = s^2, \qquad b = s(s+1), \qquad r = s+1, \qquad k = s, \qquad \lambda = 1$$

is nonexistent if $s \equiv 1$ or 2 (mod 4) and the square free part of s contains a prime $\equiv 3$ (mod 4).

Proof. For a BIB design with the parameters given in (12.3.9)

$$(12.3.10) \qquad \begin{aligned} v\beta^{t-2}(r - \lambda)^{v-1} &= s^2 \times s^{s-1} \times s^{s^2-1} \\ &= s^{s(s+1)} \sim 1, \end{aligned}$$

and hence condition (12.3.7) is always satisfied. The left-hand side of (12.3.8) reduces to $(-1, s)^{s(s^3+1)/2}$. Now if $s \equiv 0$ or 3 (mod 4), the index is even, and the necessary condition (12.3.8) is satisfied. However, if $s \equiv 1$ or 2 (mod 4), then the left-hand side of (12.3.8) is $(-1, s)_p$, which is -1, if the square free part of s contains a prime $\equiv 3$ (mod 4), and in such cases the design is nonexistent. Thus the theorem is established.

This result is equivalent to that given by Bruck and Ryser (1949), which was included in Chapter 1.

From Theorem 12.3.2 affine resolvable BIB designs with the parameters given in (12.3.9) are nonexistent for $s = 6, 14, 21, 22$, etc.

We have seen in Chapter 5 that a BIB design with the parameters

$$(12.3.11) \qquad v = 6, \qquad b = 10, \qquad r = 5, \qquad k = 3, \qquad \lambda = 2$$

cannot be resolvable and hence affine resolvable. This can also be seen from (12.3.8), which will be violated for $p = 3$.

12.4. NONEXISTENCE OF SYMMETRICAL, REGULAR PBIB DESIGNS

In this section we derive the necessary conditions for the existence of regular†, symmetrical PBIB designs with m associate classes. Let N be the incidence matrix of a regular, symmetrical PBIB design with m associate classes and let NN' have distinct, rational, characteristic roots $\rho_0 = r^2$, $\rho_1, \rho_2, \cdots, \rho_s$, with the respective multiplicities $\alpha_0 = 1$, $\alpha_1, \alpha_2, \cdots, \alpha_s$. Let us now calculate $C_p(NN')$. Clearly $E_{v,1}$ is a characteristic vector of NN' corresponding to the root ρ_0. Let $\mathbf{a_1}^i, \mathbf{a_2}^i, \cdots, \mathbf{a}_{\alpha_i}^i$ be a set of rational, linearly independent, characteristic vectors corresponding to the root ρ_i ($i = 1, 2, \cdots, s$). Put

(12.4.1) $\qquad S = [E_{v,1}, \mathbf{a_1}^1, \cdots, \mathbf{a}_{\alpha_1}^1, \cdots, \mathbf{a_1}^s, \cdots, \mathbf{a}_{\alpha_s}^s].$

Then we can easily verify that

(12.4.2) $\qquad S'NN'S = \operatorname{diag}[\rho_0 v, \rho_1 Q_1, \rho_2 Q_2, \cdots, \rho_s Q_s],$

where Q_i is the gramian of the vectors $\mathbf{a_1}^i, \mathbf{a_2}^i, \cdots, \mathbf{a}_{\alpha_i}^i$ ($i = 1, 2, \cdots, s$). Thus

(12.4.3) $\qquad NN' \sim \operatorname{diag}[\rho_0 v, \rho_1 Q_1, \rho_2 Q_2, \cdots, \rho_s Q_s],$

and hence

$$
C_p(NN') = (-1, -1)_p^{s+1}\left(\rho_0 v, -\left\{\prod_{i=1}^s \rho_i^{\alpha_i}\right\}\left\{\prod_{i=1}^s |Q_i|\right\}\right)_p
$$

$$
\times \left\{\prod_{\substack{i,j=1 \\ i<j}}^s (\rho_i^{\alpha_i}, \rho_j^{\alpha_j})_p\right\}\left\{\prod_{\substack{i,j=1 \\ i<j}}^s (\rho_i^{\alpha_i}, |Q_j|)_p\right\}
$$

(12.4.4) $\qquad \times \left\{\prod_{\substack{i,j=1 \\ i<j}}^s (\rho_j^{\alpha_j}, |Q_i|)_p\right\}\left\{\prod_{\substack{i,j=1 \\ i<j}}^s (|Q_i|, |Q_j|)_p\right\}$

$$
\times \left\{\prod_{i=1}^s (-1, \rho_i)_p^{\alpha_i(\alpha_i+1)/2}\right\}\left\{\prod_{i=1}^s (\rho_i, |Q_i|)_p^{\alpha_i-1}\right\}
$$

$$
\times \left\{\prod_{i=1}^s C_p(Q_i)\right\}.
$$

Again

(12.4.5) $\qquad S'S = \operatorname{diag}[v, Q_1, Q_2, \cdots, Q_s].$

But, as $S'S \sim I_v$, using Theorem A.6.4 of Appendix A, we have

(12.4.6) $\qquad |S'S| = v|Q_1||Q_2| \cdots |Q_s| \sim 1$

---

† A regular design is one whose incidence matrix N satisfies $|NN'| > 0$.

and

$$
\text{(12.4.7)} \quad
\begin{aligned}
C_p(S'S) &= (-1, -1)_p^{s+1}(v, -v)_p\left(\prod_{\substack{i,j=1\\i<j}}^{s}(|Q_i|, |Q_j|)_p\right)\left\{\prod_{i=1}^{s}C_p(Q_i)\right\}\\
&= (-1, -1)_p.
\end{aligned}
$$

Making use of (12.4.6) and (12.4.7) in (12.4.4), we get

$$
\text{(12.4.8)} \quad
\begin{aligned}
C_p(NN') &= (-1, -1)_p\left(v, \prod_{i=1}^{s}\rho_i^{\alpha_i}\right)_p\left\{\prod_{\substack{i,j=1\\i<j}}^{s}(\rho_i^{\alpha_i}, \rho_j^{\alpha_j})_p\right\}\\
&\times \left\{\prod_{\substack{i,j=1\\i<j}}^{s}(\rho_i^{\alpha_i}, |Q_j|)_p\right\}\left\{\prod_{\substack{i,j=1\\i<j}}^{s}(\rho_j^{\alpha_j}, |Q_i|)_p\right\}\\
&\times \left\{\prod_{i=1}^{s}(-1, \rho_i)_p^{\alpha_i(\alpha_i+1)/2}\right\}\left\{\prod_{i=1}^{s}(\rho_i, |Q_i|)_p^{\alpha_i-1}\right\}.
\end{aligned}
$$

Since N is the incidence matrix of a regular, symmetrical PBIB design, clearly $NN' \sim I_v$, and using Theorem A.6.4, we get the following theorem:

Theorem 12.4.1. The necessary conditions for the existence of regular, symmetrical PBIB design are that

$$
\text{(12.4.9)} \quad |NN'| = \rho_0\rho_1^{\alpha_1}\rho_2^{\alpha_2}\cdots\rho_s^{\alpha_s} \sim 1
$$

and, if (12.4.9) is satisfied, then

$$
\text{(12.4.10)} \quad C_p(NN') = (-1, -1)_p
$$

for all primes p, where $C_p(NN')$ is given by (12.4.8), where $\rho_0 = r^2, \rho_1, \cdots, \rho_s$ are the distinct, rational roots of NN', with the respective multiplicities $\alpha_0 = 1, \alpha_1, \cdots, \alpha_s$.

For two- and three-associate-class PBIB designs these necessary conditions are given by the following two theorems:

Theorem 12.4.2. The necessary conditions for the existence of symmetrical, regular PBIB designs with two associate classes are that

$$
\text{(12.4.11)} \quad \rho_1^{\alpha_1}\rho_2^{\alpha_2} \sim 1
$$

and, if (12.4.11) is satisfied, then

$$
\text{(12.4.12)} \quad \left\{\prod_{i=1}^{2}(-1, \rho_i)_p^{\alpha_i(\alpha_i+1)/2}\right\}(\rho_1, \rho_2)_p^{\alpha_1\alpha_2}(\rho_1\rho_2, |Q_1|)_p(\rho_2, v)_p = 1
$$

for all odd primes p, where ρ_1 and ρ_2 are the rational roots, other than r^2, of NN', with the respective multiplicities α_1 and α_2, and Q_1 is the gramian of a complete set of rational, linearly independent vectors corresponding to the root ρ_1 of NN'.

Theorem 12.4.3. The necessary conditions for the existence of symmetrical, regular PBIB designs with three associate classes are that

$$(12.4.13) \qquad \rho_1^{\alpha_1}\rho_2^{\alpha_2}\rho_3^{\alpha_3} \sim 1$$

and, if (12.4.13) is satisfied, then

$$(12.4.14) \quad \left\{\prod_{i=1}^{3}(-1, \rho_i)_p^{\alpha_i(\alpha_i+1)/2}\right\}(\rho_1, \rho_2)_p^{\alpha_1\alpha_2}(\rho_1, \rho_3)_p^{\alpha_1\alpha_3}(\rho_2, \rho_3)_p^{\alpha_2\alpha_3}$$

$$\times (\rho_1\rho_3, |Q_1|)_p(\rho_2\rho_3, |Q_2|)_p(\rho_3, v)_p = 1$$

for all odd primes p, where ρ_1, ρ_2, and ρ_3 are the rational roots, other than r^2, of NN', with the respective multiplicities α_1, α_2, and α_3, and Q_1, Q_2 are the gramians of complete sets of rational, linearly independent vectors corresponding to the roots ρ_1 and ρ_2 of NN'.

Thus the problem of finding the necessary conditions for the existence of designs reduces to the problem of finding the determinants of gramians of complete sets of rational, linearly independent, characteristics vectors of the different roots. In the subsequent theorems we prove the conditions for the existence of symmetrical, regular PBIB designs with two associate classes having group-divisible, triangular, and L_i association schemes.

Theorem 12.4.4. The necessary conditions for the existence of symmetrical, regular, group-divisible designs are that

$$(12.4.15) \qquad (r^2 - v\lambda_2)^{m-1}(r - \lambda_1)^{m(n-1)} \sim 1$$

and, if (12.4.15) is satisfied, then

$$(12.4.16) \quad \begin{aligned} &(-1, r^2 - v\lambda_2)_p^{m(m-1)/2}(-1, r - \lambda_1)_p^{m(n-1)[m(n-1)+1]/2} \\ &\qquad \times (r^2 - v\lambda_2, vn^m)_p(r - \lambda_1, n)_p^m = +1 \end{aligned}$$

for all odd primes p.

Proof. Let the $v = mn$ symbols be divided into m groups of n symbols each and let the symbols in the ith group be numbered $(i-1)n + 1$, $(i-1)n + 2, \cdots, in$, where $i = 1, 2, \cdots, m$. We have seen in Chapter 8 that the characteristic roots of NN', N being the incidence matrix of the design, are $\rho_0 = r^2$, $\rho_1 = r^2 - v\lambda_2$, and $\rho_2 = r - \lambda_1$, with the respective multiplicities $\alpha_0 = 1$, $\alpha_1 = m - 1$, and $\alpha_2 = m(n-1)$. Let us form the vth order vectors ξ_i ($i = 1, 2, \cdots, m$), where ξ_i has n unit entries in the positions $(i-1)n + 1$, $(i-1)n + 2, \cdots, in$ and zeros elsewhere. We can easily see that $E_{v, 1}$ lies in the vector space generated by the vectors ξ_i. Consider the vector space V generated by the vectors ξ_i and orthogonal to $E_{v, 1}$. Any vector $\xi \in V$ is of the form $\xi = a_1\xi_1 + a_2\xi_2 + \cdots + a_m\xi_m$, where $a_1 + a_2 + \cdots + a_m = 0$. We can easily see that $\xi \in V$ is a characteristic vector corresponding to the root ρ_1 of NN'. Thus the vector space generated by $\xi_1, \xi_2, \cdots, \xi_m$ are identical to the vector space generated by $E_{v,1}$ and a complete set of characteristic vectors corresponding to root ρ_1 of NN'. Hence

(12.4.17) $$\text{diag } [v, Q_1] \sim nI_m,$$

and hence

(12.4.18) $$|Q_1| \sim vn^m.$$

Now, applying Theorem 12.4.2, we get our present theorem.

The following group-divisible designs are nonexistent, as they violate the necessary conditions shown in parentheses:

$v = 40 = b$, $m = 8$, $n = 5$, $r = 10 = k$, $\lambda_1 = 5$, $\lambda_2 = 2$ (12.4.15);

$v = 33 = b$, $m = 11$, $n = 3$, $r = 6 = k$, $\lambda_1 = 0$, $\lambda_2 = 1$ (12.4.16);

$v = 35 = b$, $m = 7$, $n = 5$, $r = 6 = k$, $\lambda_1 = 0$, $\lambda_2 = 1$ (12.4.16).

Theorem 12.4.5. The necessary conditions for the existence of regular, symmetrical, triangular designs are that

(12.4.19) $$\rho_1^{n-1} \rho_2^{n(n-3)/2} \sim 1$$

and, if (12.4.19) is satisfied, then

(12.4.20)
$$(-1, \rho_1)_p^{(n-1)(n-2)/2}(\rho_1, n)_p(\rho_1, n-2)_p^{n-1}(-1, \rho_2)_p^{n(n-1)(n-2)(n-3)/8}$$
$$\times (\rho_2, 2)_p(\rho_2, n-1)_p(\rho_2, n-2)_p^{n-1} = +1$$

for all primes p, where $\rho_1 = r + (n-4)\lambda_1 - (n-3)\lambda_2$ and $\rho_2 = r - 2\lambda_1 + \lambda_2$.

Proof. Let the $v = n(n-1)/2$ symbols be arranged in an $n \times n$ square symmetrically with a blank diagonal. We have seen in Chapter 8 that the characteristic roots of NN', N being the incidence matrix of the triangular design, are $\rho_0 = r^2$, $\rho_1 = r + (n-4)\lambda_1 - (n-3)\lambda_2$, and $\rho_2 = r - 2\lambda_1 + \lambda_2$, with the respective multiplicities $\alpha_0 = 1$, $\alpha_1 = n-1$, and $\alpha_2 = n(n-3)/2$. Let us form the vth order vectors ξ_i $(i = 1, 2, \cdots, n)$, where ξ_i has $n-1$ unit entries in the positions corresponding to the $n-1$ symbols occurring in the ith row of the association scheme and zeros elsewhere. We can easily see that $E_{v,1}$ lies in the vector space generated by the vectors ξ_i. Consider the vector space V generated by the vectors ξ_i and orthogonal to $E_{v,1}$. Any vector $\xi \in V$ is of the form $\xi = a_1\xi_1 + a_2\xi_2 + \cdots + a_n\xi_n$, where $a_1 + a_2 + \cdots + a_n = 0$. We can easily see that if the ith symbol occurs in the pth row and the qth column of the association scheme, then the ith element of ξ is $a_p + a_q$, whereas the ith element of $NN'\xi$ is

(12.4.21)
$$\left\{ \sum_{\substack{i=1 \\ i \neq p, q}}^{n} a_i \right\}(n-3)\lambda_2 + \left\{ \sum_{\substack{i=1 \\ i \neq p, q}}^{n} a_i \right\}2\lambda_1 + (a_p + a_q)r$$
$$+ (a_p + a_q)(n-2)\lambda_1 = (a_p + a_q)[r + (n-4)\lambda_1 - (n-3)\lambda_2],$$

which is ρ_1 times the ith element of ξ. Hence $\xi \in V$ is a characteristic vector corresponding to the root ρ_1 of NN'. Thus the vector space generated by $E_{v,1}$ and a complete set of rational, linearly independent, characteristic vectors corresponding to the root ρ_1 of NN' are identical with the vector space generated by $\xi_1, \xi_2, \cdots, \xi_n$. Thus

$$(12.4.22) \qquad \text{diag}[v, Q_1] \sim (n-2)I_n + E_{n,n}$$

and hence

$$(12.4.23) \qquad |Q_1| \sim n(n-2)^{n-1}.$$

Our theorem follows by making an appeal to Theorem 12.4.2, while substituting (12.4.23) and using (12.4.19) and the properties of the Hilbert symbol.

The following triangular designs are nonexistent, as they violate the necessary conditions, shown in parentheses:

$$
\begin{array}{lllll}
v = 15 = b, & r = 5 = k, & \lambda_1 = 1, & \lambda_2 = 2 & (12.4.19); \\
v = 36 = b, & r = 8 = k, & \lambda_1 = 1, & \lambda_2 = 2 & (12.4.19); \\
v = 21 = b, & r = 6 = k, & \lambda_1 = 1, & \lambda_2 = 2 & (12.4.20); \\
v = 55 = b, & r = 10 = k, & \lambda_1 = 1, & \lambda_2 = 2 & (12.4.20).
\end{array}
$$

Theorem 12.4.6. The necessary conditions for the existence of regular, symmetrical L_i designs are that

$$(12.4.24) \qquad \rho_1{}^{\alpha_1} \rho_2{}^{\alpha_2} \sim 1$$

and, if (12.4.24) is satisfied, then

$$(12.4.25) \qquad \left\{ \prod_{i=1}^{2} (-1, \rho_i)_p^{\alpha_i(\alpha_i+1)/2} \right\} (\rho_1, \rho_2)_p^{\alpha_1 \alpha_2} (\rho_1 \rho_2, s)_p^{is} = 1$$

for all primes p, where $\rho_1 = r + (s-i)\lambda_1 - (s+1-i)\lambda_2$, $\rho_2 = r - i\lambda_1 + (i-1)\lambda_2$, $\alpha_1 = i(s-1)$, and $\alpha_2 = (s+1-i)(s-1)$.

Proof. Let us consider the characterization of the L_i association scheme in terms of orthogonal arrays, as given in Chapter 8. Let $(s^2, i, s, 2)$ be an orthogonal array in s integers $1, 2, \cdots, s$. Let us number the columns of the array $1, 2, \cdots, s^2$ and identify the $v = s^2$ symbols with the columns. Let ξ_{kj} $(k = 1, 2, \cdots, i; j = 1, 2, \cdots, s)$ be the vth order column vectors having s unit entries in the positions corresponding to the columns where the jth integer occurs in the kth row and zeros elsewhere. We can easily verify that for every k, $\sum_{j=1}^{s} \xi_{kj} = E_{v,1}$ and that the set of si vectors thus formed is of rank $i(s-1) + 1$. Let us denote the vector space generated by these vectors by U. Then U can also be generated by $E_{v,1}$ and $s-1$ vectors selected from each of the i sets (ξ_{kj}) $(j = 1, 2, \cdots, s)$, which form a rational basis of U.

Consider the subspace V of U, which is orthogonal to $E_{v,1}$. Any vector $\xi \in V$ is of the form $\xi = \sum_{k,j} a_{kj} \xi_{kj}$, where $\sum_{k,j} a_{kj} = 0$. We can easily verify that $\xi \in V$ will be a characteristic vector of NN' corresponding to the root ρ_1. Thus the vector space generated by $E_{v,1}$ and a complete set of rational, linearly independent, characteristic vectors corresponding to the root ρ_1 of NN' are identical with U. Noting the properties of ξ_{kj}, we can easily see that

$$\text{diag } [v, Q_1] \sim \begin{bmatrix} s^2 & sE_{1,s-1} & sE_{1,s-1} & \cdots & sE_{1,s-1} \\ sE_{s-1,1} & sI_{s-1} & E_{s-1,s-1} & \cdots & E_{s-1,s-1} \\ sE_{s-1,1} & E_{s-1,s-1} & sI_{s-1} & \cdots & E_{s-1,s-1} \\ \cdot & \cdot & \cdot & \cdots & \cdot \\ sE_{s-1,1} & E_{s-1,s-1} & E_{s-1,s-1} & \cdots & sI_{s-1} \end{bmatrix}$$

and hence

$$(12.4.26) \qquad\qquad |Q_1| \sim s^{is}.$$

Now applying Theorem 12.4.2, we get our present theorem.

The following L_i designs are nonexistent, as they violate the necessary conditions, shown in parentheses:

$v = 100 = b,$	$i = 5,$	$r = 18 = k,$	$\lambda_1 = 2,$	$\lambda_2 = 4$	(12.4.24);
$v = 144 = b,$	$i = 7,$	$r = 45 = k,$	$\lambda_1 = 12,$	$\lambda_2 = 16$	(12.4.24);
$v = 25 = b,$	$i = 3,$	$r = 4 = k,$	$\lambda_1 = 0,$	$\lambda_2 = 1$	(12.4.25);
$v = 49 = b,$	$i = 5,$	$r = 24 = k,$	$\lambda_1 = 10,$	$\lambda_2 = 14$	(12.4.25).

Theorems 12.4.4, 12.4.5, and 12.4.6 were first proved by Bose and Connor (1952), Ogawa (1959), and Shrikhande and Jain (1962), respectively. The method of calculating the gramian for a complete set of rational, independent vectors corresponding to the root ρ_1 of NN' in the case of triangular designs is due to Corsten (1960). It is to be noted that the gramians used in Theorems 12.4.5 and 12.4.6 were based on the association schemes and not merely the parameters. However, Bhagwandas and Shrikhande (1968) showed that the gramians of complete sets of rational, linearly independent vectors corresponding to the roots of NN' are rationally congruent for pseudotriangular (or pseudo L_2) and triangular (or L_2) designs. In the light of this result Theorems 12.4.5 and 12.4.6 hold for pseudotriangular and L_2 designs also. By using our fundamental Theorem 12.4.1, the necessary conditions for the existence of various regular, symmetrical PBIB designs with known association schemes can be obtained, and in this connection we refer the reader to the bibliography cited at the end of this chapter. Some of these results are also given as exercises at the end of this volume.

In conclusion we remark that, since NN' is positive at least semidefinite, none of its characteristic roots can be negative. Hence the following theorem:

Theorem 12.4.7. The necessary conditions for the existence of designs are that the characteristic roots of NN' be nonnegative.

12.5. NONEXISTENCE OF CERTAIN ASYMMETRICAL PBIB DESIGNS

In this section we try to obtain the necessary conditions for the existence of certain asymmetrical PBIB designs by using the Hasse–Minkowski invariant. This technique was devised by Shrikhande, Raghavarao, and Tharthare (1963) and was later improved by Raghavarao (1966).

Let N be the incidence matrix of an asymmetrical, connected PBIB design with m associate classes, where $b = v - \alpha$. Then, $\rho_0 = rk$ being a simple root of NN', let zero be a root of NN' of multiplicity α. Let the remaining s positive, rational, distinct roots of NN' be $\rho_1, \rho_2, \cdots, \rho_s$, with the multiplicities $\alpha_1, \alpha_2, \cdots, \alpha_s$, respectively.

Let

$$(12.5.1) \qquad X = (\mathbf{x_1}, \mathbf{x_2}, \cdots, \mathbf{x_\alpha})$$

be a $v \times \alpha$ matrix whose columns represent a set of mutually orthogonal, rational vectors corresponding to the zero root of NN'.

Let

$$(12.5.2) \qquad \mathbf{x_i'}\mathbf{x_i} = c_i \qquad i = 1, 2, \cdots, \alpha.$$

Put

$$(12.5.3) \qquad Y = X \operatorname{diag}(c_1^{-1/2}, c_2^{-1/2}, \cdots, c_\alpha^{-1/2}).$$

The spectral decomposition of NN' is

$$(12.5.4) \qquad NN' = \sum_{i=0}^{s} \rho_i A_i,$$

where the matrices A_i are rational, symmetrical, and idempotent, and further

$$(12.5.5) \qquad A_i A_j = 0_{v,v} \qquad i \neq j, i, j = 0, 1, \cdots, s.$$

In particular $A_0 = v^{-1} E_{v,v}$.

Let us now consider the rational, symmetrical matrix

$$(12.5.6) \qquad L = NN' + YY'.$$

Clearly

$$(12.5.7) \qquad LA_i = \rho_i A_i; \qquad LYY' = YY'.$$

Hence A_i and YY' generate vector spaces corresponding to the roots ρ_i $(i = 0, 1, \cdots, s)$ and 1 of L, and

$$(12.5.8) \qquad |L| = \prod_{i=0}^{s} \rho_i^{\alpha_i}.$$

As in the preceding section, we can easily show that

$$(12.5.9) \qquad L \sim \text{diag} \, [\rho_0 v, \rho_1 Q_1, \rho_2 Q_2, \cdots, \rho_s Q_s, Q],$$

where Q_i and Q are the gramians of complete sets of rational vectors corresponding to the roots ρ_i $(i = 1, 2, \cdots, s)$ and zero, respectively. The term $C_p(L)$ can be calculated to be

$$
\begin{aligned}
C_p(L) = {} & \left(\rho_0 v, \, -v \prod_{i=1}^{s} \rho_i^{\alpha_i}\right) \left\{\prod_{i=1}^{s} (\rho_i^{\alpha_i}, |Q|)_p\right\} \\
& \times \left\{\prod_{\substack{i,j=1 \\ i<j}}^{s} (\rho_i^{\alpha_i}, \rho_j^{\alpha_j})_p\right\} \left\{\prod_{\substack{i,j=1 \\ i<j}}^{s} (\rho_i^{\alpha_i}, |Q_j|)_p\right\} \\
& \times \left\{\prod_{\substack{i,j=1 \\ i<j}}^{s} (\rho_j^{\alpha_j}, |Q_i|)_p\right\} \left\{\prod_{i=1}^{s} (-1, \rho_i)_p^{\alpha_i(\alpha_i+1)/2}\right\} \\
& \times \left\{\prod_{i=1}^{s} (\rho_i, |Q_i|)_p^{\alpha_i-1}\right\} (-1, -1)_p .
\end{aligned}
$$
(12.5.10)

Again

$$(12.5.11) \qquad
\begin{aligned}
L = {} & NN' + YY' \\
= {} & (N|X)\{\text{diag} \, [I_{v-\alpha}, c_1^{-1}, c_2^{-1}, \cdots, c_\alpha^{-1}]\}(N|X)'
\end{aligned}
$$

and therefore

$$(12.5.12) \qquad
\begin{aligned}
L \sim {} & \text{diag} \, [I_{v-\alpha}, c_1^{-1}, c_2^{-1}, \cdots, c_\alpha^{-1}] \\
\sim {} & \text{diag} \, [I_{v-\alpha}, Q].
\end{aligned}
$$

Thus

$$(12.5.13) \qquad |L| \sim |Q|$$

and

$$(12.5.14) \qquad C_p(L) = C_p\{\text{diag} \, [I_{v-\alpha}, Q]\} = C_p(Q).$$

Equating (12.5.8) and (12.5.13) and also (12.5.10) and (12.5.14), we get the following theorem:

Theorem 12.5.1. For a PBIB design with $b = v - \alpha$ let NN' have positive, rational roots $\rho_0 = rk$, $\rho_1, \rho_2, \cdots, \rho_s$, with the respective multiplicities $\alpha_0 = 1, \alpha_1, \alpha_2, \cdots, \alpha_s$, and let zero be a root with the multiplicity α. Let

$Q, Q_1, Q_2, \cdots, Q_s$ be the gramians of complete sets of rational, linearly independent vectors corresponding to the roots $0, \rho_1, \rho_2, \cdots, \rho_s$, respectively, of NN'. Then the necessary conditions for the existence of the design are that

$$(12.5.15) \qquad \left\{ \prod_{i=0}^{s} \rho_i^{\alpha_i} \right\} |Q| \sim 1$$

and, if (12.5.15) is satisfied, then

$$(\rho_0, -v)_p \left(v, \prod_{i=1}^{s} \rho_i^{\alpha_i} \right) \left\{ \prod_{i=1}^{s} (-1, \rho_i)_p^{\alpha_i(\alpha_i+3)/2} \right\} (-1, -1)_p$$

$$12.5.16 \qquad \times \left\{ \prod_{\substack{i,j=1 \\ i<j}}^{s} (\rho_i^{\alpha_i}, \rho_j^{\alpha_j})_p \right\} \left\{ \prod_{\substack{i,j=1 \\ i<j}}^{s} (\rho_i^{\alpha_i}, |Q_j|)_p \right\} \left\{ \prod_{\substack{i,j=1 \\ i<j}}^{s} (\rho_j^{\alpha_j}, |Q_i|)_p \right\}$$

$$\times \left\{ \prod_{i=1}^{s} (\rho_i, |Q_i|)_p^{\alpha_i-1} \right\} C_p(Q) = 1$$

for all primes p.

In particular we have the following theorem:

Theorem 12.5.2. For a connected PBIB design with two associate classes and $b = v - \alpha$ let NN' have the roots $\rho_0 = rk$, necessarily of multiplicity 1, and zero, with the multiplicity α, the remaining root being a positive, rational number ρ_1 of multiplicity $b - 1$. Then the necessary conditions for the existence of the design are that

$$(12.5.17) \qquad \rho_0 \rho_1^{\alpha_1} |Q| \sim 1$$

and, if (12.5.17) is satisfied, then

$$(12.5.18) \quad (-1, -1)_p (\rho_0, -v\rho_1^{\alpha_1+1})_p (v, \rho_1)_p (-1, \rho_1)_p^{\alpha_1(\alpha_1+3)/2} C_p(Q) = 1$$

for all primes p, Q being the gramian corresponding to a complete set of rational, linearly independent vectors corresponding to the zero root of NN'.

We give an alternative proof for Theorem 12.5.2 based on the ideas developed in Chapter 10. Clearly the designs under consideration are linked block designs and hence their duals are BIB designs. Now, let Q and Q_1 be the gramians of complete sets of rational, linearly independent vectors corresponding to the roots 0 and ρ_1, respectively, of NN'. Then

$$(12.5.19) \qquad |Q| |Q_1| \sim v$$

and

$$(12.5.20) \qquad (-1, -1)_p (|Q|, |Q_1|)_p C_p(Q) C_p(Q_1) = 1,$$

from which we get

$$(12.5.21) \qquad C_p(Q_1) = (-1, -1)_p (|Q|, v)_p (|Q|, -1)_p C_p(Q).$$

If Q_1^* is the gramian of a complete set of rational, linearly independent vectors corresponding to the root ρ_1 of $N'N$, we get

$$(12.5.22) \qquad Q_1^* = \rho_1 Q_1$$

and hence

$$(12.5.23) \qquad |Q_1^*| = \rho_1^{\alpha_1} |Q_1|$$

and

$$(12.5.24) \qquad C_p(Q_1^*) = (-1, \rho_1)_p^{\alpha_1(\alpha_1+1)/2}(\rho_1, |Q_1|)_p^{\alpha_1-1} C_p(Q_1).$$

But, as the design is a linked block design, $Q_1^* \sim P_1^*$, where P_1^* is the gramian of a complete set of linearly independent, rational vectors corresponding to the multiple root of $N'N$ and satisfying

$$(12.5.25) \qquad |P_1^*| \sim b$$

and

$$(12.5.26) \qquad C_p(P_1^*) = 1.$$

Now applying Theorem A.6.4 for the rational congruence of $Q_1^* \sim P_1^*$ and making use of (12.5.19) and (12.5.20), we get the results of Theorem 12.5.2.

Making use of the gramians obtained in the proofs of Theorems 12.4.4, 12.4.5, and 12.4.6 and the relation (12.4.6), and making an appeal to Theorem 12.5.2, the following results can be obtained for group-divisible, triangular, and L_i designs:

Theorem 12.5.3. The necessary conditions for the existence of semiregular group-divisible designs with $b = v - m + 1$ are that

$$(12.5.27) \qquad \lambda_2 n^m (r - \lambda_1)^{m(n-1)} \sim 1$$

and, if (12.5.27) is satisfied, then

$$(12.5.28) \quad (\rho_0, -v\rho_1^{\alpha_1+1})_p (v, \rho_1)_p (-1, \rho_1)_p^{\alpha_1(\alpha_1+3)/2}(v, n^m)_p (-1, n)_p^{m(m+1)/2} = 1$$

for all primes p, where $\rho_1 = r - \lambda_1$ and $\alpha_1 = m(n-1)$.

Theorem 12.5.4. The necessary conditions for the existence of triangular designs with $r + (n-4)\lambda_1 - (n-3)\lambda_2 = 0$ and $b = (n-1)(n-2)/2$ are that

$$(12.5.29) \qquad \rho_0 \rho_1^{n(n-3)/2} n(n-2)^{n-1} \sim 1$$

and, if (12.5.29) is satisfied, then

$$(12.5.30) \quad \begin{aligned} (\rho_0, &-v\rho_1^{\alpha_1+1})_p (v, \rho_1)_p (-1, \rho_1)_p^{\alpha_1(\alpha_1+3)/2}(-1, n-2)_p^{n(n-1)/2} \\ &\times (-1, n-1)_p^n (-1, n)_p (n-2, n)_p (-1, v)_p (2v, \rho_0 \rho_1^{\alpha_1})_p = 1 \end{aligned}$$

for all primes p, where $\rho_1 = r - 2\lambda_1 + \lambda_2$ and $\alpha_1 = n(n-3)/2$.

Theorem 12.5.5. The necessary conditions for the existence of triangular designs with $r - 2\lambda_1 + \lambda_2 = 0$ and $b = n$ are that

$$(12.5.31) \qquad 2\rho_0 \rho_1^{\alpha_1}(n-1)(n-2)^{n-1} \sim 1$$

and, if (12.5.31) is satisfied, then

$$
\begin{aligned}
(12.5.32) \qquad & (\rho_0, \ -v\rho_1^{\alpha_1+1})_p(v, \rho_1)_p(-1, \rho_1)_p^{\alpha_1(\alpha_1+3)/2}(-1, n-2)_p^{n(n-1)/2} \\
& \times (-1, n-1)_p^n(-1, n)_p(n-2, n)_p(-1, v)_p \\
& \times (2v, n(n-2)^{n-1})_p(n(n-2)^{n-1}, \rho_0\rho_1^{\alpha_1})_p = 1
\end{aligned}
$$

for all primes p, where $\rho_1 = r + (n-4)\lambda_1 - (n-3)\lambda_2$ and $\alpha_1 = n-1$.

Theorem 12.5.6. The necessary conditions for the existence of L_i designs with $r + (s-i)\lambda_1 - (s+1-i)\lambda_2 = 0$ and $b = s(s-i) + i$ are that

$$(12.5.33) \qquad \rho_0 \rho_1^{\alpha_1} s^{is} \sim 1$$

and, if (12.5.33) is satisfied, then

$$(12.5.34) \qquad (\rho_0, \ -\rho_1^{\alpha_1+1})_p(-1, \rho_1)_p^{\alpha_1(\alpha_1+3)/2}(-1, s)_p^{is(s+i)/2} = 1$$

for all primes p, where $\rho_1 = r - i\lambda_1 + (i-1)\lambda_2$ and $\alpha_1 = (s-1)(s+1-i)$.

Theorem 12.5.7. The necessary conditions for the existence of L_i designs with $r - i\lambda_1 + (i-1)\lambda_2 = 0$ and $b = i(s-1) + 1$ are that

$$(12.5.35) \qquad \rho_0 \rho_1^{\alpha_1} s^{is} \sim 1$$

and, if (12.5.35) is satisfied, then

$$(12.5.36) \qquad (\rho_0, \ -\rho_1^{\alpha_1+1})_p(-1, \rho_1)_p^{\alpha_1(\alpha_1+3)/2}(-1, s)_p^{is(i+s+2)/2} = 1$$

for all primes p, where $\rho_1 = r + (s-i)\lambda_1 - (s+1-i)\lambda_2$ and $\alpha_1 = i(s-1)$.

12.6. NONEXISTENCE OF AFFINE α-RESOLVABLE PBIB DESIGNS WITH TWO ASSOCIATE CLASSES

Let N^* be the incidence matrix of an affine α-resolvable PBIB design with two associate classes and with the parameters v, b, r, k, λ_1, λ_2. Let the b sets be divided into t classes of β sets each, such that every symbol occurs exactly α times in each class. Let $\rho_0 = rk$, $\rho_1 \neq 0$, and 0 be the three distinct rational roots of $N^*N^{*\prime}$, with the respective multiplicities 1, α_1, and μ, respectively, where $1 + \alpha_1 + \mu = v$. We know from Chapter 8 that $b - t = \alpha_1$. Let $\mathbf{x}_1, \mathbf{x}_2, \cdots, \mathbf{x}_\mu$ be a complete set of rational, orthogonal, characteristic vectors corresponding to the zero root of $N^*N^{*\prime}$. Put

$$(12.6.1) \qquad X = [\mathbf{x}_1, \mathbf{x}_2, \cdots, \mathbf{x}_\mu].$$

Let

(12.6.2) $$c_i = \mathbf{x}_i' \mathbf{x}_i \qquad i = 1, 2, \cdots, \mu$$

and

(12.6.3) $$Y = X \operatorname{diag} [c_1^{-\frac{1}{2}}, c_2^{-\frac{1}{2}}, \cdots, c_\mu^{-\frac{1}{2}}].$$

Put

(12.6.4) $$N_1 = \begin{bmatrix} N^* & Y \\ L(\beta, t) & 0_{t-1, \mu} \end{bmatrix}$$

and

(12.6.5) $$N_2 = \begin{bmatrix} N^* & X \\ L(\beta, t) & 0_{t-1, \mu} \end{bmatrix},$$

where

(12.6.6) $$L(\beta, t) = \begin{bmatrix} E_{1,\beta} & -E_{1,\beta} & 0_{1,\beta} & \cdots & 0_{1,\beta} \\ E_{1,\beta} & 0_{1,\beta} & -E_{1,\beta} & \cdots & 0_{1,\beta} \\ \cdot & & & \cdots & \cdot \\ E_{1,\beta} & 0_{1,\beta} & 0_{1,\beta} & \cdots & -E_{1,\beta} \end{bmatrix}.$$

We note that N_2 is a rational matrix. Now

(12.6.7) $$N_1 = N_2 \begin{bmatrix} I_b & 0_{b, \mu} \\ 0_{\mu, b} & \operatorname{diag} [c_1^{-\frac{1}{2}}, c_2^{-\frac{1}{2}}, \cdots, c_\mu^{-\frac{1}{2}}] \end{bmatrix}$$

and hence

(12.6.8) $$N_1 N_1' \sim \begin{bmatrix} I_b & 0_{b, \mu} \\ 0_{\mu, b} & \operatorname{diag} [c_1, c_2, \cdots, c_\mu] \end{bmatrix} \sim \begin{bmatrix} I_b & 0_{b, \mu} \\ 0_{\mu, b} & Q \end{bmatrix},$$

where Q is the gramian of a complete set of rational, linearly independent, characteristic vectors corresponding to the zero root of $N^* N^{*'}$. Now, applying Theorem A.6.4 to (12.6.8), we get

(12.6.9) $$|N_1 N_1'| \sim |Q|$$

and

(12.6.10) $$C_p(N_1 N_1') = C_p(Q).$$

Making use of the results of the last section, we can easily show that

(12.6.11) $$|N_1 N_1'| = \rho_0 \rho_1^{\alpha_1} t \beta^{t-1}$$

and

(12.6.12) $$\begin{aligned} C_p(N_1 N_1') = &(\rho_0 \rho_1^{\alpha_1}, t\beta^{t-1})_p (t, \beta)_p^{t-2} (\rho_0, -v\rho_1^{\alpha_1})_p (v, \rho_1)_p \\ &\times (\rho_1, |Q|)_p (-1, \rho_1)_p^{\alpha_1(\alpha_1+1)/2} (-1, \beta)_p^{t(t-1)/2}. \end{aligned}$$

Equating (12.6.9) with (12.6.11) and (12.6.10) with (12.6.12), we get the following theorem:

Theorem 12.6.1. The necessary conditions for the existence of affine α-resolvable PBIB designs with two associate classes are that

$$(12.6.13) \qquad \rho_0 \rho_1{}^{\alpha_1} t \beta^{t-1} |Q| \sim 1$$

and, if (12.6.13) is satisfied, then

$$
\begin{aligned}
(12.6.14) \quad &(-1, \beta)_p^{t(t-1)/2}(t, \beta)_p^{t-2}(\rho_0, v\rho_1{}^{\alpha_1})_p(v, \rho_1)_p \\
&\qquad \times (\rho_0 \rho_1^{\alpha_1+1}, |Q|)_p(-1, \rho_1)_p^{\alpha_1(\alpha_1+3)/2} C_p(Q) = 1
\end{aligned}
$$

for all primes p, where $\rho_0 = rk$ and ρ_1 are the nonzero roots of $N^*N^{*\prime}$, with the multiplicities 1 and α_1, and Q is the gramian of a complete set of rational, linearly independent vectors corresponding to the zero root of $N^*N^{*\prime}$.

Using the gramians of the rational vectors of $N^*N^{*\prime}$ as obtained earlier for group-divisible, triangular, and L_i designs, and making an appeal to the above theorem, we get the following theorems:

Theorem 12.6.2. The necessary conditions for the existence of an affine α-resolvable semiregular group-divisible design are that

$$(12.6.15) \qquad \lambda_2 t \beta^{t-1} n^m \rho_1{}^{\alpha_1} \sim 1$$

and, if (12.6.15) is satisfied, then

$$
\begin{aligned}
(12.6.16) \quad &(-1, \beta)_p^{t(t-1)/2}(t, \beta)_p^{t-2}(\rho_0, v\rho_1{}^{\alpha_1})_p(v, \rho_1)_p(-1, \rho_1)_p^{\alpha_1(\alpha_1+3)/2} \\
&\qquad \times (\rho_0 \rho_1^{\alpha_1+1}, vn^m)_p(v, n^m)_p(-1, n)_p^{m(m+1)/2} = 1
\end{aligned}
$$

for all primes p, where $\rho_1 = r - \lambda_1$ and $\alpha_1 = m(n - 1)$.

Theorem 12.6.3. The necessary conditions for the existence of affine α-resolvable triangular designs satisfying $r + (n - 4)\lambda_1 - (n - 3)\lambda_2 = 0$ are that

$$(12.6.17) \qquad t \beta^{t-1} n(n - 2)^{n-1} \rho_0 \rho_1{}^{\alpha_1} \sim 1$$

and, if (12.6.17) is satisfied, then

$$
\begin{aligned}
(12.6.18) \quad &(-1, \beta)_p^{t(t-1)/2}(t, \beta)_p^{t-2}(\rho_0, v\rho_1{}^{\alpha_1})_p(v, \rho_1)_p(-1, \rho_1)_p^{\alpha_1(\alpha_1+3)/2} \\
&\qquad \times (\rho_0 \rho_1^{\alpha_1+1}, n(n-2)^{n-1})_p(-1, n-2)_p^{n(n-1)/2}(-1, n-1)_p^n \\
&\qquad \times (2, n(n-2)^{n-1})_p(-1, n)_p(n, n-2)_p(-1, v)_p \\
&\qquad \times (v, n(n-2)^{n-1})_p = 1
\end{aligned}
$$

for all primes p, where $\rho_1 = r - 2\lambda_1 + \lambda_2$ and $\alpha_1 = n(n - 3)/2$.

Theorem 12.6.4. The necessary conditions for the existence of affine α-resolvable triangular designs satisfying $r - 2\lambda_1 + \lambda_2 = 0$ are that

$$(12.6.19) \qquad 2t\beta^{t-1}(n-1)(n-2)^{n-1}\rho_0\rho_1{}^{\alpha_1} \sim 1$$

and, if (12.6.19) is satisfied, then

$$
\begin{aligned}
(12.6.20) \quad & (-1, \beta)_p^{t(t-1)/2}(t, \beta)_p^{t-2}(\rho_0, v\rho_1{}^{\alpha_1})_p(v, \rho_1)_p(-1, \rho_1)_p^{\alpha_1(\alpha_1+3)/2} \\
& \times (\rho_0\rho_1^{\alpha_1+1}, 2(n-1)(n-2)^{n-1})_p \\
& \times (n(n-2)^{n-1}, 2(n-1)(n-2)^{n-1})_p \\
& \times (-1, n-2)_p^{n(n-1)/2}(-1, n-1)_p^n \\
& \times (2, n(n-2)^{n-1})_p(-1, n)_p(n, n-2)_p(-1, v)_p \\
& \times (v, n(n-2)^{n-1})_p = 1
\end{aligned}
$$

for all primes p, where $\rho_1 = r + (n-4)\lambda_1 - (n-3)\lambda_2$ and $\alpha_1 = n-1$.

Theorem 12.6.5. The necessary conditions for the existence of affine α-resolvable L_i designs satisfying $r + (s-i)\lambda_1 - (s-i+1)\lambda_2 = 0$, are that

$$(12.6.21) \qquad t\beta^{t-1}s^{is}\rho_0\rho_1{}^{\alpha_1} \sim 1$$

and, if (12.6.21) is satisfied, then

$$
\begin{aligned}
(12.6.22) \quad & (-1, \beta)_p^{t(t-1)/2}(t, \beta)_p^{t-2}(\rho_0, v\rho_1{}^{\alpha_1})_p(v, \rho_1)_p(\rho_0\rho_1^{\alpha_1+1}, s^{is})_p \\
& \times (-1, \rho_1)_p^{\alpha_1(\alpha_1+3)/2}(-1, s)_p^{is(s+i)/2} = 1
\end{aligned}
$$

for all primes p, where $\rho_1 = r - i\lambda_1 + (i-1)\lambda_2$ and $\alpha_1 = (s-1)(s-i+1)$.

Theorem 12.6.6. The necessary conditions for the existence of affine α-resolvable L_i designs satisfying $r - i\lambda_1 + (i-1)\lambda_2 = 0$ are that

$$(12.6.23) \qquad t\beta^{t-1}s^{is}\rho_0\rho_1{}^{\alpha_1} \sim 1$$

and, if (12.6.23) is satisfied, then

$$
\begin{aligned}
(12.6.24) \quad & (-1, \beta)_p^{t(t-1)/2}(t, \beta)_p^{t-2}(\rho_0, v\rho_1{}^{\alpha_1})_p(v, \rho_1)_p \\
& \times (\rho_0\rho_1^{\alpha_1+1}, s^{is})_p(-1, \rho_1)_p^{\alpha_1(\alpha_1+3)/2}(-1, s)_p^{is(s+i+2)/2} = 1
\end{aligned}
$$

for all primes p, where $\rho_1 = r + (s-i)\lambda_1 - (s-i+1)\lambda_2$ and $\alpha_1 = i(s-1)$.

These results were first obtained by Shrikhande and Raghavarao (1963). For an alternative simple proof based on the ideas of Chapter 10 we refer to Raghavarao (1966).

12.7. NONEXISTENCE OF BIB DESIGNS IN THE USEFUL RANGE

We close this chapter by listing all the known nonexisting BIB designs in the range $v, b \leq 100, r, k \leq 15$ and indicating the reasons for their nonexistence in Table 12.7.1.

Table 12.7.1. *Nonexisting BIB designs satisfying*
$v, b \leq 100, r, k \leq 15$

Series	Parameters					Classification†
	v	b	r	k	λ	
1	15	21	7	5	2	c
2	22	22	7	7	2	a
3	21	28	8	6	2	d
4	29	29	8	8	2	b
5	34	34	12	12	4	a
6	36	45	10	8	2	c
7	43	43	7	7	1	b
8	43	43	15	15	3	b
9	46	46	10	10	2	a
10	53	53	13	13	3	b
11	55	66	12	10	2	d
12	67	67	12	12	2	b
13	78	91	14	12	2	c
14	92	92	14	14	2	a

† These designs are classified as follows:

a. Nonexistent, violating (12.2.1).

b. Nonexistent, violating (12.2.2).

c. Its dual is a triangular design satisfying the requirements of Theorem 12.5.4; it is nonexistent, violating (12.5.29).

d. Its dual is a triangular design satisfying the requirements of Theorem 12.5.4; it is nonexistent, violating (12.5.30).

We remark that the lists of parameters in Tables 5.9.1 and 12.7.1 do not exhaust all the possible parameter combinations in the useful range $v, b \leq 100$ and $r, k \leq 15$. There are certain other parameter combinations whose solutions are not known and which may presumably be nonexisting. However, the methods developed in this chapter do not rule out their existence.

REFERENCES

1. Bhagwandas and Shrikhande, S. S. (1968). Seidel equivalence of strongly regular graphs. *Sankhyā*, **A30**, 359–368.
2. Bose, R. C., and Connor, W. S. (1952). Combinatorial properties of group-divisible incomplete block designs. *Ann. Math. Stat.*, **23**, 367–383.
3. Bruck, R. H., and Ryser, H. J. (1949). The non-existence of certain finite projective planes. *Can. J. Math.*, **1**, 88–93.
4. Chowla, S., and Ryser, H. J. (1950). Combinatorial problems. *Can. J. Math.*, **2**, 93–99.
5. Corsten, L. C. A. (1960). Proper spaces related to triangular partially balanced incomplete block designs. *Ann. Math. Stat.*, **31**, 498–501.
6. Ogawa, J. (1959). A necessary condition for existence of regular symmetrical experimental designs of triangular type with partially balanced incomplete blocks. *Ann. Math. Stat.*, **30**, 1063–1071.
7. Raghavarao, D. (1966). Duals of partially balanced incomplete block designs and some nonexistence theorems. *Ann. Math. Stat.*, **37**, 1048–1052.
8. Shrikhande, S. S. (1950). The impossibility of certain symmetrical balanced incomplete block designs. *Ann. Math. Stat.*, **21**, 106–111.
9. Shrikhande, S. S. (1953). The non-existence of certain affine resolvable balanced incomplete block designs. *Can. J. Math.*, **5**, 413–420.
10. Shrikhande, S. S., and Jain, N. C. (1962). The non-existence of some partially balanced incomplete block designs with latin square type association schemes. *Sankhyā*, **A24**, 259–268.
11. Shrikhande, S. S., and Raghavarao, D. (1963). Affine α-resolvable incomplete block designs. *Contributions to Statistics*. Presented to Prof. P. C. Mahalanobis on his 70th birthday. Pergamon Press, New York, pp. 471–480.
12. Shrikhande, S. S., Raghavarao, D., and Tharthare, S. K. (1963). Non-existence of some unsymmetrical partially balanced incomplete block designs. *Can. J. Math.*, **15**, 686–701.

BIBLIOGRAPHY

Adhikary, B. Some types of *m*-associate PBIB association schemes. *Calcutta Stat. Assn. Bull.*, **15**, 47–74 (1966).

Connor, W. S., Jr. On the structure of balanced incomplete block designs. *Ann. Math. Stat.*, **23**, 57–71 (1952).

Hinkelmann, K. Extended group divisible partially balanced incomplete block designs. *Ann. Math. Stat.*, **35**, 681–695 (1964).

Hussain, Q. M. Impossibility of the symmetrical incomplete block design with $\lambda = 2$, $k = 7$. *Sankhyā*, **7**, 317–322 (1946).

Hussain, Q. M. Alternative proof of the impossibility of the symmetrical design with $\lambda = 2$, $k = 7$. *Sankhyā*, **8**, 384 (1948).

Kusumoto, K. A necessary condition for the existence of regular, symmetrical PBIB designs of T_3 type. *Ann. Inst. Stat. Math.*, **17**, 149–165 (1965).

Kusomoto, K. Association schemes of new types and necessary conditions for existence of regular and symmetrical PBIB designs with those association schemes. *Ann. Inst. Stat. Math.*, **19**, 73–100 (1967).

Ogawa, J. On a unified method of deriving necessary conditions for existence of symmetrically partially balanced incomplete block designs of certain types. *Bull. Intern. Inst. Stat.*, **38**, 43–57 (1960).

Ogawa, J. On the non-existence of certain block designs. *Sûgaku*, **17**, 65–72 (1965).

Raghavarao, D. A generalization of group divisible designs. *Ann. Math. Stat.*, **31**, 756–771 (1960).

Raghavarao, D. Some results on tactical configurations and nonexistence of difference set solutions for certain symmetrical PBIB designs. *Ann. Inst. Stat. Math.*, **22**, 501–506 (1970).

Raghavarao, D., and Chandrasekhararao, K. Cubic designs. *Ann. Math. Stat.*, **35**, 389–397 (1964).

Saraf, W. S. On the structure and combinatorial properties of certain semiregular group divisible designs. *Sankhyā*, **A23**, 287–296 (1961).

Schutzenberger, M. P. A non-existence theorem for an infinite family of symmetrical block designs. *Ann. Eugenics*, **14**, 286–287 (1949).

Seiden, E. On necessary conditions for the existence of some symmetrical and unsymmetrical triangular PBIB designs and BIB designs. *Ann. Math. Stat.*, **34**, 348–351 (1963).

Shrikhande, S. S. On the non-existence of certain difference sets for incomplete group designs. *Sankhyā*, **11**, 183–184 (1951).

Shrikhande, S. S. On the non-existence of affine resolvable balanced incomplete block designs. *Sankhyā*, **11**, 185–186 (1951).

Shrikhande, S. S., and Raghavarao, D. A note on the non-existence of symmetric balanced incomplete block designs. *Sankhyā*, **A26**, 91–92 (1964).

Singh, N. K., and Shukla, G. C. Non-existence of some PBIBD. *J. Indian Stat. Assn.*, **1**, 71–78 (1963).

Singh, N. K., and Singh, K. N. The non-existence of some partially balanced incomplete block designs with three associate classes. *Sankhyā*, **26**, 239–250 (1964).

Tharthare, S. K. Right angular designs. *Ann. Math. Stat.*, **34**, 1057–1067 (1963).

Tharthare, S. K. Generalized right angular designs. *Ann. Math. Stat.*, **36**, 1535–1553 (1965).

Vartak, M. N. The nonexistence of certain PBIB designs. *Ann. Math. Stat.*, **30**, 1051–1062 (1959).

Yamamoto, K. On an orthogonal basis of the eigenspace associated with partially balanced incomplete block designs of a latin square type association scheme. *Mem. Fac. Sci. Kyushya Univ.*, **19**, 99–104 (1965).

CHAPTER 13

Confounding in Symmetrical Factorial Experiments

13.1. INTRODUCTION

When different factors influence a character under study, it is always desirable to test different combinations of the factors at various levels. Such experiments are called factorial experiments and are widely used because of the wider inductive basis of the conclusions drawn from them and of estimating the interactions between different factors.

Consider a factorial experiment with n factors, the ith factor being experimented with at s_i levels ($i = 1, 2, \cdots, n$). Clearly there are $s_1 s_2 \cdots s_n$ different treatment combinations carrying $s_1 s_2 \cdots s_n - 1$ degrees of freedom between them.† In factorial experiments we can partition the treatment sum of squares into orthogonal contrasts possessing single degrees of freedom corresponding to the main effects and interactions of all the n factors.

A contrast may be said to represent the main effect of the ith factor if the coefficients in the linear function constituting the contrast are independent of the levels of factors other than the ith factor's. We can easily see that there are $s_i - 1$ contrasts representing the ith factor's main effect.

A contrast may be said to represent a two-factor (or first-order) interaction of the ith and jth factors if (a) the coefficients in the linear function constituting the contrast are independent of the levels of the factors other than the ith and jth factors; (b) the contrast is orthogonal to any contrast representing the main effect of the ith and jth factors. Clearly there are $(s_i - 1)(s_j - 1)$ contrasts representing the two-factor interaction of the ith and jth factors.

By induction, after defining when a contrast belongs to the $(k - 1)$th order interaction of any k factors, for $k = 2, 3, \cdots, r - 1$ we say that a contrast

---

† In this and the next three chapters we use the terms "treatment combinations" and "blocks" for symbols and sets, respectively, as most of the results of these chapters are mainly related to the combinatorial problems of the analysis of experimental designs.

belongs to the $(r - 1)$th order interaction of r factors, say i_1th, i_2th, $\cdots$, i_rth if (a) the coefficients in the linear function constituting the contrast are independent of the levels of the factors other than the i_1th, i_2th, $\cdots$, i_rth factors; (b) the contrast is orthogonal to all contrasts belonging to all possible main effects and interactions of the i_1th, i_2th, $\cdots$, i_rth factors. We can see that there are $(s_{i_1} - 1)(s_{i_2} - 1) \cdots (s_{i_r} - 1)$ contrasts representing the r-factor interaction of the i_1th, i_2th, $\cdots$, i_rth factors.

For another method of analyzing factorial experiments—a method based on a special calculus—we refer to Kurkjian and Zelen (1962, 1963).

The factorial experiment is said to be symmetrical if $s_1 = s_2 = \cdots = s_n = s$ and is called an s^n factorial experiment. Otherwise it is said to be an asymmetrical experiment and is called an $s_1 \times s_2 \times \cdots \times s_n$ experiment.

Let us consider the simplest 2^n factorial experiment and denote the first level of a factor by unity and the second level of the factor by a lowercase latin letter. We use capital latin letters for the main effects and interactions. With n factors a, b, c, d, e, etc., the effects and interactions may be represented by

$$(13.1.1) \qquad X = r^{-1} 2^{-(n-1)}(a \pm 1)(b \pm 1)(c \pm 1) \cdots,$$

where r is the number of replications of each treatment combination and the sign in the parentheses is positive if the corresponding capital letter is not contained in X and negative if it is contained in X, the multiplication being completed in the algebraic way and yields substituted for the symbols. Conversely the effect of treatment combination $a_i b_j c_k \cdots$, where absence is denoted by the subscript taking the value 0 and presence by the subscript taking the value 1, is given by

$$2a_i b_j c_k \cdots = 2\mu \pm A \pm B \pm AB \pm C \pm AC \text{ etc.,}$$

where μ is the general mean; the sign of A is -1 if $i = 0$ and $+1$ if $i = 1$, and the sign of B is -1 if $j = 0$ and $+1$ if $j = 1$, etc., and the sign of a term involving several letters is the product of the signs of the individual letters. The variance of a main effect or interaction can be seen with the usual assumptions to be $\sigma^2 / r2^{n-2}$, and the sum of squares with one degree of freedom corresponding to X is $X^2 r2^{n-2}$.

In the case of an s^n factorial experiment, where $s(\geq 2)$ is a prime or a prime power, let us identify the s^n treatment combinations with the points $(x_1, x_2, \cdots, x_n)$ of $EG(n, s)$.† Any $(n - 1)$-flat of $EG(n, s)$ has an equation of the form

$$(13.1.2) \qquad a_0 + a_1 x_1 + a_2 x_2 + \cdots + a_n x_n = 0 \qquad a_i \in GF(s)$$

---

† Though the geometrical characterization described in this paragraph can be used for $s = 2$, it is more simple to use the notation developed earlier in the case of $s = 2$.

and contains s^{n-1} points. By keeping $a_1, a_2, \cdots, a_n$ constant and varying a_0 over the elements of GF(s), we generate s parallel $(n-1)$-flats that have no common points and constitute a pencil, called $P(a_1, a_2, \cdots, a_n)$, of $(n-1)$-flats. This pencil divides the s^n treatment combinations into s sets, which give rise to $s-1$ independent comparisons. Hence the pencil $P(a_1, a_2, \cdots, a_n)$ is said to carry $s-1$ degrees of freedom. We observe that pencils $P(a_1, a_2, \cdots, a_n)$ and $P(\rho a_1, \rho a_2, \cdots, \rho a_n)$, where ρ is a nonzero element of GF(s), represent the same comparison and hence are identical. Without loss of generality, we assume the first nonzero coordinate in $P(a_1, a_2, \cdots, a_n)$ to be unity.

By using the definition of main effects and interactions, we can see that the pencil $P(a_1, a_2, \cdots, a_n)$ represents the interaction of the i_1th, i_2th, $\cdots, i_r$th factors if and only if $a_{i_1}, a_{i_2}, \cdots, a_{i_r}$ are nonzero and the other coordinates in the pencil $P(a_1, a_2, \cdots, a_n)$ are zero.

When a large number of treatment combinations are to be tested, it is undesirable to use complete blocks because in that case the plots of each block may not be homogeneous in their fertility. In such a case the block size will be reduced by the concept of *confounding*. We form the blocks of each replicate in such a way that the block comparisons are identical with certain interaction comparisons. Such interactions are said to be confounded in that particular replication. If the same interaction is confounded in all replications, it is said to be totally confounded, and if there is at least one replication in which it is unconfounded, it is said to be partially confounded. No information will be available on totally confounded interactions, and partial information will be available on partially confounded interactions. If an interaction is confounded in r_1 replications and is unconfounded in r_2 replications, the loss of information on that interaction can be seen to be $r_1/(r_1 + r_2)$, where information on an effect or interaction is the reciprocal of the variance of its estimate.

In this chapter we study the confounding of symmetrical factorial experiments and related results, postponing the problems related to confounding in asymmetrical factorial experiments to the next chapter. In an s^n factorial experiment of constant block size the block size must be necessarily a power of s, and we say an (s^n, s^k) experiment for an s^n factorial experiment, where each replicate contains s^k blocks each of size s^{n-k}.

13.2. CONFOUNDING IN 2^n EXPERIMENTS

To construct a $(2^n, 2^k)$ class of design we first choose k independent interactions $X_1, X_2, \cdots, X_k$ such that none of them is a product of any others. Then we form the set of treatment combinations having an even number of symbols in common with each interaction $X_1, X_2, \cdots, X_k$. Such a set can be easily shown to form a group, called *intrablock subgroup*, under the binary operation of multiplication, subject to the restriction that the square of any symbol will be replaced by 1. The intrablock subgroup is also known in the literature as the *key block* or the *initial block*. Let $(x_1, x_2, \cdots, x_{2^{n-k}})$ be the intrablock subgroup that will be taken as block 1. If y is a treatment combination not belonging to block 1, then block 2 will be obtained as $(x_1 y, x_2 y, \cdots, x_{2^{n-k}} y)$, where the square of any symbol is replaced by 1. If z is a treatment combination not occurring in blocks 1 and 2, then the third block will be $(x_1 z, x_2 z, \cdots, x_{2^{n-k}} z)$, where again the squares of the symbols are replaced by 1. Continuing similarly, we can generate the 2^k blocks of the replication.† In the design thus obtained not only $X_1, X_2, \cdots, X_k$ will be confounded but the other $2^k - 1 - k$ effects or interactions obtained by multiplying $X_1, X_2, \cdots, X_k$ in all possible ways, subject to the condition of replacing the square of any letter by 1, will also be confounded. The confounded interactions can also be seen to form the nonidentity elements of a group with similar generators as the intrablock subgroup.

For example, let us construct a $(2^6, 2^3)$ experiment in the factors $a, b, c, d, e,$ and f. The three interactions ACE, BDE, and BCF are independent since none is a product of the others. When we confound them, automatically the interactions $ACEBDE = ABCD$, $ACEBCF = ABEF$, $BDEBCF = CDEF$, and $ACEBDEBCF = ADF$ will also be confounded. The intrablock subgroup in this case is

(13.2.1)   (1, acf, $abef$, bce, bdf, $abcd$, ade, $cdef$).

Treating (13.2.1) as block 1, the remaining seven blocks are obtained by multiplying it by $a, b, c, d, e, f,$ and ab, respectively, while replacing the square of any symbol by 1. The plan is as follows:

Block	Constituents of the Block
1	(1, acf, $abef$, bce, bdf, $abcd$, ade, $cdef$)
2	(a, cf, bef, $abce$, $abdf$, bcd, de, $acdef$)
3	(b, $abcf$, aef, ce, df, acd, $abde$, $bcdef$)
4	(c, af, $abcef$, be, $bcdf$, abd, $acde$, def)
5	(d, $acdf$, $abdef$, $bcde$, bf, abc, ae, cef)
6	(e, $acef$, abf, bc, $bdef$, $abcde$, ad, cdf)
7	(f, ac, abe, $bcef$, bd, $abcdf$, $adef$, cde)
8	(ab, bcf, ef, ace, adf, cd, bde, $abcdef$)

† If the 2^n treatment combinations are taken to form a group, G, then the 2^k blocks are the cosets of the intrablock subgroup in G.

Much work on the confounding of 2^n designs has been done by Fisher (1942, 1945) and Yates (1935, 1937). For a detailed account of the results of this section we refer the reader to Kempthorne (1952). The plans of 2^n confounded designs are given by Cochran and Cox (1957) and Kitagawa and Mitome (1953).

13.3. CONFOUNDING IN s^n EXPERIMENTS THROUGH PENCILS

To construct an (s^n, s^k) experiment we first choose k independent pencils $P_i = P(a_{i1}, a_{i2}, \cdots, a_{in})$ $(i = 1, 2, \cdots, k)$. Let $\sum_{i_1}, \sum_{i_2}, \cdots, \sum_{i_k}$ be flats belonging to the independent pencils $P_1, P_2, \cdots, P_k$, respectively. These flats pass through a common $(n - k)$-flat $\sum_{i_1, i_2, \cdots, i_k}$. We denote the treatment combinations on the $(n - k)$-flat by $(\sum_{i_1, i_2, \cdots, i_k})$. The totality of s^n treatment combinations will be divided in this way into s^k sets of the type $(\sum_{i_1, i_2, \cdots, i_k})$. We form the $(i_1 i_2 \cdots i_k)$th block with the treatment combinations $(\sum_{i_1, i_2, \cdots, i_k})$ for $i_1, i_2, \cdots, i_k = 0, 1, \cdots, s - 1$. Then the $s^k - 1$ degrees of freedom carried by the $(s^k - 1)/(s - 1)$ pencils

$$(13.3.1) \qquad P\left(\sum_{i=1}^{k} \lambda_i a_{i1}, \sum_{i=1}^{k} \lambda_i a_{i2}, \cdots, \sum_{i=1}^{k} \lambda_i a_{in} \right) \qquad \lambda_i \in \mathrm{GF}(s)$$

can also be verified to be confounded with the s^k blocks. The block where the treatment combination $(0, 0, \cdots, 0)$ occurs can be shown to form a group, called the intrablock subgroup, as in the case $s = 2$ of the last section, and the whole plan can be generated from the intrablock subgroup. If $(x_1, x_2, \cdots, x_{s^{n-k}})$ is the intrablock subgroup and y is a treatment combination not belonging to the intrablock subgroup, then the second block can be taken as $(x_1 + y, x_2 + y, \cdots, x_{s^{n-k}} + y)$, where the addition of treatment combinations will be vector addition where each component is in $\mathrm{GF}(s)$. If z is a treatment combination not in the first and second blocks, the third block will be taken as $(x_1 + z, x_2 + z, \cdots, x_{s^{n-k}} + z)$. Continuing in this way, the whole replication of an (s^n, s^k) plan will be constructed.

As an example, let us construct a $(3^4, 3^3)$ experiment confounding the degrees of freedom carried by the pencils $P(1, 0, 0, 0)$, $P(0, 1, 1, 0)$, and $P(0, 1, 0, 1)$. Then automatically the degrees of freedom carried by the pencils $P(1, 1, 1, 0)$, $P(1, 2, 2, 0)$, $P(1, 1, 0, 1)$, $P(1, 2, 0, 2)$, $P(0, 0, 1, 2)$, $P(0, 1, 2, 2)$, $P(1, 2, 1, 1)$, $P(1, 0, 1, 2)$, $P(1, 0, 2, 1)$, and $P(1, 1, 2, 2)$ will also be confounded with the blocks. It is remarked that, though the confounding of main effects is undesirable in experiments, in constructing a $(3^4, 3^3)$ experiment the main effects cannot be kept free from confounding. To save space

Table 13.3.1

Block	x_1	$x_2 + x_3$	$x_2 + x_4$	Constituents of the Block
000	0	0	0	(0000, 0122, 0211)
001	0	0	1	(0120, 0212, 0001)
002	0	0	2	(0121, 0210, 0002)
010	0	1	0	(0010, 0221, 0102)
011	0	1	1	(0011, 0222, 0100)
012	0	1	2	(0012, 0220, 0101)
020	0	2	0	(0020, 0201, 0112)
021	0	2	1	(0021, 0202, 0110)
022	0	2	2	(0022, 0200, 0111)
100	1	0	0	(1000, 1122, 1211)
101	1	0	1	(1120, 1212, 1001)
102	1	0	2	(1121, 1210, 1002)
110	1	1	0	(1010, 1221, 1102)
111	1	1	1	(1011, 1222, 1100)
112	1	1	2	(1012, 1220, 1101)
120	1	2	0	(1020, 1201, 1112)
121	1	2	1	(1021, 1202, 1110)
122	1	2	2	(1022, 1200, 1111)
200	2	0	0	(2000, 2122, 2211)
201	2	0	1	(2120, 2212, 2001)
202	2	0	2	(2121, 2210, 2002)
210	2	1	0	(2010, 2221, 2102)
211	2	1	1	(2011, 2222, 2100)
212	2	1	2	(2012, 2220, 2101)
220	2	2	0	(2020, 2201, 2112)
221	2	2	1	(2021, 2202, 2110)
222	2	2	2	(2022, 2200, 2111)

we present the experiment in Table 13.3.1, in which columns 2, 3, and 4 list the right members of the three equations of each 1-flat.

The ideas introduced so far in this section were first presented by Bose and Kishen (1940) and were later improved by Bose (1947). Constructed designs are given by Cochran and Cox (1957) and Kitagawa and Mitome (1953).

An important aspect of confounded factorial experiments is to retain the main effects and lower order interactions unconfounded. The intrablock subgroup of an (s^n, s^k) experiment in which t-factor or lower factor interactions and main effects are unconfounded was shown by Rao (1946) to be a hypercube $[n, s, n - k, t]$ of strength t or equivalently an orthogonal array (s^{n-k}, n, s, t) of index s^{n-k-t}. [Also see Plackett and Burman (1946) for further results in this connection.]

13.4. PRINCIPLE OF GENERALIZED INTERACTION

Consider n independent pencils $P_i = P(a_{i1}, a_{i2}, \cdots, a_{in})$ ($i = 1, 2, \cdots, n$), and let the degrees of freedom carried by $P_1, P_2, \cdots, P_n$ be regarded as belonging to the main effects of certain conceptual factors $\phi_1, \phi_2, \cdots, \phi_n$. Then it is easy to verify that the treatment combination $(x_1, x_2, \cdots, x_n)$ transforms to $(y_1, y_2, \cdots, y_n)$, where

(13.4.1) $y_i = a_{i1} x_1 + a_{i2} x_2 + \cdots + a_{in} x_n$ $i = 1, 2, \cdots, n$,

and the pencil $P(\lambda_1, \lambda_2, \cdots, \lambda_n)$ transforms to $P'(\mu_1, \mu_2, \cdots, \mu_n)$, where

(13.4.2) $\lambda_j = a_{1j} \mu_1 + a_{2j} \mu_2 + \cdots + a_{nj} \mu_n$ $j = 1, 2, \cdots, n$.

For the set of treatment combinations corresponding to any particular flat of the pencil P_i, the general factor ϕ_i remains at a constant level. Hence the sets $(\sum_{i_1, i_2, \cdots, i_k})$ considered in the preceding section are just those that are obtained by associating a particular level of each of the general factors $\phi_1, \phi_2, \cdots, \phi_k$ with all possible combinations of the levels of the remaining factors $\phi_{k+1}, \phi_{k+2}, \cdots, \phi_n$. The degrees of freedom carried by the pencils (13.3.1) clearly belong to all possible generalized interactions of one or more of the pseudofactors.

A detailed illustration of the principle of generalized interaction was given by Bose (1947).

13.5. CONFOUNDING WITH THE HELP OF PSEUDOFACTORS

Consider a symmetrical factorial experiment s^n where $s = p^m$, p is a prime, and $m > 1$. Though this case was dealt with in Section 13.3 with the help of projective geometries and Galois fields, in this section we present a different approach, which will help us to construct new types of confounded design.

Let us consider a 4^2 experiment in factors nitrogen (n) and phosphorus (p). We are interested in forming two blocks of size 8 for each replicate. This design does not fit in the frame of confounded designs discussed in Section 13.3. However, we can construct this design with the help of certain pseudofactors. Let the four levels n_0, n_1, n_2, and n_3 of the factor n be identified with the four treatment combinations 1, a, b, and ab of a 2^2 factorial experiment in pseudofactors a and b, and let the four levels p_0, p_1, p_2, and p_3 of the factor p be identified with the four treatment combinations 1, c, d, and cd of another 2^2 factorial experiment in pseudofactors c and d. Then the 16 treatment combinations in the two factors n and p have a bijection with the 16 treatment combinations of the 2^4 experiment in pseudofactors a, b, c, and d. In pseudofactors we can confound the interaction $ABCD$ to get a design in two blocks of size 8. The required plan is as follows:

Block	Constituents of the Block
1	$(1, ab, ac, ad, bc, bd, cd, abcd)$
2	$(a, b, c, d, abc, abd, acd, bcd)$

Replacing these treatment combinations by the treatment combinations in the original factors n and p, we get the following:

Block	Constituents of the Block
1	$(n_0 p_0, n_3 p_0, n_1 p_1, n_2 p_1, n_1 p_2, n_2 p_2, n_0 p_3, n_3 p_3)$
2	$(n_1 p_0, n_2 p_0, n_0 p_1, n_0 p_2, n_3 p_1, n_3 p_2, n_1 p_3, n_2 p_3)$

In general the s^n experiment where $s = p^m$, p is a prime, and $m > 1$ can be considered as a p^{mn} experiment in mn pseudofactors. Hence confounding in such experiments can be made to accommodate the experiment in p^k blocks of p^{mn-k} plots.

13.6. ALTERNATIVE METHOD OF CONFOUNDING

Das (1964) provided a somewhat alternative method of constructing confounded symmetrical factorial designs—a method that ultimately leads to all the results of Bose (1947). It may, however, be noted that the method proposed by Das had been described in various ways by several other writers dating back to 1950 [cf. Review No. 8464, *Mathematical Reviews*, **32** (1966)]. In this section we follow the version given by Das (1964).

In this method the independent treatment combinations are first written and multiplied in all possible combinations with an added control treatment combination to generate the remaining treatment combinations of the key block. After the initial block is obtained, the remaining blocks can be constructed by the standard procedure of our Sections 13.2 and 13.3.

Let us construct an (s^n, s^k) experiment. Let the n factors be $F_1, F_2, \cdots, F_n$. Since the block size is s^{n-k}, $n - k$ independent treatment combinations generate all the s^{n-k} treatment combinations. We take $n - k$ basic factors $F_1, F_2, \cdots, F_{n-k}$ and write $n - k$ treatment combinations, where the ith treatment combination contains the ith factor F_i at the first level and all other factors at the 0th level. We then add new factors $F_{n-k+1}, F_{n-k+2}, \cdots, F_n$, where each added factor contains any of the s symbols of GF(s) in the $n - k$ independent treatment combinations. If the factor F_{n-k+j} contains a_{ij} in the ith treatment combination, we can easily see that the k independent confounded pencils are

(13.6.1)
$$P(a_{11}, a_{21}, \cdots, a_{n-k, 1}, \quad -1, \quad 0, \cdots, \quad 0);$$
$$P(a_{12}, a_{22}, \cdots, a_{n-k, 2}, \quad 0, \quad -1, \cdots, \quad 0);$$
$$\cdots$$
$$P(a_{1k}, a_{2k}, \cdots, a_{n-k, k}, \quad 0, \quad 0, \cdots, \quad -1).$$

We illustrate this method with two constructions. Consider a $(2^6, 2^2)$ experiment. Then there will be four basic factors A, B, C, D and two added factors E, F. Let the four independent treatment combinations be as follows:

Independent Treatment Combinations	Basic Factors				Added Factors	
	A	B	C	D	E	F
1	1	0	0	0	1	0
2	0	1	0	0	1	1
3	0	0	1	0	0	1
4	0	0	0	1	1	0

Since the factor E contains the element 1 in the 1, 2, and 4 treatment combinations, the interaction $ABDE$ will be confounded. Similarly, since the factor F contains the element 1 in the treatment combinations 2 and 3, the interaction BCF will be confounded. By the theory of generalized interaction, the interaction $ACDEF$ will also be confounded. The developed key block is

(13.6.2)
$$(1, \; ae, \; bef, \; cf, \; de, \; abf, \; acef, \; bcef, \; abc,$$
$$ad, \; bdf, \; abdef, \; cdef, \; acdf, \; bcdf, \; abcde).$$

As another illustration, let us construct a $(3^4, 3^2)$ experiment. Let F_1, F_2 be the basic factors, F_3, F_4 the added factors, and the independent treatment combinations be as follows:

Independent Treatment Combinations	Basic Factors		Added Factors	
	F_1	F_2	F_3	F_4
1	1	0	1	1
2	0	1	1	2

Since F_3 has entries 1, 1 in the treatment combinations 1 and 2, the pencil $P(1, 1, 2, 0)$ will be confounded in this plan. Since F_4 has entries 1, 2 in the treatment combinations 1 and 2, the pencil $P(1, 2, 0, 2)$ will also be confounded. By the theory of generalized interaction, the pencils $P(1, 0, 1, 1)$ and $P(0, 1, 1, 2)$ will automatically be confounded in the plan. The completed initial block is

(13.6.3)   (1011, 0112, 2022, 0221, 1120, 1202, 2101, 2210, 0000).

13.7. PACKING PROBLEM

We have seen in the design $(3^4, 3^3)$ of Section 13.3 that it is impossible to keep the main effects free from confounding. We can easily realize that we cannot have any arbitrary number of factors at s levels in an experiment for a constant block size preserving t-factor or lower order interactions: there is definitely a ceiling on this number. Let $m_t(r, s)$ denote the maximum number of factors, each at s levels, that can be accommodated in a confounded design in blocks of size s^r so that no t or lower factor interactions are confounded. Clearly the geometric interpretation of $m_t(r, s)$ is that it is the maximum number of points that can be chosen in $\mathrm{PG}(r - 1, s)$ such that no t of them are linearly dependent. The problem of finding an upper bound for $m_t(r, s)$ and forming the key block without confounding t or lower factor interactions is of considerable interest not only in experimental designs but also in information theory.

For the interrelationship between the theory of confounding as developed in this chapter and the theory of error-correcting codes due to Hamming (1950) and Slepian (1956), we refer to Bose (1960). In the notation of error-correcting codes, if s denotes the number of distinct symbols to be transmitted, t is the Hamming distance of the code, being the measure of the code's error-correcting capability, and r is the number of redundant parity check symbols included in each block of transmitted symbols, then $m_t(r, s)$ denotes the maximum length of the block in a linear code with given t, r, s. This problem may be called the packing problem. Its complete solution is yet unknown, but partial solutions have been discussed by Barlotti (1957), Bose (1947, 1960), Bose and Srivastava (1964), Fisher (1942, 1945), Peterson (1961), Qvist (1952), Seiden (1950), and Tallini (1956). We give some of their results here.

Theorem 13.7.1. [Fisher (1942, 1945)]

(13.7.1) $$m_2(r, s) = \frac{s^r - 1}{s - 1}.$$

Proof. We prove this result by the method of construction described in Section 13.6. Any main effect will be confounded if the column corresponding to any added factor has zeros in each of the r initial treatment combinations.

Similarly any two-factor interaction will be confounded if any two columns of the basic or additional factors are identical. Thus the total number of factors equals the number of points in $PG(r-1, s)$ geometry. Since there are $Q_{r-1} = (s^r - 1)/(s - 1)$ points in $PG(r-1, s)$, the required equality (13.7.1) follows.

Theorem 13.7.2. [Bose, (1947)]

$$(13.7.2) \qquad m_3(r, 2) = 2^{r-1}.$$

Proof. We can easily see that the three-factor interaction will be unconfounded if the independent treatment combinations of the key block with basic and added factors are such that no column is the sum of any other two columns (Cf. Section 13.6). It is clear that the columns with an odd number of ones without replications will be the independent treatment combinations of the key block without confounding one-, two-, or three-factor interactions. Since we can form

$$(13.7.3) \qquad \binom{r}{1} + \binom{r}{3} + \cdots = 2^{r-1},$$

columns of dimension r with an odd number of ones, we have (13.7.2).

Theorem 13.7.3. [Bose (1947)]

$$(13.7.4) \qquad \begin{aligned} m_3(3, s) &= s + 1 && \text{when } s \text{ is odd}; \\ &= s + 2 && \text{when } s \text{ is even}. \end{aligned}$$

Proof. We can easily observe that the key block of the design with s^3 treatment combinations without confounding one-, two-, or three-factor interactions is an orthogonal array $(s^3, k, s, 3)$ of index unity where the rows correspond to factors and the columns to treatment combinations. Construction of the orthogonal arrays of Section 2.4 translated in the present language gives (13.7.4).

In the following theorems we give some other bounds for $m_t(r, s)$, omitting the proofs:

Theorem 13.7.4. [Bose (1947), Qvist (1952)]

$$(13.7.5) \qquad m_3(4, s) = s^2 + 1 \qquad s > 2.$$

The result of this theorem was derived by Bose for the case when s is odd, and Qvist proved it for the case when s is even. Seiden (1950) obtained a particular case of the result when $s = 4$.

Theorem 13.7.5. [Tallini (1956)]

$$(13.7.6) \qquad m_3(r, s) < s^{r-2} + 1 \qquad s > 2, r \geqq 4.$$

The bounds of Tallini hold for even and odd s. Barlotti (1957) improved some of Tallini's bounds. He established the following theorem:

Theorem 13.7.6. [Barlotti (1957)]

$$(13.7.7) \qquad m_3(r, s) \leq s^{r-2}(s - 5) \sum_{i=1}^{r-5} s^i + 1 \qquad r \geq 5, s \geq 7 \text{ and odd};$$

$$(13.7.8) \qquad m_3(5, 5) \leq 124;$$

$$(13.7.9) \qquad m_3(r, 5) \leq 5^{r-2} - 10 \sum_{i=0}^{r-6} (5^i - 1) \qquad r \geq 6;$$

$$(13.7.10) \qquad m_3(5, s) \leq s^3 \qquad s \text{ even};$$

and

$$(13.7.11) \qquad m_3(r, s) \leq s^{r-2} - s \sum_{i=0}^{r-6} s^i \qquad r \geq 6, s \text{ even}.$$

Bose and Srivastava (1964) improved the bounds of Tallini and Barlotti for the case when $s = 3$ or when s is even. They proved the following:

Theorem 13.7.7. [Bose and Srivastava (1964)]
When $s > 2$, $r \geq 4$, $m_3(r, s)$ cannot exceed the positive root of the equation

$$(13.7.12) \quad x^2(s^2 - s - 1) - x[(s^2 - 2s - 1) + Q_{r-1}(s - 2)] - 2Q_{r-1} = 0,$$

where $Q_{r-1} = (s^r - 1)/(s - 1)$.

13.8. BALANCING IN FACTORIAL EXPERIMENTS

Bose (1947) introduced the concept of balancing in symmetrical factorial experiments. He defined the following:

Definition 13.8.1. In a partially confounded symmetrical factorial experiment, if each of the $(s - 1)^{k-1}$ pencils of $(s - 1)$ degrees of freedom carried by the $(k - 1)$th order interaction between the factors $F_{i_1}, F_{i_2}, \cdots, F_{i_k}$ is confounded in r_1 replications and remains unconfounded in r_2 replications, then we say that the interaction $F_{i_1} F_{i_2} \cdots F_{i_k}$ has been balanced.

We note that if the interaction $F_{i_1} F_{i_2} \cdots F_{i_k}$ is balanced, there is a uniform loss of information equal to $r_1/(r_1 + r_2)$ on every degree of freedom belonging to this interaction.

Bose (1947) further defined the following:

Definition 13.8.2. Of the $\binom{n}{k}$, $(k - 1)$th order interactions carrying in all $\binom{n}{k}(s - 1)^{k-1}$ pencils of $s - 1$ degrees of freedom, if each of the pencils is confounded in r_1 replications and unconfounded in r_2 replications, we say that the $(k - 1)$th order interaction is completely balanced.

Bose (1947) established the following theorem:

Theorem 13.8.1. For an (s^n, s^{n-1}) design we can find $(s - 1)^{n-1}$ replications in which the main effects remain unconfounded, and a complete balance is achieved over interactions of all orders, from the first to the $(n - 1)$th. The loss of information on the $(k - 1)$th order interaction is given by

(13.8.1) $$\frac{(s - 1)^{k-1} - (-1)^{k-1}}{s(s - 1)^{k-1}}.$$

Proof. From the relation between Euclidean and projective geometries, any $(n - 1)$-flat

(13.8.2) $$\alpha_0 + a_1 x_1 + a_2 x_2 + \cdots a_n x_n = 0$$

of the pencil $P(a_1, a_2, \cdots, a_n)$ can be embedded in the $(n - 1)$-flat

(13.8.3) $$\alpha_0 x_0 + a_1 x_1 + a_2 x_2 + \cdots + a_n x_n = 0$$

of PG(n, s). In fact the points of (13.8.2) are the same as the finite points of (13.8.3), but the latter contains points at infinity lying on the $(n - 2)$-flat

(13.8.4) $$x_0 = 0, a_1 x_1 + a_2 x_2 + \cdots + a_n x_n = 0.$$

We observe that each of the $(n - 1)$-flats of the pencil $P(a_1, a_2, \cdots, a_n)$, when embedded in PG(n, s), pass through (13.8.4). This flat may be termed the vertex of the parallel pencil $P(a_1, a_2, \cdots, a_n)$.

The idea of the parallel pencil may be extended to what may be called a parallel bundle. If S is any $(n - k - 1)$-flat at infinity, there are exactly $(s^{k+1} - 1)/(s - 1)$, $(n - k)$-flats of PG(n, s) that pass through S. Of these, $(s^k - 1)/(s - 1)$ lie wholly at infinity, whereas the remaining s^k have also finite points. Thus there are exactly s^k, $(n - k)$-flats of EG(n, s) that, when extended to infinity, pass through the $(n - k - 1)$-flat S. They may be said to form a parallel bundle of the kth order interaction with vertex S. A parallel pencil is only a parallel bundle of the first order.

The nature of confounding done by the parallel bundle of kth order in forming the design (s^n, s^k) depends on the nature of its vertex.

A fundamental simplex is one formed from the points $X_1, X_2, \cdots, X_n$ at infinity of PG(n, s), where the coordinates of X_i are all zero except $x_i = 1$ $(i = 1, 2, \cdots, n)$. The $(k - 1)$th dimensional flats formed by any k of the n points $X_1, X_2, \cdots, X_n$ $(k < n)$ may be called the $(k - 1)$th cell of the fundamental simplex. We can easily verify that the pencil $P(a_1, a_2, \cdots, a_n)$ belongs to the k-factor interaction of the factors $F_{i_1}, F_{i_2}, \cdots, F_{i_k}$ if and only if the vertex of the pencil passes through all the vertices of the fundamental simplex other than $X_{i_1}, X_{i_2}, \cdots, X_{i_k}$ but does not pass through any of these.

In forming an (s^n, s^{n-1}) design the vertex K of the parallel bundle will be a point of $PG(n, s)$, and no main effect will be confounded if the vertex K has the coordinates $(0, \xi_1, \xi_2, \cdots, \xi_n)$ $(\xi_i \neq 0; i = 1, 2, \cdots, n)$, and we call such points clear points. If an (s^n, s^{n-1}) design is formed with the vertex $(0, \xi_1, \xi_2, \cdots, \xi_n)$, the pencils $P(a_1, a_2, \cdots, a_n)$, satisfying

$$(13.8.5) \qquad a_1 \xi_1 + a_2 \xi_2 + \cdots + a_n \xi_n = 0,$$

will be confounded in that replicate. Let us now form $(s-1)^{n-1}$ replications corresponding to different clear points as vertices.

The pencil $P(a_1, a_2, \cdots, a_k, 0, 0, \cdots, 0)$ $(a_i \neq 0; i = 1, 2, \cdots, k)$ carrying interaction degrees of freedom of the factors $F_1, F_2, \cdots, F_k$ will be confounded in the replications whose vertices $(0, \xi_1, \xi_2, \cdots, \xi_n)$ satisfy

$$(13.8.6) \qquad a_1 \xi_1 + a_2 \xi_2 + \cdots + a_k \xi_k = 0.$$

We can enumerate

$$(13.8.7) \qquad q = (s-1) \frac{(s-1)^{k-1} - (-1)^{k-1}}{s}$$

solutions of (13.8.6) with none of the coordinates $\xi_1, \xi_2, \cdots, \xi_k$ taking the zero value. Hence the number of vertices $(0, \xi_1, \xi_2, \cdots, \xi_n)$ satisfying (13.8.6) for which no ξ vanishes is $q(s-1)^{n-k}$. Since to each clear point there correspond $(s-1)$ such sets, the number of replications confounding the pencil $P(a_1, a_2, \cdots, a_k, 0, 0, \cdots, 0)$ is

$$(13.8.8) \qquad r_k = (s-1)^{n-k} \frac{(s-1)^{k-1} - (-1)^{k-1}}{s}.$$

Similarly every set of $(s-1)$ degrees of freedom belonging to any k-factor interaction is confounded in r_k replications. Hence complete balance is achieved over the k-factor interaction with a loss of information (13.8.1).

The definition of balance provided by Bose fails when the block size is not a power of s, or when s is not a prime or a prime power, or in asymmetrical factorial designs. Hence there is a need of a broad definition of balance in factorial experiments. Similar to the definition of balancing in a single-factor design given in Chapter 4, Shah (1958) provided two definitions of complete balance as follows:

Definition 13.8.3. Complete balance is achieved over a set of n normalized orthogonal contrasts if the dispersion matrix of their estimates is a scalar multiple of the identity matrix of order n.

Definition 13.8.4. Complete balance is achieved over a set of contrasts if every linear combination of these contrasts giving a normalized contrast is estimated with the same variance.

Shah observed that both his definitions are equivalent.

13.9. BALANCED FACTORIAL EXPERIMENTS

Shah (1958) defined the following:

Definition 13.9.1. A factorial experiment will be called a balanced factorial experiment (BFE) if the following conditions are satisfied:

1. Each of the treatment combinations is replicated the same number of times, say r.
2. Each of the blocks is of the same size, say k.
3. Estimates of contrasts belonging to different interactions are uncorrelated with each other.
4. "Complete balance" is achieved over each of the interactions.

For the s^m treatment combinations $(x_1, x_2, \cdots, x_m)$ $(x_i = 0, 1, \cdots, s - 1;$ $i = 1, 2, \cdots, m)$, Shah defined an m-associate-class hypercubic association scheme (see Chapter 8) by calling two treatment combinations ith associates $(i = 1, 2, \cdots, m)$ if the levels of $(m - i)$ factors in their representations are identical but the levels of the remaining i factors are different in their representation. With a little algebraic manipulation Shah showed that the parameters of the association scheme are given by

$$(13.9.1) \qquad n_i = \binom{m}{i} (s - 1)^i \qquad i = 1, 2, \cdots, m$$

and

$$(13.9.2) \quad p_{jk}{}^i = \sum_u \binom{m - i}{m - u}\binom{i}{u - j}\binom{i + j - u}{u - k} (s - 1)^{u - i}(s - 2)^{i + j + k - 2u},$$

where $\sum_u$ is the summation over all the values of u such that

$$(13.9.3) \qquad \max(i, j, k) \le u \le \tfrac{1}{2}(i + j + k).$$

By using certain results on the analysis of incomplete block designs, Shah (1958) proved the following:

Theorem 13.9.1. If in an s^m balanced factorial experiment any normalized contrast belonging to any k-factor interaction $(k \le m)$ is estimated with the same variance σ^2/θ_k, then it is a PBIB design with m associate classes, with a hypercubic association scheme with the parameters given in (13.9.1) and (13.9.2), and with the λ-parameters given by

$$(13.9.4) \qquad -\frac{1}{k} \begin{bmatrix} \lambda_1 \\ \lambda_2 \\ \cdot \\ \lambda_m \end{bmatrix} = \begin{bmatrix} f_1{}^1 & f_1{}^2 & \cdots & f_1{}^m \\ f_2{}^1 & f_2{}^2 & \cdots & f_2{}^m \\ \cdot & \cdot & \cdots & \cdot \\ f_m{}^1 & f_m{}^2 & \cdots & f_m{}^m \end{bmatrix} \begin{bmatrix} \theta_1 \\ \theta_2 \\ \cdot \\ \theta_m \end{bmatrix},$$

where

$$(13.9.5) \qquad f_p{}^q = \sum_j{}^* \binom{p}{j} \binom{m-p}{q-j} (-1)^j (s-1)^{q-j} \qquad p, q = 1, 2, \cdots, m,$$

the summation $\sum_j^*$ being extended over all the values of j such that

$$(13.9.6) \qquad \max(0, p + q - m) \leqq j \leqq \min(p, q).$$

Other results that will hold for symmetrical as well as asymmetrical confounded factorial experiments have been given by Kshirsagar (1966) and Kurkjian and Zelen (1963). We discuss these results in Chapter 14, as they will be more useful with respect to asymmetrical factorial experiments.

REFERENCES

1. Barlotti, A. (1957). Una limitazione superiore per il numero di punti appartenenti a una calotta $\mathscr{C}(k, 0)$ di uno spazio lineare finito. *Boll. Un. Math. Ital.*, **12**, 67–70.
2. Bose, R. C. (1947). Mathematical theory of the symmetrical factorial design. *Sankhyā*, **8**, 107–166.
3. Bose, R. C. (1960). *On some Connections between the Design of Experiments and Information Theory.* Report No. 1022, Case Institute of Technology, Ohio.
4. Bose, R. C., and Kishen, K. (1940). On the problem of confounding in general symmetrical factorial design. *Sankhyā*, **5**, 21–36.
5. Bose, R. C., and Srivastava, J. N. (1964). On a bound useful in the theory of factorial designs and error correcting codes. *Ann. Math. Stat.*, **35**, 408–414.
6. Cochran, W. G., and Cox, G. M. (1957). *Experimental Designs*, 2nd edition. Wiley, New York.
7. Das, M. N. (1964). A somewhat alternative approach for construction of symmetrical factorial designs and obtaining maximum number of factors. *Calcutta Stat. Assoc. Bull.*, **13**, 1–17.
8. Fisher, R. A. (1942). The theory of confounding in factorial experiments in relation to the theory of groups. *Ann. Eugenics*, **11**, 341–353.
9. Fisher, R. A. (1945). A system of confounding for factors with more than two alternatives giving completely orthogonal cubes and higher powers. *Ann. Eugenics*, **12**, 283–290.

10. Hamming, R. W. (1950). Error detecting and error correcting codes. *Bell System Tech. J.*, **29**, 147–160.

11. Kempthorne, O. (1952). *The Design and Analysis of Experiments.* Wiley, New York.

12. Kitagawa, T., and Mitome, M. (1953). *Tables for the Design of Factorial Experiments.* Baifukan Co., Ltd., Tokyo.

13. Kshirsagar, A. M. (1966). Balanced factorial designs. *J. Roy. Stat. Soc.*, **B28**, 559–567.

14. Kurkjian, B., and Zelen, M. (1962). A calculus of factorial arrangements. *Ann. Math. Stat.*, **33**, 600–619.

15. Kurkjian, B., and Zelen, M. (1963). Applications of the calculus of factorial arrangements. I. Block and direct product designs. *Biometrika*, **50**, 63–73.

16. Peterson, W. W. (1961). *Error Correcting Codes.* MIT Press, Cambridge, Mass.

17. Plackett, R. L., and Burman, J. P. (1946). The design of optimum multifactorial experiments. *Biometrika*, **33**, 305–325.

18. Qvist, B. (1952). Some remarks concerning curves of the second degree in a finite plane. *Ann. Acad. Sci. Fenn.*, *Ser. AI*, No. 134.

19. Rao, C. R. (1946). Hypercubes of strength d leading to confounded designs in factorial experiments. *Bull. Calcutta Math. Soc.*, **38**, 67–78.

20. Seiden, E. (1950). A theorem in finite projective geometry and an application to statistics. *Proc. Am. Math. Soc.*, **1**, 282–286.

21. Shah, B. V. (1958). On balancing in factorial experiments. *Ann. Math. Stat.*, **29**, 766–779.

22. Slepian, D. (1956). A class of binary signalling alphabets. *Bell System Tech. J.*, **35**, 203–234.

23. Tallini, G. (1956). Sulla k-calotta di uno spazio lineare finito. *Ann. Math.*, **42**, 119–164.

24. Yates, F. (1935). Complex experiments. *J. Roy. Stat. Soc. Suppl.*, **2**, 181–247.

25. Yates, F. (1937). *The Design and Analysis of Factorial Experiments.* Imperial Bureau of Soil Sciences, Technical Communication No. 35.

BIBLIOGRAPHY

Barnard, M. M. An enumeration of confounded arrangement in the $2 \times 2 \times 2 \times \cdots$ factorial designs. *Suppl. J. Roy. Stat. Soc.*, **3**, 195–202 (1936).

Kishen, K. A note on the construction of the $(2^{16}, 2^{11})$ and other associated confounded designs keeping up to second order interactions unconfounded. *J. Indian Soc. Agric. Stat.*, **11**, 180–186 (1959).

Nair, K. R. On a method of getting confounded arrangements in the general symmetrical type of experiments. *Sankhyā*, **4**, 121–138 (1938).

Rao, C. R. Factorial experiments derivable from combinatorial arrangement of arrays. *J. Roy. Stat. Soc. Suppl.*, **9**, 128–139 (1947).

Shah, B. V. Balanced factorial experiments. *Ann. Math. Stat.*, **31**, 502–514 (1960).

CHAPTER 14

Confounding in Asymmetrical Factorial Experiments

14.1. INTRODUCTION

In the preceding chapter we studied thoroughly the problem of confounding in symmetrical factorial experiments. Whereas confounding in symmetrical factorial experiments was almost put on a firm footing in 1940 through the work of Bose and Kishen, till recently not much work had been done regarding confounding in asymmetrical factorial experiments. Yates (1937) was the first to tackle the asymmetrical factorial designs, and following his lead, the papers cited in the references and the bibliography appeared on constructions in particular cases. Nair and Rao (1941, 1942, 1948) have developed a set of sufficient conditions that lead to the construction of confounded asymmetrical factorial designs. Kishen and Srivastava (1959) gave some general methods of constructing these designs through Galois fields and finite geometries. So far no general technique for constructing confounded asymmetrical factorial designs has been evolved.

The possibility of confounding in asymmetrical factorial experiments poses more complicated problems than the construction of symmetrical factorial experiments. In symmetrical factorial experiments we can confound higher order interactions without losing any information on main effects. In mixed factorials the situation is different. In the case of experiments of the form $p^m \times q^n$ we can confound r independent effects and interactions of factors at p levels and s independent effects and interactions of factors at q levels in blocks of size $p^{m-r}q^{n-s}$.

In the literature on asymmetrical factorial experiments the concept of "balance" is being used to show that the relative loss of information on any affected interaction is the same. The concept of balanced factorial experiments introduced by Shah (1958, 1960a,c) encompasses a wider sphere of this concept and is more general.

14.2. TOTAL RELATIVE LOSS OF INFORMATION IN CONFOUNDED EXPERIMENTS

Consider a factorial experiment in $v = s_1 s_2 \cdots s_n$ treatment combinations and let l_i' be the mutually orthogonal, normalized contrasts representing main effects and all order interactions ($i = 1, 2, \cdots, v - 1$). Of these effects and interactions, let α interactions be totally confounded in a design formed in b blocks of sizes $k_1, k_2, \cdots, k_b$, where the jth treatment combination is replicated r times. Let N be the incidence matrix of the design. We assume that the last α interactions are totally confounded. Then the C-matrix of the design has $v - 1 - \alpha$ nonzero characteristic roots. If t is the column vector of the v treatment effects and if $l_i't$ is estimated by $l_i'\hat{t}$ with a variance σ^2/θ_i for $i = 1, 2, \cdots, v - 1 - \alpha$, then following the work of Shah (1958), we can express C in the spectral decomposition form

$$(14.2.1) \qquad C = \sum_{i=1}^{v-1-\alpha} \theta_i l_i l_i'.$$

In fact, let

$$(14.2.2) \qquad \begin{aligned} P &= [l_1 l_2 \cdots l_{v-1-\alpha} m_1 m_2 \cdots m_{\alpha+1}] \\ &= [L \quad M] \end{aligned}$$

be an orthogonal matrix, where L is a $v \times (v - 1 - \alpha)$ matrix and M is a $v \times (\alpha + 1)$ matrix. Clearly $m_{\alpha+1} = E_{v,1}$. Since $PP' = P'P = I_v$, we have

$$(14.2.3) \qquad Q = CPP'\hat{t},$$

and hence

$$(14.2.4) \qquad P'Q = (P'CP)(P'\hat{t}).$$

Since $m_j't$ ($j = 1, 2, \cdots, \alpha + 1$) are nonestimable,

$$(14.2.5) \qquad m_j'Q = 0 \quad \text{and} \quad m_j'C = 0 \qquad j = 1, 2, \cdots, \alpha + 1.$$

Thus (14.2.4) reduces to

$$(14.2.6) \qquad L'Q = (L'CL)(L'\hat{t}).$$

Since $L'CL$ is of rank $(v - 1 - \alpha)$, equivalently we have

$$(14.2.7) \qquad (L'\hat{t}) = (L'CL)^{-1}(L'Q),$$

and hence

$$(14.2.8) \qquad \text{var}(L'\hat{t}) = (L'CL)^{-1}\sigma^2.$$

From hypothesis

(14.2.9) $$\text{var}(L'\hat{t}) = \text{diag}\left[\frac{1}{\theta_1}, \frac{1}{\theta_2}, \cdots, \frac{1}{\theta_{v-1-\alpha}}\right]\sigma^2.$$

Equating the right-hand sides of (14.2.8) and (14.2.9) and simplifying, we get (14.2.1).

If none of the interactions or main effects is confounded, not even partially, the variance of the estimate of $l_i' t$ could be σ^2/r. Thus, recollecting that the reciprocal of variance is the available amount of information on an estimate, we find that the relative loss of information on the estimate of $l_i' t$ is

(14.2.10) $$\frac{r - \theta_i}{r}$$

for a given $i = 1, 2, \cdots, v - 1 - \alpha$. Hence the total relative loss of information on these $v - 1 - \alpha$ partially confounded or unconfounded interactions is

(14.2.11) $$\sum_{i=1}^{v-1-\alpha} \frac{(r - \theta_i)}{r} = v - 1 - \alpha - \frac{\text{tr}(C)}{r}$$

$$= v - 1 - \alpha - \frac{vr - b}{r} = \frac{b}{r} - 1 - \alpha.$$

Hence the following theorem:

Theorem 14.2.1. The total relative loss of information in a confounded experiment is as given by (14.2.11).

When no interaction is confounded, we have the following corollary:

Corollary 14.2.1.1. In a confounded experiment, if no interaction is totally confounded, the total relative loss of information is $(b/r) - 1$.

The above results have been proved by Kshirsagar (1958).

14.3. CONFOUNDED ASYMMETRICAL FACTORIAL EXPERIMENTS AS PBIB DESIGNS

In the preceding chapter we saw that an m-associate-class hypercubic association scheme can be defined for the treatment combinations of a symmetrical factorial experiment. But in the case of treatment combinations in asymmetrical factorial experiments the m associate classes will not suffice to represent the association scheme. Let us consider an asymmetrical experiment of the type $s_1^{m_1} \times s_2^{m_2} \times \cdots \times s_h^{m_h}$, where m_i factors are at s_i levels

$(i = 1, 2, \cdots, h)$. We consider the representation of the treatment combinations in an ordered m-ple where $m = \sum_{i=1}^{h} m_i$ and the order corresponds to the order in which the factors appear in the representation

$$s_1{}^{m_1} \times s_2{}^{m_2} \times \cdots \times s_h{}^{m_h}.$$

Two treatments will be called $c_1 c_2 \cdots c_h$th associates, not all c_i factors being zero, if in their representations the levels of $m_i - c_i$ factors out of the m_i factors at s_i levels are the same, while the levels of the remaining c_i factors among the m_i factors at s_i levels are distinct. Shah (1958) showed that such an association scheme has the parameters

$$(14.3.1) \qquad n_{c_1 c_2 \cdots c_h} = \prod_{i=1}^{h} \binom{m_i}{c_i} (s_i - 1)^{c_i}$$

and

$$(14.3.2) \qquad p_{d_1 d_2 \cdots d_h, \, e_1 e_2 \cdots e_h}^{c_1 c_2 \cdots c_h} = \prod_{i=1}^{h} p_{d_i, \, e_i}^{c_i}(m_i, s_i),$$

where

$$(14.3.3) \quad p_{d_i, \, e_i}^{c_i}(m_i, s_i) = \sum_u \binom{m_i - c_i}{m_i - u}\binom{c_i}{u - d_i}\binom{c_i + d_i - u}{u - e_i}$$
$$\times (s_i - 1)^{u - c_i}(s_i - 2)^{c_i + d_i + e_i - 2u},$$

the summation being taken over all the values of u such that

$$(14.3.4) \qquad \max(c_i, d_i, e_i) \le u \le \tfrac{1}{2}(c_i + d_i + e_i)$$

and it has $\prod_{i=1}^{h} (m_i + 1) - 1$ associate classes.

The above association scheme reduces to the $2^h - 1$ extended group-divisible association scheme of Kempthorne and Hinkelman, introduced in Chapter 8, if $m_1 = m_2 = \cdots = m_h = 1$.

Shah (1958) and Kshirsagar (1966) showed that a balanced factorial experiment in h factors of the type $s_1 \times s_2 \times \cdots \times s_h$ is a PBIB design with an extended group-divisible association scheme and λ-parameters given by

$$(14.3.5) \qquad \lambda_{d_1 d_2 \cdots d_h} = \frac{-k}{s_1 s_2 \cdots s_h} \sum \prod_{i=1}^{h} G_i(c_i, d_i) \theta_{c_1 c_2 \cdots c_h},$$

where c_i and d_i take the values 0 or 1; the summation is over all the values of $(c_1 c_2 \cdots c_h)$,

$$(14.3.6) \quad G_i(1, 1) = s_i - 1, \quad G_i(1, 0) = -1, \quad G_i(0, 0) = G_i(0, 1) = 1,$$

and $\theta_{c_1 c_2 \cdots c_h}$ are the characteristic roots of the C-matrix of the balanced factorial experiment subject to $\theta_{00\cdots 0} = 0$. It may be noted that when $h = 2$ and $\lambda_{00} = \lambda_{01}$ or λ_{10}, the balanced factorial experiment becomes a group-divisible design.

The converse of the above result was established by Kurkjian and Zelen (1963), who introduced the following definition:

Definition 14.3.1. If N is the incidence matrix of a design in $v = s_1 s_2 \cdots s_n$ treatment combinations in b blocks, then N is said to possess the property (A) if

$$
\begin{aligned}
NN' = \sum_{\alpha=0}^{t} \Bigg\{ & \sum_{\delta_1+\delta_2+\cdots \delta_n = \alpha} h(\delta_1, \delta_2, \cdots, \delta_n) \\
& \times [D_1^{\delta_1} \times D_2^{\delta_2} \times \cdots \times D_n^{\delta_n}] \Bigg\},
\end{aligned}
$$

(14.3.7)

where $\times$ is the Kronecker-product sign, $\delta_i = 0$ or 1, the $h(\delta_1, \delta_2, \cdots, \delta_n)$ are constants depending on δ_i, and $D_i^{\delta_i}$ is an $s_i \times s_i$ matrix defined by $D_i^{\delta_i} = I_{s_i}$ if $\delta_i = 0$ and $D_i^{\delta_i} = E_{s_i, s_i}$ if $\delta_i = 1$.

The incidence matrix of randomized block, BIB, group-divisible, and L_i designs possesses the property (A). Also the direct product of two incidence matrices possessing the property (A) will also possess the property (A). Kurkjian and Zelen (1963) showed that factorial designs whose incidence matrix possesses the property (A) are balanced factorial experiments.

For constructions of balanced factorial experiments not covered in this chapter we refer to Shah (1960b), Muller (1966), and Kishen and Tyagi (1964). The constructions of 117 balanced asymmetrical factorial experiments were given by Shah (1960c) in his unpublished Ph.D thesis.

The balanced asymmetrical factorial experiments can be generalized to partially balanced asymmetrical factorial experiments to reduce the number of replications; for their constructions we refer to Kishen and Tyagi (1963).

14.4. CONFOUNDING IN ASYMMETRICAL FACTORIALS WHERE THE LEVELS OF FACTORS ARE DIFFERENT POWERS OF THE SAME PRIME

Consider a $t^m \times s^n$ factorial experiment, where $t = p^\alpha$, $s = p^\beta$, p being a prime, and α, β are positive integers. The t levels of a factor can be identified with all treatment combinations of α pseudofactors, and the s levels of a factor can be identified with all treatment combinations of β pseudofactors. Thus the $t^m \times s^n$ experiment can be treated as a symmetrical factorial experiment $p^{m\alpha + n\beta}$ in $m\alpha + n\beta$ pseudofactors each at p levels, and confounding can be done through the well-known techniques of symmetrical factorial experiments. In order to prevent any main effect of the mixed factorial from being confounded, it is essential that confounding be not made with pseudofactors corresponding to a given factor's level alone of the original experiment.

As an example, let us consider a 4×2^2 experiment in three factors a, b, c, the levels of a being a_0, a_1, a_2, a_3, the levels of b being b_0, b_1, and those of c being c_0 and c_1. We identify four levels of a with the treatment combinations $d_0 e_0$, $d_0 e_1$, $d_1 e_0$, $d_1 e_1$, respectively, of a 2^2 experiment in pseudofactors d and e. Confounding the interaction $BCDE$ in the identified 2^4 experiment and writing the plan in original factors, we obtain the following arrangement:

Block	Constituents of the Block
1	$(a_0 b_0 c_0, a_0 b_1 c_1, a_1 b_1 c_0, a_1 b_0 c_1, a_2 b_1 c_0, a_2 b_0 c_1, a_3 b_0 c_0, a_3 b_1 c_1)$
2	$(a_0 b_0 c_1, a_0 b_1 c_0, a_1 b_0 c_0, a_1 b_1 c_1, a_2 b_0 c_0, a_2 b_1 c_1, a_3 b_0 c_1, a_3 b_1 c_0)$

Noting that with four equally spaced levels we can form three orthogonal comparisons,

$$A' = a_3 + a_2 - a_1 - a_0,$$

(14.4.1) $$A'' = a_3 - a_2 - a_1 + a_0,$$

$$A''' = a_3 - a_2 + a_1 - a_0,$$

we observe that they represent respectively E, DE, and D main effects and interaction of pseudofactors d and e. Hence the confounded interaction in the plan given above is $A''BC$. Balancing over the three-factor interaction can be achieved by taking three replications and confounding $A'BC$, $A''BC$, and $A'''BC$, one in each replication.

14.5. CONFOUNDING IN $3^m \times 2^n$ EXPERIMENTS

Let there be m factors $F_1, F_2, \cdots, F_m$, each at three levels, and n factors $G_1, G_2, \cdots, G_n$, each at two levels. Choose a pencil, say $P(1, 1, \cdots, 1)$ for a 3^m experiment on factors $F_1, F_2, \cdots, F_m$ and let the three $(m-1)$-flats be denoted by X_1, X_2, and X_3. Choose the two fractions Y_1 and Y_2 into which the 2^n treatment combinations will be divided by the $(n-1)$th order interaction $G_1 G_2 \cdots G_n$. Table 14.5.1 shows the plan to be formed in three replications.

The plan of Table 14.5.1 can be verified to confound the interaction $G_1 G_2 \cdots G_n$, carrying one degree of freedom, and the generalized interaction of the pencil $P(1, 1, \cdots, 1)$ of the factors $F_1, F_2, \cdots, F_m$ with the interaction $G_1 G_2 \cdots G_n$, carrying two degrees of freedom. The relative loss of

Table 14.5.1. Confounding of $3^m \times 2^n$ experiments in blocks of $3^m \times 2^{n-1}$ plots

Replication	I		II		III	
Block	1	2	1	2	1	2
Levels of Factors $F_1, F_2, \cdots, F_m$	Levels of Factors $G_1, G_2, \cdots, G_n$					
X_1	Y_1	Y_2	Y_1	Y_2	Y_2	Y_1
X_2	Y_1	Y_2	Y_2	Y_1	Y_1	Y_2
X_3	Y_2	Y_1	Y_1	Y_2	Y_1	Y_2

information on the former being 1/9 and on each degree of freedom of the latter being 4/9, the total relative loss of information is 1.

With the same notation as above we can also construct the design in blocks of $3^{m-1} \times 2^n$ plots in two replications, as shown in Table 14.5.2.

Table 14.5.2. Confounding of $3^m \times 2^n$ experiments in blocks of $3^{m-1} \times 2^n$ plots

Replication	I			II		
Block	1	2	3	1	2	3
Treatment combinations	$X_2 Y_1$ $X_3 Y_2$	$X_3 Y_1$ $X_1 Y_2$	$X_1 Y_1$ $X_2 Y_2$	$X_3 Y_1$ $X_2 Y_2$	$X_1 Y_1$ $X_3 Y_2$	$X_2 Y_1$ $X_1 Y_2$

In the plan of Table 14.5.2 we can verify that the interaction of $F_1 F_2 \cdots F_m$ carried by the pencil $P(1, 1, \cdots, 1)$ with two degrees of freedom and its generalized interaction with $G_1 G_2 \cdots G_n$ is confounded. The relative loss of information on each degree of freedom of the former can be seen to be 1/4 and on each degree of freedom of the latter, 3/4. The total relative loss of information on confounded interactions is thus 2.

In a similar manner various confoundings can be made on $3^m \times 2^n$ experiments and in general $p^m \times q^n$; we refer the reader to Chakrabarti (1962), Kempthorne (1952).

14.6. CONFOUNDING IN $v \times s^m$ EXPERIMENTS IN BLOCKS OF vs^{m-1} PLOTS

Let the factor F be at v levels and the factors $G_1, G_2, \cdots, G_m$ be each at s levels, s being a prime. Choose a high-order interaction pencil Z for the factors $G_1, G_2, \cdots, G_m$ and let $X_1, X_2, \cdots, X_s$ be the $(m-1)$-flats based on that pencil. Let there be a BIB design with the parameters v, b, r, k, and λ and let the levels of the factor F be identified with the treatments. From each block of the BIB design we generate $s(s-1)$ blocks by associating X_i combinations with the levels of the factor F occurring in that block and combinations $X_{i'}$ with the levels not occurring in that block ($i \neq i'$; $i, i' = 1, 2, \cdots, s$). The design thus obtained was shown by Tharthare (1965) to be a balanced design confounding the interaction Z and AZ. In fact the design thus obtained is a generalized right-angular design with the parameters

$$(14.6.1) \quad \begin{aligned} v^* &= vs^m, & b^* &= bs(s-1), & r^* &= b(s-1), \\ k^* &= vs^{m-1}, & \lambda_1^* &= r^*, & \lambda_2^* &= 0, \\ \lambda_3^* &= (s-1)[b - 2(r - \lambda)], & \lambda_4^* &= 2(r - \lambda). \end{aligned}$$

Tharthare showed that the relative loss of information on each degree of freedom of Z is $[r^* + (v-1)(\lambda_3^* - \lambda_4^*)]/vr^*$, and the relative loss of information on each degree of freedom of AZ is $(r^* - \lambda_3^* + \lambda_4^*)/vr^*$, the total relative loss of information being $(b^*/r^*) - 1$. This design is $(s-1)$-resolvable.

The above construction method can be slightly generalized. We can use an equireplicated pairwise balanced design $(v; k_1, k_2, \cdots, k_n)$ of index λ in place of a BIB design in the above construction and get a confounded asymmetrical design $v \times s^n$ to form a generalized right-angular design with the same set of parameters (14.6.1). Kishen and Tyagi (1964) have given $v \times 2^2$ and $v \times 3^2$ series of confounded designs, with slight modifications in the construction.

In particular, Tharthare constructed a 5×2^2 design, as shown in Table 14.6.1, based on the interaction $G_1 G_2$; the relative loss of information on the $G_1 G_2$ interaction is 9/25 and on each degree of freedom of the $FG_1 G_2$ interaction, 4/25.

Shah (1960b) and Kishen and Tyagi (1963) also constructed a plan of this design in 10 nonresolvable blocks, as shown in Table 14.6.2. This plan is

Table 14.6.1. Plan of 5×2^2 resolvable experiment of Tharthare

Replication	I		II		III		IV		V	
Block	1	2	1	2	1	2	1	2	1	2
Levels of Factor F	Levels of Factors G_1 and G_2									
0	X_1	X_2	X_1	X_2	X_1	X_2	X_1	X_2	X_2	X_1
1	X_1	X_2	X_1	X_2	X_1	X_2	X_2	X_1	X_1	X_2
2	X_1	X_2	X_1	X_2	X_2	X_1	X_1	X_2	X_1	X_2
3	X_1	X_2	X_2	X_1	X_1	X_2	X_1	X_2	X_1	X_2
4	X_2	X_1	X_1	X_2	X_1	X_2	X_1	X_2	X_1	X_2

Table 14.6.2. Plan of 5×2^2 nonresolvable plan of Shah

Block	1	2	3	4	5	6	7	8	9	10
Levels of Factor F				Levels of Factors G_1 and G_2						
0	X_1	X_2	X_2	X_2	X_1	X_2	X_1	X_1	X_2	X_1
1	X_1	X_1	X_2	X_2	X_2	X_1	X_2	X_1	X_1	X_2
2	X_2	X_1	X_1	X_2	X_2	X_2	X_1	X_2	X_1	X_1
3	X_2	X_2	X_1	X_1	X_2	X_1	X_2	X_1	X_2	X_1
4	X_2	X_2	X_2	X_1	X_1	X_1	X_1	X_2	X_1	X_2

balanced and has a smaller relative loss of information on the two-factor interaction than the plan of Table 14.6.1 but more loss of information on the three-factor interaction. For this design the loss of information on each degree of freedom of two- and three-factor interactions is 1/25 and 6/25, respectively.

It may be added that Li (1944) was the first to construct a 5×2^2 design in 10 blocks, but his design was only partially balanced.

14.7. USE OF FINITE GEOMETRIES IN THE CONSTRUCTION OF CONFOUNDED ASYMMETRICAL DESIGNS

We let s_1 be a prime and are interested in constructing a confounded asymmetrical factorial experiment $s_1 \times s_2 \times \cdots \times s_n$ with s_1 blocks in each replication $(s_1 \geqq s_2 \geqq \cdots \geqq s_n)$.   Let $\alpha_0 = 0, \alpha_1 = x, \alpha_2 = x^2, \cdots, \alpha_{s_1 - 1}$ $= x^{s_1 - 1} = 1$ be the elements of $GF(s_1)$, where x is a primitive root of $GF(s_1)$. Let us identify these elements with the s_1 levels of the first factor F_1. The s_i levels of the ith factor F_i $(i = 2, 3, \cdots, n)$ can be selected as any s_i elements of the elements of $GF(s_1)$. In this connection it may be noted that Kishen and Srivastava (1959a, b) described a very nice way of constructing a polynomial over $GF(s_1)$ that takes s_i specified values. Though the polynomials constructed by them are of theoretical importance and provide a natural way to indicate the levels of the different factors, for practical constructions it is adequate to restrict the levels of F_i to any s_i elements of $GF(s_1)$ in an arbitrary manner. After suitably choosing the levels of the n factors, let the $s_1 s_2 \cdots s_n$ treatment combinations be denoted by $(x_1, x_2, \cdots, x_n)$, where x_i takes the s_i suitably selected elements of $GF(s_1)$.

Now to confound a k-factor interaction involving F_1, we form s_1 blocks according to the s_1 flats of the pencil

$$(14.7.1) \qquad \begin{aligned} x_1 + (a_{i_2} x_{i_2} + \cdots + a_{i_{k-1}} x_{i_{k-1}}) &= \alpha, \qquad \alpha \in GF(s_1), \\ a_{i_r} &\in GF(s_1), \qquad r = 2, 3, \cdots, k - 1. \end{aligned}$$

In the plan thus formed the interaction of the factors $F_1 F_{i_2} \cdots F_{i_{k-1}}$ carried by the pencil (14.7.1) is deliberately confounded. But the main effect of F_1 and all the interactions of F_1 with $F_{i_2}, F_{i_3}, \cdots, F_{i_{k-1}}$ will automatically unintentionally get confounded. When there are at least two factors at s_1 levels each, no main effect will be partially confounded.

Let us take an illustrative example. Consider a $3^2 \times 2$ experiment. The elements of GF(3) are 0, 1, 2, and let us assume that the three factors of the experiment are $A(0, 1, 2)$, $B(0, 1, 2)$ and $C(0, 1)$. Let the 18 treatment combinations be denoted by (x_1, x_2, x_3), where $x_1, x_2 = 0, 1, 2$, and $x_3 = 0, 1$. We can obtain balance in four replications, confounding the following four pencils, one in each replication:

(14.7.2)
$$\begin{aligned}
x_1 + x_2 + x_3 &= 0, 1, 2; \\
x_1 + x_2 + 2x_3 &= 0, 1, 2; \\
x_1 + 2x_2 + x_3 &= 0, 1, 2; \\
x_1 + 2x_2 + 2x_3 &= 0, 1, 2.
\end{aligned}$$

The plan of the design is shown in Table 14.7.1. The relative loss of information on each degree of freedom on $F_1 F_2$ can be seen to be 1/8 and on each degree of freedom of the $F_1 F_2 F_3$ interaction, 3/8; hence the total relative loss of information is 2.

For other methods of this type of construction we refer the reader to the paper of Kishen and Srivastava (1959b).

White and Hultquist (1965) extended the use of finite fields in the construction of asymmetrical plans by defining the addition and multiplication of elements from distinct finite fields after mapping them on a finite commutative subring containing subrings isomorphic to each of the fields in question. They then applied the standard procedure of constructing confounded symmetrical factorial experiments.

Table 14.7.1. Plan of a $3^2 \times 2$ design in six plot blocks

Replication	I			II			III			IV		
Block	1	2	3	1	2	3	1	2	3	1	2	3
Treatment combi-nations	000	100	200	000	100	200	110	100	200	000	100	200
	111	010	020	120	010	020	220	020	010	110	020	010
	120	220	110	210	220	110	000	210	120	220	210	120
	210	001	011	011	111	001	201	001	101	101	201	001
	201	121	101	101	201	121	011	111	021	021	011	111
	021	211	221	221	021	211	121	221	211	211	121	221

14.8. CONFOUNDED ASYMMETRICAL $3^m \times 2^n$ EXPERIMENTS IN BLOCKS OF 2^k PLOTS

Let $F_1, F_2, \cdots, F_m$ be m factors, each at three levels, and $G_1, G_2, \cdots, G_n$ be n factors, each at two levels. Let us consider a symmetrical factorial 2^{2m+n} experiment in factors $F_1', F_1'', F_2', F_2'', \cdots, F_m', F_m'', G_1, G_2, \cdots, G_n$, the levels being denoted by -1 and $+1$. Choose a suitable set of $2^{2m+n-k} - 1$ interactions and form the confounded design in blocks of 2^k plots. Then add the levels of F_i', F_i'' to get -2, 0, or $+2$ as levels of the factor F_i. This design has a drawback in that all the treatment combinations do not occur with the same frequency, but it has an added advantage in that repeated treatment combinations in a block provide some degree of freedom for error. A 3×2^2 design constructed by this method in blocks of eight plots was given by Das and Rao (1967).

REFERENCES

1. Chakrabarti, M. C. (1962). *Mathematics of Design and Analysis of Experiments.* Asia Publishing House, Bombay.
2. Das, M. N., and Rao, P. S. (1967). Construction and analysis of some new series of confounded asymmetrical factorial designs. *Biometrics,* **23**, 813–822.
3. Kempthorne, O. (1952). *The Design and Analysis of Experiments.* Wiley, New York.
4. Kishen, K., and Srivastava, J. N. (1959a). Confounding in asymmetrical factorial designs in relation to finite geometries. *Current Science,* **28**, 98–100.
5. Kishen, K., and Srivastava, J. N. (1959b). Mathematical theory of confounding in asymmetrical and symmetrical factorial designs. *J. Indian Soc. Agric. Stat.,* **11**, 73–110.
6. Kishen, K., and Tyagi, B. N. (1963). Partially balanced asymmetrical factorial designs. *Contributions to Statistics.* Presented to Professor P. C. Mahalanobis on his 70th birthday. Pergamon Press, New York, pp. 147–158.
7. Kishen, K., and Tyagi, B. N. (1964). On the construction and analysis of some balanced aysmmetrical factorial designs. *Calcutta Stat. Assn. Bull.,* **13**, 123–149.
8. Kshirsagar, A. M. (1958). A note on the total relative loss of information in any design. *Calcutta Stat. Assn. Bull.,* **7**, 78–81.
9. Kshirsagar, A. M. (1966). Balanced factorial designs. *J. Roy. Stat. Soc.,* **B28**, 559–567.
10. Kurkjian, B., and Zelen, M. (1963). Applications of the calculus of factorial arrangements. I. Block and direct product designs. *Biometrika,* **50**, 63–73.
11. Li, J. C. R. (1944). Design and statistical analysis of some confounded factorial experiments. Research Bulletin 333, Iowa State College of Agriculture.
12. Muller, E. R. (1966). Balanced confounding of factorial experiments. *Biometrika,* **53**, 507–524.

13. Nair, K. R., and Rao, C. R. (1941). Confounded designs for asymmetrical factorial experiments. *Science and Culture*, **7**, 313–314.
14. Nair, K. R., and Rao, C. R. (1942). Confounded designs for $k \times p^m \times q^n \times \cdots$ type of factorial experiment. *Science and Culture*, **7**, 361–362.
15. Nair, K. R., and Rao, C. R. (1948). Confounding in asymmetrical factorial experiments. *J. Roy. Stat. Soc.*, **B10**, 109–131.
16. Shah, B. V. (1958). On balancing in factorial experiments. *Ann. Math. Stat.*, **29**, 766–779.
17. Shah, B. V. (1960a). Balanced factorial experiments. *Ann. Math. Stat.*, **31**, 502–514.
18. Shah, B. V. (1960b). On a 5×2^2 factorial design. *Biometrics*, **16**, 115–118.
19. Shah, B. V. (1960c). Some aspects of the construction and analysis of incomplete block designs. Unpublished Ph.D. thesis submitted to Bombay University.
20. Tharthare, S. K. (1965). Generalized right angular designs. *Ann. Math. Stat.*, **36**, 1535–1553.
21. White, D., and Hultquist, R. A. (1965). Construction of confounding plans for mixed factorial designs. *Ann. Math. Stat.*, **36**, 1256–1271.
22. Yates, F. (1937). *The Design and Analysis of Factorial Experiments*. Imperial Bureau of Soil Science, Technical Communication No. 35.

BIBLIOGRAPHY

Das, M. N. Fractional replicates as asymmetrical factorial designs. *J. Indian Soc. Agric. Stat.*, **12**, 159–174 (1960).

Inkson, R. H. E. The analysis of a $3^2 \times 2^2$ factorial experiment with confounding. *J. Applied Stat.*, **10**, 98–107 (1961).

Kishen, K. Recent developments in experimental design. Presidential address to the Statistics Section, Indian Science Congress, Madras, 1958.

Kishen, K. On a class of asymmetrical factorial designs. *Current Science*, **29**, 465–466 (1960).

Kishen, K., and Tyagi, B. N. On some methods of construction of asymmetrical factorial designs. *Current Science*, **30**, 407–409 (1961).

Kramer, C. Y., and Bradley, R. A. Intrablock analysis for factorial in two associate class group divisible designs. *Ann. Math. Stat.*, **28**, 349–361 (1957).

Rao, C. R. A general class of quasi factorial and related designs. *Sankhyā*, **17**, 165–174 (1956).

Sardana, M. G., and Das, M. N. On the construction and analysis of some confounded asymmetrical factorial designs. *Biometrics*, **21**, 948–956 (1965).

Thompson, H. R., and Dick, I. D. Factorial designs in small blocks derivable from orthogonal latin squares. *J. Roy. Stat. Soc.*, **B13**, 126–130 (1951).

Yates, F. Complex experiments. *Suppl. J. Roy. Stat. Soc.*, **2**, 181–247 (1935).

Zelen, M. The use of group divisible designs for confounded asymmetrical factorial arrangements. *Ann. Math. Stat.*, **29**, 22–40 (1958).

CHAPTER 15

Fractional Replication

15.1. INTRODUCTION

In a factorial experiment, when the number of treatment combinations is very large, it will be beyond the resources of the investigator to experiment with all of them. For such cases Finney (1945) proposed a method in which only a fraction of the treatment combinations will be experimented with. Plackett and Burman (1946) studied the problem in more detail and gave fractional designs with the minimum possible size of s^n factorial experiments for $s = 2, 3, 4, 5, 6,$ and 7. In fractional replication, though the size of the experiment is reduced, information on certain higher order interactions is sacrificed. The crucial part of the specification of the fractionally replicated design is the suitable choice of the defining or identity relationship. The nonestimable effects or interactions for the selected fraction of treatment combinations, when equated with I, are called the identity relation. After selecting a fraction of treatment combinations, one can easily note that any contrast of the selected treatment combinations represents more than one effect or interaction, and all effects or interactions represented by the same treatment combinations are called aliases. In aliases, by assuming that other interactions are negligible when compared with one of them, in which he is interested, the experimenter can estimate it by the corresponding contrast of the selected treatment combinations. We illustrate these ideas in Sections 15.2 and 15.3 with respect to fractions of 2^n and s^n experiments and study other results on fractional replication in the succeeding sections of this chapter. A good survey on the work done on the technique of fractional replicate designs through 1962 has been made by Addelman (1963).

The reader is requested to observe the resemblances between the methods of constructing compounded plans and the methods of constructing fractional replicate plans and also note that both methods can be together employed advantageously in some circumstances.

15.2. $1/2^k$ REPLICATE OF 2^n FACTORIAL EXPERIMENTS

Let us consider a 2^3 factorial experiment in three factors n, p, k, each at two levels. The eight complete treatment combinations will be 1, n, p, np, k, nk, pk, and npk. Instead of experimenting with all eight treatment combinations, let us consider the four treatment combinations 1, np, nk, and pk. We observe that these four treatment combinations occur with the negative sign in the NPK interaction. If $[np]$ denotes the total yield of r plots receiving the treatment combination np, etc., we note that

$$(4r)^{-1}\{[1] + [np] + [nk] + [pk]\} \quad \text{represents } \mu - \tfrac{1}{2}NPK,$$

$$(4r)^{-1}\{-[1] + [np] + [nk] - [pk]\} \text{ represents } \tfrac{1}{2}(N - PK),$$

$$(4r)^{-1}\{-[1] + [np] - [nk] + [pk]\} \text{ represents } \tfrac{1}{2}(P - NK),$$

$$(4r)^{-1}\{-[1] - [np] + [nk] + [pk]\} \text{ represents } \tfrac{1}{2}(K - NP),$$

μ being the general mean. If we assume that the two- and three-factor interactions are negligible, the above four orthogonal functions of the treatment combinations can be used to estimate the general mean μ and the main effects N, P, and K, respectively. This state of affairs will be represented by

$$(15.2.1) \qquad\qquad I = NPK,$$

which is known as the *identity relation*, and

$$(15.2.2) \qquad N = PK, \qquad P = NK, \qquad K = NP,$$

which are called *alias sets*. We see that the alias sets are obtained from the identity relation by multiplying both sides with the main-effect symbols and denoting the square of a symbol by unity. We remark that the experiment could have been conducted with the treatment combinations n, p, k, and npk, and in that case also it is impossible to separate the mean μ from the NPK interaction, the N effect from the PK interaction, the P effect from the NK interaction, and the K effect from the NP interaction. Hence we have the identity relationship and alias sets given in (15.2.1) and (15.2.2). Thus we note that the treatment combinations with positive sign in the NPK interaction or the treatment combinations with negative sign in the NPK interaction have given rise to the same state of affairs. Conventionally we select the set of treatment combinations including the control 1.

In general let us consider the problem of constructing a $1/2^k$ fraction of a 2^n factorial experiment. Such an experiment will be denoted by 2^{n-k}. Of the total of $(2^n - 1)$ effects and interactions in the full factorial experiment, $2^k - 1$ will be inseparable from the mean, and the remaining $2^n - 2^k$ will be mutually inseparable in sets of 2^k, there being $(2^{n-k} - 1)$ such sets. The treatment combinations will be selected to be of the same sign as the control in the interactions $X_1, X_2, \cdots, X_k$, none of which is obtainable from the others by multiplication and will not result in having two effects or interactions of interest to the experimenter in the same alias set. Then the generalized interactions of $X_1, X_2, \cdots, X_k$ will also be inseparable from the mean, and the identity relationship will become

$$(15.2.3) \qquad I = X_1 = X_2 = X_1 X_2 = X_3 = X_1 X_3 = X_2 X_3 = X_1 X_2 X_3 = \text{etc.}$$

The alias sets of an effect or interaction Y are the generalized interactions of Y with $X_1, X_2, X_1 X_2$, etc.

As a further illustration, the treatment combinations

$$1,\ ab,\ cd,\ abcd,\ ef,\ abef,\ cdef,\ abcdef,\ bce,\ ace,\ bde,\ ade,\ bcf,\ acf,\ bdf,\ adf,$$

can be shown to have been formed on the defining relation

$$(15.2.4) \qquad\qquad I = ABCD = ABEF = CDEF,$$

and the alias sets in this case are

$$(15.2.5)
\begin{aligned}
A &= BCD = BEF = ACDEF, & B &= ACD = AEF = BCDEF, \\
C &= ABD = ABCEF = DEF, & D &= ABC = ABDEF = CEF, \\
E &= ABCDE = ABF = CDF, & F &= ABCDF = ABE = CDE, \\
AB &= CD = EF = ABCDEF, & AC &= BD = BCEF = ADEF, \\
AD &= BC = BDEF = ACEF, & AE &= BCDE = BF = ACDF, \\
AF &= BCDF = BE = ACDE, & CE &= ABDE = ABCF = DF, \\
CF &= ABDF = ABCE = DE, & ACE &= BDE = BCF = ADF, \\
ACF &= BDF = BCE = ADE.
\end{aligned}$$

15.3. $1/s^k$ REPLICATE OF s^n FACTORIAL EXPERIMENTS

Let s be a prime or a prime power. To select a $1/s^k$ replicate of an s^n experiment, which is denoted by s^{n-k}, we first select k independent pencils $P(a_{i1}, a_{i2}, \cdots, a_{in})$ $(i = 1, 2, \cdots, k)$ such that none is a generalized interaction of the others and will not result in two pencils of interest to the experimenter being in the same alias set. Then we select the treatment combinations that occur on the same flat as $(0, 0, \cdots, 0)$ in the pencils $P(a_{i1}, a_{i2}, \cdots, a_{ik})$. Evidently they occur on the same flat in each of the generalized interactions of $P(a_{i1}, a_{i2}, \cdots, a_{in})$ $(i = 1, 2, \cdots, k)$. The identity relation will then be

(15.3.1) $I = P\left(\sum_i \lambda_i a_{i1}, \sum_i \lambda_i a_{i2}, \cdots, \sum_i \lambda_i a_{in}\right)$ $\lambda_i \in GF(s)$,

and there will be $(s^k - 1)/(s - 1)$ pencils in the identity relationship. The remaining $(s^n - s^k)/(s - 1)$ pencils will be divided into $(s^{n-k} - 1)/(s - 1)$ alias sets of s^k pencils. The alias set of the pencil $P(b_1, b_2, \cdots, b_n)$ consists of pencils representing generalized interactions of it with each of the $(s^k - 1)/(s - 1)$ pencils of the identity relationship.

As an example, we construct a 1/3 replicate of a 3^3 experiment based on the identity relationship

(15.3.2) $I = P(1, 1, 1)$.

The treatment combinations forming the design are

000, 012, 102, 120, 021, 201, 210, 111, 222

and give rise to the alias sets

$$P(1, 0, 0) = P(1, 2, 2) = P(0, 1, 1);$$
$$P(0, 1, 0) = P(1, 2, 1) = P(1, 0, 1);$$
(15.3.3)
$$P(0, 0, 1) = P(1, 1, 2) = P(1, 1, 0);$$
$$P(1, 2, 0) = P(1, 0, 2) = P(0, 1, 2).$$

For plans of standard fractional replicate designs we refer the reader to Brownlee, Kelly, and Loraine (1948); Connor and Zelen (1959); Cochran and Cox (1957); Kitagawa and Mitome (1953); and a publication by the National Bureau of Standards.

From the last and present sections we can clearly see that two main effects will not be entangled in an alias set if the number of factors involved in the interactions of the defining relation is at least 3, and two first-order interactions will not be entangled in an alias set if the number of factors involved in the interactions of the defining relations is at least 5.

Confounding can be resorted to even in fractionally replicated factorial experiments, and in such cases the information on the confounded interactions and their aliases will be lost while the rest of the factorial effects are undisturbed and estimated in the usual way [cf. Addelman (1965), Cochran and Cox (1957), Patel (1962), and Youden (1961)].

The reader might have realized the resemblance between the confounding and the fractional replication concepts. Clearly the selected fraction of treatment combinations of an s^n experiment based on the identity relationship $I = X_1 = X_2 = X_1 X_2 = X_3 = X_1 X_3 = X_2 X_3 = X_1 X_2 X_3 = \cdots$ forms the intrablock subgroup of a confounded s^n experiment confounding X_1, X_2,

$X_3, \cdots$ and their generalized interactions. A slightly different way of seeing the formal equivalence of fractional replication and confounding was proposed by Kempthorne (1947, 1952). Consider a 5^2 factorial experiment, where each treatment combination is represented by $(x_1, x_2)(x_1, x_2 = 0, 1, 2, 3, 4)$, confounding the pencil $P(1, 1)$. To the treatment combinations of the ith block, satisfying $x_1 + x_2 = i$, we add a new factor c at the ith level $(i = 0, 1, 2, 3, 4)$. Then the confounded design can clearly be verified to be a $1/5$ replication of a 5^3 experiment based on the identity relation $P(1, 1, 4)$. This method of equivalence will sometimes be useful in constructing a $1/s^{k-1}$ replicate of an s^{n+1} experiment with s^{r-1} blocks, when a $1/s^k$ replicate of an s^n experiment in s^r blocks is given.

A fraction of an s^n experiment will be called an orthogonal main-effect plan if all the main effects are estimable orthogonally from the selected treatment combinations. However, if the main effects are estimable with possible correlations between their estimates based on a fractional replication, the selected treatment combinations will be called nonorthogonal main-effect plans.

15.4. FRACTIONAL REPLICATE PLANS WITH THE HELP OF MOLS

Tippett (1934) used complete sets of MOLS of order 5 to construct a fractional replicate plan of a 5^6 experiment in 5^2 treatment combinations. In fact a complete set of MOLS of order s can be utilized to construct a fractional replicate design for an s^{s+1} factorial experiment in s^2 treatment combinations. Let the rows $0, 1, 2, \cdots, s - 1$ correspond to the s levels of the first factor, the columns $0, 1, 2, \cdots, s - 1$ to the s levels of the second factor, the symbols $0, 1, 2, \cdots, s - 1$ of the first MOLS to the s levels of the third factor, $\cdots$, the symbols $0, 1, 2, \cdots, s - 1$ of the $(s - 1)$th MOLS to the s levels of the $(s + 1)$th factor. Then the treatment combinations corresponding to the s^2 cells of a hypergraecolatin square give the required fractional replicate plan for estimating all the main effects orthogonally.

For a 3^4 experiment a pair of MOLS

(15.4.1)
$$\begin{array}{ccc} 00 & 12 & 21 \\ 11 & 20 & 02 \\ 22 & 01 & 10 \end{array}$$

results in the selection of the nine treatment combinations 0000, 0112, 0221, 1011, 1120, 1202, 2022, 2101, and 2210 as an orthogonal main-effect plan for estimating the main effects orthogonally, assuming all interactions to be negligible.

15.5. USE OF HADAMARD MATRICES IN THE CONSTRUCTION OF FRACTIONAL REPLICATE PLANS

A square matrix of order n, H_n, is called a Hadamard matrix if it has the elements $+1$ and -1 such that $H_n' H_n = nI_n$; we study this in detail in Section 17.4. We can assume, without loss of generality, that the first row and column of a Hadamard matrix of order $4t$ can be so arranged that they contain $+1$ everywhere. By omitting the first column and identifying the columns with $4t - 1$ factors and interpreting -1 as the first level and $+1$ as the second level of each factor, we get an orthogonal main-effect plan of 2^{4t-1} factorial experiments in $4t$ treatment combinations [cf. Plackett and Burman (1946)].

15.6. USE OF ORTHOGONAL ARRAYS IN THE CONSTRUCTION OF FRACTIONAL REPLICATE PLANS

The orthogonal arrays of Chapter 2 can be used for constructing fractional replicate plans of symmetrical factorial experiments. In an orthogonal array (N, k, s, t) we identify the k rows with k factors and the s symbols of the ith row with the s levels of the ith factor. Then the N assemblies (or runs or treatment combinations) give the required fractional replicate plan. Rao (1947) has proved that an orthogonal array $(N, k, s, t + m - 1)$ of index λ yields a fractionally replicated design from which all main effects and m-factor interactions can be measured when interactions of t or more factors are assumed to be absent. In particular orthogonal arrays $(s^n, (s^n - 1)/(s - 1), s, 2)$ and $(2s^n, 2(s^n - 1)/(s - 1) - 1, s, 2)$ can be used to construct fractional replicate plans of $s^{(s^n-1)/(s-1)}$ and $s^{2(s^n-1)/(s-1)-1}$ factorial experiments in s^n and $2s^n$ selected treatment combinations such that the main effects are orthogonally estimated. Since complete sets of MOLS form a particular case of orthogonal arrays, the fractional replicate plans described in Section 15.4 form a special case of the present method of this section.

Partially balanced arrays (N, k, s, t) with any λ-parameters of Chakravarti (1956) can also be used to get fractional replicate plans of an s^k experiment in N treatment combinations. However, they will not be orthogonal main-effect plans. Furthermore in such plans the estimability of all main effects depends on N and the λ-parameters.

15.7. SPECIAL TYPES OF 2^n FRACTIONAL REPLICATE PLANS

Box and Hunter (1961a) divided the fractional replicate plans into various types, as follows:

1. A fractional replicate plan is said to be of resolution III type if no main

effect is mixed up with any other main effect but main effects are confounded with two-factor interactions and two-factor interactions with one another.

2. A fractional replicate plan is said to be of resolution IV type if no main effect is mixed up with any other main effect or two-factor interaction but two-factor interactions are confounded with one another.

3. A fractional replicate plan is said to be of resolution V type if no main effect is mixed up with any other main effect or two-factor interaction but two-factor interactions are confounded with three-factor interactions.

In general we can say that a fractional replicate plan is of resolution R type if no p-factor effect is confounded with any other effect containing less than $R - p$ factors. A resolution R type 2^{n-k} experiment will be denoted by 2_R^{n-k}. Clearly R denotes the minimum number of factors involved in the interactions of the defining relations.

Fractional replicate plans derivable from the Hadamard matrices of Section 15.5 are of resolution III type. These plans exist for 2^{4t-1} factorial experiments in $4t$ treatment combinations for many values of t. These designs are also called *saturated designs* since no degree of freedom is left for estimating error. If the number of factors is not congruent to 3 (mod 4), fractional replicate plans of resolution III type are obtained by omitting some factors from the resolution III design of the next higher order.

The treatment combinations that form a fractional replicate plan of resolution IV type form an orthogonal array of strength 3. These plans are slightly advantageous over resolution III designs because the main-effect estimates are not contaminated by the two-factor interactions, whether the latter are negligible or not. Box and Wilson (1951) gave an interesting method of constructing resolution IV plans. If H_n is a Hadamard matrix of order n, then

$$(15.7.1) \qquad \begin{bmatrix} H_n \\ -H_n \end{bmatrix}$$

leads to a resolution IV plan of a 2^n factorial experiment in $2n$ treatment combinations by identifying the columns and rows of (15.7.1) with the n factors and $2n$ fractional treatment combinations and interpreting 1 (or -1) in the (i, j)th position to the second level (or first level) of the jth factor in the ith treatment combination. For an intermediate number of factors we obtain the required plan by omitting factors from the next available order resolution IV plan.

If one is interested in estimating two-factor interactions clearly, resolution V type fractional replicate plans are helpful. The selected treatment combinations will form an orthogonal array of strength 4. In general resolution V type fractional replicate plans require a large number of runs, and for such cases certain compromise plans were evolved by Addelman (1962b), and Whitwell and Morbey (1961). Addelman developed three classes of compromise plans, which we describe below.

In the class I compromise plan of a fraction of an s^k experiment consisting of s^n runs we consider the problem of estimating all main effects and two-factor interactions from among d specific factors. We consider the orthogonal main-effect plan with $(s^n - 1)/(s - 1)$ factors, represented by $X_1, X_2, \cdots, X_n$, and their generalized interactions in s^n treatment combinations. From these factors we omit those that are two-factor interactions of any two of the given d factors. The columns corresponding to the retained factors give the required class 1 compromise plan.

As an illustration let us consider the orthogonal main-effect plan in 2^3 treatment combinations consisting of the factors $X_1, X_2, X_3, X_1 X_2, X_1 X_3, X_2 X_3$, and $X_1 X_2 X_3$. Let us be interested in studying the two-factor interactions between the $d = 3$ factors X_1, X_2, X_3. The plan accommodates only four factors with the representations X_1, X_2, X_3, and $X_1 X_2 X_3$. We observe that the interactions $X_1 X_2, X_1 X_3$, and $X_2 X_3$ are omitted, as they are two-factor interactions of the three selected factors X_1, X_2, X_3. Thus the plan for the four factors X_1, X_2, X_3, and $X_1 X_2 X_3$, which can be designated by A, B, C, and D, is

(15.7.2)

A	B	C	D
0	0	0	0
0	0	1	1
0	1	0	1
0	1	1	0
1	0	0	1
1	0	1	0
1	1	0	0
1	1	1	1

In the class 2 compromise plans of fractional replicate plans of s^k experiments in s^n runs we consider the problem of orthogonally estimating main effects, two-factor interactions of d factors of one set, and two-factor interactions of $k - d$ factors of the remaining set. We again consider the orthogonal main-effect plan with $(s^n - 1)/(s - 1)$ factors, represented by $X_1, X_2, \cdots, X_n$, and their generalized interactions in s^n runs. We choose d-factor representations from this plan in such a manner that the interaction of any two or three of these representations is not among those chosen. These factors then represent the d factors of the first set. Excluding the d factors of the first set and their two-factor interactions, we choose $k - d$ factor representations such that (a) the interaction of no two or three of these representations is among them and (b) the interaction of no two of these representations is one of the factor representations or the interactions of two of the representations of the first set.

As an illustration let us consider the orthogonal main-effect plan in 2^5 treatment combinations with the factors X_1, X_2, X_3, X_4, X_5 and their generalized interactions. The factors X_1, X_2, X_3 are independent, and we select them to be our $d = 3$ factors of the first set. The factor representations X_4, X_5, $X_1 X_2 X_3$, $X_1 X_4 X_5$ are such that no two- or three-factor interactions are the main effects or two-factor interactions of X_1, X_2, and X_3. Hence these four factors qualify to be the factors of the second set. If A, B, C, D, E, F, G are used to designate these factors, the plan can be easily completed as in (15.7.2).

In class 3 compromise plans of s^k experiments in s^n treatment combinations we consider the problem of orthogonally estimating all main effects and all two-factor interactions that contain any one of d specified factors. We once again consider the orthogonal main-effect plan in $(s^n - 1)/(s - 1)$ factors, represented by X_1, X_2, $\cdots$, X_n, and their generalized interactions in s^n treatment combinations. We choose the k factors in such a manner that none of them includes two- or three-factor interactions of any two or three of the d given factors. In order to obtain uncorrelated estimates of the remaining $k - d$ factors and all the interactions of each of the d factors with the remaining $k - d$ factors, it is necessary that none of the k factors be represented by the interactions of any one or two of the d interacting factors with any of the remaining $k - d$ factors.

As an illustration, let us consider the orthogonal main-effect plan for 15 factors represented by X_1, X_2, X_3, X_4 and their generalized interactions in 16 runs. Let two factors X_1, X_2 be interacting and let them be the given $d = 2$ factors. The remaining factors X_3, X_4, and $X_3 X_4$ satisfy the requirement of the above construction; choosing the five factors X_1, X_2, X_3, X_4, and $X_3 X_4$, the plan can be completed as in (15.7.2).

Tables of compromise plans constructible by these methods were given by Addelman (1962b).

For the literature on other resolution-type designs we refer to Addelman (1965) and Draper and Mitchell (1967, 1968).

15.8. IRREGULAR FRACTIONS OF 2^n DESIGNS

Irregular fractions of 2^n factorial experiments have been studied by Addelman (1961), Banerjee (1950), Banerjee and Federer (1966), Cochran and Cox (1957), Davies (1945), Dykstra (1959), Finney (1945), Fry (1961), and Rao (1947).

We have already seen that there are s^m different fractions of the $1/s^m$

replicate of an s^n experiment. Of these, if we select k sets, we get a k/s^m replicate of an s^n experiment, which can also be denoted by $k(s^{n-m})$. Addelman (1961) investigated the properties of $k(2^{n-m})$ irregular fractions and concluded that the most useful irregular fractional plans are those for which $k = 3$. He also observed that when $k > 3$ and is odd, the plans are either inefficient or there are plans with fewer than $k2^{n-m}$ trials that permit uncorrelated estimates of the desired effects and interactions.

15.9. FRACTIONAL PLANS FOR ASYMMETRICAL FACTORIAL EXPERIMENTS

Though fractions of symmetrical factorial experiments were evolved in 1945, fractions of asymmetrical factorial experiments were considerably developed only in the last decade. Work on fractions of asymmetrical factorial experiments has been described by Addelman (1962a,b), Box and Hunter (1961a,b), Box and Wilson (1951), Connor and Zelen (1959), and John (1962).

Before we proceed with the study of fractions of asymmetrical factorial plans, we state the following theorem, whose proof is left as an exercise to the reader:

Theorem 15.9.1. In a factorial experiment with N treatment combinations if the ith level of factor A, the jth level of factor B, and the ith level of factor A, jth level of factor B occur in $N_{i.}$ $N_{.j}$ and n_{ij} runs, then the main effects of factors A and B are orthogonally estimated if the frequencies are proportional, that is,

$$n_{ij} = \frac{N_{i.} N_{.j}}{N}.$$

Generalizing the above result, we have the following:

Theorem 15.9.2. In an experiment with N runs the AB interaction will be orthogonally estimated with the main effect of C if

$$(15.9.2) \qquad n_{ijk} = \frac{N_{ij.} N_{..k}}{N},$$

where n_{ijk} is the number of runs in which the factors A, B, and C occur at the i, j, and k levels, respectively; $N_{ij.}$ is the number of runs in which the factors A and B occur at the i and j levels, and $N_{..k}$ is the number of runs in which the factor C occurs at the k level.

Chakravarti (1956) constructed fractional replicate plans for $s_1^m \times s_2^n$ designs. Let orthogonal main-effect plans exist for s_1^m and s_2^n experiments in M and N runs. Then by putting side by side for each run of the first plan every run of the second plan, we get a fractionally replicated design for an $s_1^m \times s_2^n$ experiment in MN runs that estimates all main effects and two-factor interactions between one s_1- and another s_2-level factors orthogonally.

Let us illustrate this method of construction by constructing a plan for a $2^7 \times 3^4$ experiment in 72 treatment combinations. The orthogonal main-effect plan for a 2^7 experiment with the factors A, B, C, D, E, F, and G in eight treatment combinations is

	A	B	C	D	E	F	G
	1	1	1	1	1	1	1
	0	1	0	1	0	1	0
	1	0	0	1	1	0	0
(15.9.3)	0	0	1	1	0	0	1
	1	1	1	0	0	0	0
	0	1	0	0	1	0	1
	1	0	0	0	0	1	1
	0	0	1	0	1	1	0

The orthogonal main-effect plan for a 3^4 experiment with the factors H, I, J, and K in nine treatment combinations is

	H	I	J	K
	0	0	0	0
	0	1	1	2
	0	2	2	1
	1	0	1	1
(15.9.4)	1	1	2	0
	1	2	0	2
	2	0	2	2
	2	1	0	1
	2	2	1	0

Chakravarti's method gives the following 72 runs for a $2^7 \times 3^4$ experiment that will estimate main effects and the two-factor interaction of one of the factors A, B, C, D, E, F, and G with one of the factors H, I, J, and K orthogonally:

A	B	C	D	E	F	G	H	I	J	K
1	1	1	1	1	1	1	0	0	0	0
1	1	1	1	1	1	1	0	1	1	2
					$\cdots$					
0	0	1	0	1	1	0	2	2	1	0

The generalization of this method is straightforward to include more than two different levels of factors.

Addelman (1962a) gave an ingenious method of collapsing the levels of factors to construct fractionally replicated orthogonal main-effect plans of asymmetrical factorial experiments. Let us consider an asymmetrical factorial experiment $s_1^{n_1} \times s_2^{n_2} \times \cdots \times s_k^{n_k}$, where without loss of generality we assume $s_1 > s_2 > \cdots > s_k$. Let us construct a fraction consisting of s_1^n runs, where

$$(15.9.5) \qquad n_1 + n_2 + \cdots + n_k < \frac{s_1^n - 1}{s_1 - 1}.$$

We first form an orthogonal main-effect plan in s_1^n treatment combinations in $(s_1^n - 1)/(s_1 - 1)$ factors. Of these factors we retain $n_1 + n_2 + \cdots + n_k$ factors. The levels of the first n_1 factors will be left as they are. For each of the next n_2 factors we show a many–one mapping of s_1 levels on s_2 levels, for each of the next n_3 factors we show a many–one mapping of s_1 levels on s_3 levels, etc. The plan thus arrived at can easily be seen to be an orthogonal main-effect plan for an asymmetrical factorial experiment. It is to be noted that the efficiency of the plan depends on the nature of the correspondences made.

We illustrate this method in the construction of a $5^2 \times 3 \times 2$ experiment in 25 runs. We first construct an orthogonal main-effect plan in 25 treatment combinations in six factors as follows:

A	B	C	D	E	F		A	B	C	D	E	F
0	0	0	0	0	0		2	3	0	3	1	4
0	1	1	2	3	4		2	4	1	0	4	3
0	2	2	4	1	3		3	0	3	3	3	3
0	3	3	1	4	2		3	1	4	0	1	2
0	4	4	3	2	1		3	2	0	2	4	1
1	0	1	1	1	1		3	3	1	4	2	0
1	1	2	3	4	0		3	4	2	1	0	4
1	2	3	0	2	4		4	0	4	4	4	4
1	3	4	2	0	3		4	1	0	1	2	3
1	4	0	4	3	2		4	2	1	3	0	2
2	0	2	2	2	2		4	3	2	0	3	1
2	1	3	4	0	1		4	4	3	2	1	0
2	2	4	1	3	0							

We now retain the first four factors. For the C levels we show the correspondence

$$(15.9.6) \qquad 0 \to 0, \qquad 1 \to 1, \qquad 2 \to 2, \qquad 3 \to 0, \qquad 4 \to 1,$$

and for the D levels we show the correspondence

(15.9.7) $0 \to 0, \qquad 1 \to 1, \qquad 2 \to 0, \qquad 3 \to 1, \qquad 4 \to 1.$

The required plan is then

A	B	C	D	A	B	C	D
0	0	0	0	2	3	0	1
0	1	1	0	2	4	1	0
0	2	2	1	3	0	0	1
0	3	0	1	3	1	1	0
0	4	1	1	3	2	0	0
1	0	1	1	3	3	1	1
1	1	2	1	3	4	2	1
1	2	0	0	4	0	1	1
1	3	1	0	4	1	0	1
1	4	0	1	4	2	1	1
2	0	2	0	4	3	2	0
2	1	0	1	4	4	0	0
2	2	1	1				

The levels of factors can be collapsed in various ways to yield different orthogonal main-effect plans for an asymmetrical factorial experiment; we refer the reader in this connection to Addelman (1962a).

Morrison (1956), Connor (1960), and Tukey (1959) developed certain nonorthogonal fractional replicate plans for asymmetrical factorial experiments. In this section we describe Connor's method, which is a generalization of Morrison's method. Connor's method of constructing fractions of $2^m \times 3^n$ experiments consists of adjoining fractions of a 3^n factorial to fractions of a 2^m experiment in such a way as to preserve two-factor-interaction estimates.

As an illustration of Connor's method, let us construct a 1/4 replicate of a $2^5 \times 3^2$ experiment. Let the two-level factors be A, B, C, D, and E and the three-level factors be F and G. Let L_1, L_2, L_3 be three 1/4 replicates of a 2^5 experiment based on the identity relationship

(15.9.8) $I = ABC = ABDE = CDE.$

Let M_1, M_2, M_3 be the three 1/3 replicates of a 3^2 experiment in factors F and G, based on the identity relationship

(15.9.9) $I = P(1, 1).$

We form 24 treatment combinations by adjoining each of the three treatment combinations of M_1 to each of the eight treatment combinations of L_1, as in Chakravarti's method, described in the beginning of this section. Similarly

Table 15.9.1

A	B	C	D	E	F	G
0	0	0	0	0	0	0
0	0	0	0	0	1	2
0	0	0	0	0	2	1
0	0	0	1	1	0	0
0	0	0	1	1	1	2
0	0	0	1	1	2	1
1	0	1	1	0	0	0
1	0	1	1	0	1	2
1	0	1	0	0	2	1
1	0	1	0	1	0	0
1	0	1	0	1	1	2
1	0	1	0	1	2	1
1	1	1	0	0	0	0
1	1	0	0	0	1	2
1	1	0	1	0	2	1
1	1	0	1	1	0	0
1	1	0	1	1	1	2
1	1	0	1	1	2	1
0	1	0	1	0	0	0
0	1	1	1	0	1	2
0	1	1	0	0	2	1
0	1	1	0	1	0	0
0	1	1	0	1	1	2
0	1	1	0	1	2	1

A	B	C	D	E	F	G
0	0	1	0	0	0	1
0	1	1	0	0	1	0
0	1	1	0	0	2	2
0	1	1	1	1	0	1
0	1	1	1	1	1	0
0	1	1	1	1	2	2
1	1	0	1	0	0	1
1	1	0	1	0	1	0
1	1	0	0	0	2	2
1	0	0	0	1	0	1
1	0	0	0	1	1	0
1	0	0	0	1	2	2
1	0	0	0	0	0	1
1	0	1	0	0	1	0
1	0	1	1	0	2	2
1	0	1	1	1	0	1
1	0	1	1	1	1	0
1	0	1	1	1	2	2
0	0	1	1	0	0	1
0	0	0	1	0	1	0
0	0	0	0	0	2	2
0	0	0	0	1	0	1
0	0	0	0	1	1	0
0	0	0	0	1	2	2

A	B	C	D	E	F	G
1	0	0	0	0	0	2
1	0	0	0	0	2	0
1	0	0	0	0	1	1
1	0	0	1	1	0	2
1	0	0	1	1	2	0
1	0	0	1	1	1	1
0	0	1	1	0	0	2
0	0	1	1	0	2	0
0	0	1	0	0	1	1
0	0	1	0	1	0	2
0	0	1	0	1	2	0
0	0	1	0	1	1	1
0	0	1	0	0	0	2
0	1	0	0	0	2	0
0	1	0	1	0	1	1
0	1	0	1	1	0	2
1	1	0	1	1	2	0
1	1	0	1	1	1	1
1	1	0	1	0	0	2
1	1	1	1	0	2	0
1	1	1	0	0	1	1
1	1	1	0	1	0	2
1	1	1	0	1	2	0
1	1	1	0	1	1	2

we can combine L_2 with M_2 to generate 24 more treatment combinations. Combining L_3 and M_3, we get 24 further treatment combinations, thus generating all the 72 required treatment combinations. The completed plan is shown in Table 15.9.1.

15.10. NONORTHOGONAL FRACTIONAL PLANS

The P_n-matrices [with elements 1 and -1 such that $P_n'P_n = (n-1)I_n + E_{n,n}$, introduced by Raghavarao (1959) as the best weighing designs, which we shall study in Chapter 17] lead to efficient nonorthogonal main-effect plans of a 2^{n-1} experiment in n treatment combinations.† The following plan derived from P_5 is the efficient nonorthogonal main-effect plan of a 2^4 experiment in five runs:

$$
(15.10.1) \qquad
\begin{array}{cccc}
0 & 0 & 0 & 0 \\
0 & 1 & 1 & 1 \\
1 & 0 & 1 & 1 \\
1 & 1 & 0 & 1 \\
1 & 1 & 1 & 0
\end{array}
$$

The rows of the plan (15.10.1) can easily be verified to be the assemblies of a partially balanced array. The other best weighing designs of Chapter 17 also lead to efficient nonorthogonal main-effect plans, and we leave it to the interested reader to translate the weighing-design results in terms of fractional plans.

We close this chapter by giving a series of nonorthogonal fractional replicate plans of 3^{4n+1} experiments in $8n+4$ design points developed by Raghavarao (1965) by using the S_n-matrices introduced as best weighing designs in Section 17.6.

The S_{4n+2} matrix is a $(4n+2)$th order square matrix with elements $+1$, -1, and 0 such that

$$(15.10.2) \qquad S_{4n+2}' S_{4n+2} = (4n+1)I_{4n+2}.$$

Let us omit the first column which can be assumed to contain 0 or 1 elements, without loss of generality, of S_{4n+2} to obtain a $(4n+2) \times (4n+1)$ matrix T. Then the $8n+4$ runs given by the arrangement

$$(15.10.3) \qquad \begin{bmatrix} T \\ -T \end{bmatrix},$$

† The efficiency is in terms of minimum average variance and generalized variance for the estimated main effects. In chapter 17, these will be called A-optimality and D-optimality respectively.

where the columns correspond to $4n + 1$ factors and the rows to treatment combinations, the elements -1, 0, and 1 representing the three levels of each factor, was shown by the author to provide the plan for a 3^{4n+1} experiment, where (a) all linear effects are orthogonally estimated, (b) all linear effects are orthogonally estimated to quadratic effects, and (c) quadratic effects are not necessarily orthogonally estimated.

REFERENCES

1. Addelman, S. (1961). Irregular fractions of the 2^n factorial experiments. *Technometrics*, **3**, 479–496.
2. Addelman, S. (1962a). Orthogonal main-effect plans for asymmetrical factorial experiments. *Technometrics*, **4**, 21–46.
3. Addelman, S. (1962b). Symmetrical and asymmetrical fractional factorial plans, *Technometrics*, **4**, 47–58.
4. Addelman, S. (1963). Techniques for constructing fractional replicate plans. *J. Am. Stat. Assn.*, **58**, 45–71.
5. Addelman, S. (1965). The construction of 2^{17-9} resolution V plan in eight blocks of 32. *Technometrics*, **7**, 439–443.
6. Banerjee, K. S. (1950). A note on fractional replication of factorial arrangements. *Sankhyā*, **10**, 87–94.
7. Banerjee, K. S., and Federer, W. T. (1966). On estimation and construction in fractional replication. *Ann. Math. Stat.*, **37**, 1033–1039.
8. Box, G. E. P., and Hunter, J. S. (1961a). The 2^{k-p} fractional factorial designs, I. *Technometrics*, **3**, 311–352.
9. Box, G. E. P., and Hunter, J. S. (1961b). The 2^{k-p} fractional factorial designs, II. *Technometrics*, **3**, 449–458.
10. Box, G. E. P., and Wilson, K. B. (1951). On the experimental attainment of optimum conditions. *J. Roy. Stat. Soc.*, **B13**, 1–45.
11. Brownlee, K. A., Kelly, B. K., and Loraine, P. K. (1948). Fractional replication arrangements for factorial experiments with factors at two levels. *Biometrika*, **25**, 268–282.
12. Chakravarti, I. M. (1956). Fractional replication in asymmetrical factorial designs and partially balanced arrays. *Sankhyā*, **17**, 143–164.
13. Cochran, W. G., and Cox, G. M. (1957). *Experimental Designs*, 2nd edition. Wiley, New York.
14. Connor, W. S. (1960). Construction of fractional factorial designs of the mixed $2^m 3^n$ series. *Contributions to Probability and Statistics*. Stanford University Press.
15. Connor, W. S., and Zelen, M. (1959). *Fractional Factorial Experiment Design for Factors at Three Levels*. National Bureau of Applied Mathematics Series, 54, Washington, D.C.

16. Davies, O. L. (1945). *The Design and Analysis of Experiments.* Hafner Publishing Company, New York.
17. Draper, N. R., and Mitchell, T. J. (1967). The construction of saturated 2_R^{k-p} designs. *Ann. Math. Stat.*, **38**, 1110–1126.
18. Draper, N. R., and Mitchell, T. J. (1968). Construction of the set of 256-run designs of resolution $\geqq 5$ and the set of even 512-run designs of resolution $\geqq 6$ with special reference to the unique saturated designs. *Ann. Math. Stat.*, **39**, 246–255.
19. Dykstra, O. (1959). Partial duplication of factorial experiments. *Technometrics*, **1**, 63–70.
20. Finney, D. J. (1945). The fractional replication of factorial arrangements. *Ann. Eugenics*, **12**, 291–301.
21. Fry, R. E. (1961). Finding new fractions of factorial experimental designs. *Technometrics*, **3**, 359–370.
22. John, P. W. M. (1962). Three-quarter replicates of 2^n designs. *Biometrics*, **18**, 172–184.
23. Kempthorne, O. (1947). A simple approach to confounding and fractional replication in factorial experiments. *Biometrika*, **34**, 255–272.
24. Kempthorne, O. (1952). *The Design and Analysis of Experiments.* Wiley, New York.
25. Kitagawa, T., and Mitome, M. (1953). *Tables for the Design of Factorial Experiments.* Baifuken Co., Ltd., Japan.
26. Morrison, M. (1956). Fractional replication for mixed series. *Biometrics*, **12**, 1–19.
27. National Bureau of Standards (1957). *Fractional Factorial Experiment Designs for Factors at 2 Levels.* National Bureau of Standards Applied Mathematics Series, 48.
28. Patel, M. S. (1962). On constructing the fractional replicate of the 2^m designs with blocks. *Ann. Math. Stat.*, **33**, 1440–1449.
29. Plackett, R. L., and Burman, J. P. (1946). The design of optimum multifactorial experiments. *Biometrika*, **33**, 305–325.
30. Raghavarao, D. (1959). Some optimum weighing designs. *Ann. Math. Stat.*, **30**, 295–303.
31. Raghavarao, D. (1965). A note on fractions of 3^{4n+1} plans. *Technometrics*, **7**, 69–71.
32. Rao, C. R. (1947). Factorial experiments derivable from combinatorial arrangements of arrays. *J. Roy. Stat. Soc.*, **B9**, 128–139.
33. Tippett, L. C. H. (1934). *Applications of Statistical Methods to the Control of Quality in Industrial Production.* Manchester Statistical Society.
34. Tukey, J. W. (1959). Little pieces of mixed factorials. (abstract). *Biometrics*, **15**, 641–642.
35. Whitwell, J. C., and Morbey, G. K. (1961). Reduced designs of resolution five. *Technometrics*, **3**, 459–478.
36. Youden, W. J. (1961). Partial confounding in fractional replication. *Technometrics*, **3**, 353–358.

BIBLIOGRAPHY

Addelman, S. Augmenting factorial plans to accommodate additional two level factors. *Biometrics*, **18**, 308–322 (1962).

Addelman, S., and Kempthorne, O. Some main effect plans and orthogonal arrays of strength 2. *Ann. Math. Stat.*, **32**, 1167–1176. (1961).

Banerjee, K. S., and Federer, W. T. On a special subset giving an irregular fractional replicate of a 2^n factorial experiment. *J. Roy. Stat. Soc.*, **B29**, 292–299 (1967).

Bose, R. C., and Connor, W. S. Analysis of fractionally replicated $2^m \times 3^m$ designs. *Bull. Intern. Stat. Inst.*, **37**, 142–160 (1960).

Bose, R. C., and Srivastava, J. N. Analysis of irregular factorial fractions. *Sankhyā*, **A26**, 117–144 (1964).

Bose, R. C., and Srivastava, J. N. Mathematical theory of factorial designs. I. Analysis. II. Construction. *Bull. Intern. Stat. Inst.*, **40**, 780–794 (1964).

Connor, W. S., and Young, S. *Fractional Factorial Designs for Experiments with Factors at Two and Three Levels.* National Bureau of Standards Applied Mathematics Series, 58, Washington, D.C., 1961.

Daniel, C. Fractional replication in industrial research. *Proceedings of the Third Berkeley Symposium on Mathematical Statistics and Probability*, Vol. 5, 1956, pp. 87–98. University of California Press, Berkeley.

Daniel, C. Sequences of fractional replicates in the 2^{p-q} series. *J. Am. Stat. Assoc.*, **57**, 403–429 (1962).

Das, M. N. Fractional replication as asymmetrical factorial designs. *J. Indian Soc. Agric. Stat.*, **12**, 159–174 (1960).

Das, M. N., Shukla, G. K., and Kartha, C. P. On a method of construction of symmetrical fractional factorials and related error correcting codes. *Calcutta Stat. Assn. Bull.*, **16**, 164–179 (1967).

Davies, O. L., and Hay, W. A. The construction and uses of fractional factorial designs in industrial research. *Biometrics*, **6**, 233–249 (1950).

Dey, A., and Saha, G. M. On some irregular fractions of 2^n factorial. *J. Indian Soc. Agric. Stat.*, **20**, 45–48 (1968).

Fisher, R. A. A system of confounding for factors with more than two alternatives giving completely orthogonal cubes and higher powers. *Ann. Eugenics*, **12**, 283–290 (1945).

John, P. W. M. Three-quarter replicates of 2^4 and 2^5 designs. *Biometrics*, **17**, 319–321 (1961).

Kishen, K. On fractional replication of general symmetrical factorial designs. *J. Indian Soc. Agric. Stat.*, **1**, 91–106 (1948).

Plackett, R. L. Some generalizations in multifactorial designs. *Biometrika*, **33**, 328–332 (1946).

Rao, C. R. On hypercubes of strength d and a system of confounding in factorial experiments. *Bull. Calcutta Math. Soc.*, **38**, 67–78 (1946).

Rao, C. R. The theory of fractional replication in factorial experiments. *Sankhyā*, **10**, 81–86 (1950).

Srivastava, J. N., and Bose, R. C. Some economic partially balanced 2^m factorial fractions. *Ann. Inst. Stat. Math.*, **18**, 57–73 (1966).

Webb, S. R. Some new incomplete factorial designs (abstract). *Ann. Math. Stat.*, **33**, 296 (1962).

Westlake, W. J. Composite designs based on irregular fractions. *Biometrics*, **21**, 324–336 (1965).

Williams, E. J. Confounding and fractional replication in factorial experiments. *J. Australian Inst. Agric. Sci.*, **15**, 145–153 (1949).

Yates, F. Complex experiments. *J. Roy. Stat. Soc.*, **2**, 181–247 (1933).

Youden, W. J., and Hunter, J. S. Partially replicated latin squares. *Biometrics*, **11**, 399–405 (1955).

CHAPTER 16

Rotatable Designs

16.1. INTRODUCTION AND DEFINITION

Let k factors $F_1, F_2, \cdots, F_k$ affect the yield on a particular character and let the expected yield y satisfy the functional relation

$$(16.1.1) \qquad y = f(x_1, x_2, \cdots, x_k),$$

where $x_1, x_2, \cdots, x_k$ are the levels of the factors $F_1, F_2, \cdots, F_k$, respectively, used for getting that response. We assume that f can be represented by a polynomial of degree d. We define the following:

Definition 16.1.1.† A k-dimensional design of order d is said to be constituted of n runs of the k factors

$$(16.1.2) \qquad \begin{aligned} & (x_{11}, x_{12}, \cdots, x_{1k}) \\ & (x_{21}, x_{22}, \cdots, x_{2k}) \\ & \qquad \cdots \\ & (x_{n1}, x_{n2}, \cdots, x_{nk}) \end{aligned}$$

if from the responses recorded at the n points all the coefficients in the dth degree polynomial are estimable.

The estimates of the coefficients in the polynomial f, as well as the accuracies of the estimates, can be calculated by the method of least squares. It was observed by Box and Hunter (1957) that, instead of considering the variances of individual coefficients, the accuracy of the estimated response at a point $(x_1, x_2, \cdots, x_k)$ should provide a criterion for the selection of design. They defined the following:

---

† We are using the notation x_{ij} to denote the level of the jth factor in the ith treatment combination, whereas in most of the existing literature x_{ij} denotes the level of the ith factor in the jth treatment combination.

291

Definition 16.1.2. A k-dimensional design of order d is said to be a rotatable design if the variance of the estimated response at the point $(x_1, x_2, \cdots, x_k)$ is a function of $\rho^2 = \sum_{i=1}^{k} x_i^2$.

By using the properties of a spherical distribution, Box and Hunter (1957) showed that the following conditions are necessary for the n design points to form a second-order rotatable design:

A. $\sum x_i = 0, \sum x_i x_j = 0, \sum x_i x_j^2 = 0, \sum x_i^3 = 0, \sum x_i x_j^3 = 0, \sum x_i x_j x_k^2 = 0,$ $\sum x_i x_j x_k = 0$, and $\sum x_i x_j x_k x_l = 0$ for $i \neq j \neq k \neq l$.

B(i). $\sum x_i^2 = \text{constant} = n\lambda_2$.

B(ii). $\sum x_i^4 = \text{constant} = 3n\lambda_4$.

C. $\sum x_i^2 x_j^2 = \text{constant for } i \neq j$.

D. $\sum x_i^4 = 3\sum x_i^2 x_j^2$ for $i \neq j$.

E. $\lambda_4/\lambda_2^2 > k/(k + 2)$.

The summations in the above five relations are being taken over the n design points. For a third-order rotatable design the design points were shown by them to satisfy additionally the following:

A_1. Each of the sums of powers or products of powers of the x_i points, with at least one power odd, is zero.

B_1. $\sum x_i^6 = \text{constant} = 15n\lambda_6$.

C_1(i). $\sum x_i^2 x_j^4 = \text{constant for } i \neq j$.

C_1(ii). $\sum x_i^2 x_j^2 x_k^2 = \text{constant for } i \neq j \neq k$.

D_1(i). $\sum x_i^6 = 5\sum x_i^2 x_j^4$ for $i \neq j$.

D_1(ii). $\sum x_i^2 x_j^4 = 3\sum x_i^2 x_j^2 x_k^2$ for $i \neq j \neq k$.

E_1. $(\lambda_2 \lambda_6)/\lambda_4^2 > (k + 2)/(k + 4)$.

It may be noted that conditions E and E_1 are nonsingularity conditions necessary for all of the $\binom{k + 2}{k}$ and $\binom{k + 3}{k}$ coefficients estimable in the case of second- and third-order rotatable designs, respectively. Condition E will always be satisfied by the mere addition of central points $(0, 0, \cdots, 0)$ for second-order designs. Gardiner, Grandage, and Hader (1959) established that the mere addition of central points will not enable a third-order rotatable design to satisfy condition E_1. We can also choose $\lambda_2 = 1$, to fix a particular design so that better comparisons of designs may be obtained.

16.2. SIMPLE GEOMETRICAL DESIGNS

From the very definition of rotatable designs, one may expect that a number of component sets of points, each set consisting of equidistant points from the center, might provide solutions to rotatable designs. Sets of points at a distance ρ from the center are called equiradial sets of radius ρ. Furthermore, an equiradial set is said to be a rotatable set of order d if all moments to order $2d$ are preserved by rotation. We can see that no single equiradial set can supply a second-order rotatable design, whereas two or more sets can provide a solution. For an equiradial set in two dimensions consisting of n points equally spaced on a circle, $n > 2d$ is the condition for the rotatability conditions of dth order to be satisfied. The nonsingularity condition requires that $n_0 \geqq 1$ central points be appended to the equiradial rotatable sets of order 2. Box and Hunter (1957) showed that, by taking equiradial rotatable sets of points with the same center, the ith set consisting of n_i points with radius ρ_i and n_0 central points, one can construct rotatable designs with $n = \sum_{i=0}^{s} n_i$ points satisfying

$$(16.2.1) \qquad \lambda_4 = \frac{nk\left\{ \sum_{i=0}^{s} n_i \rho_i^4 \right\}}{(k+2)\left\{ \sum_{i=0}^{s} n_i \rho_i^2 \right\}^2}.$$

Box and Hunter (1957) gave second order rotatable designs, and Gardiner, Grandage and Hader (1959) gave third order rotatable designs that can be constructed from different geometrical configurations.

16.3. SECOND- AND THIRD-ORDER ROTATABLE DESIGNS WITH THE HELP OF A TRANSFORMATION GROUP

The technique of a transformation group was used by Bose and Draper (1959) for the construction of second-order three-dimensional rotatable designs and was continued by Draper (1960a,b,c) and Herzberg (1964) in the construction of other rotatable designs.

Given a point in k dimensions $(x_1, x_2, \cdots, x_k)$, let T be a transformation

$$T(x_1, x_2, \cdots, x_k) = (x_2, x_3, \cdots, x_k, x_1).$$

Then

$$T^2(x_1, x_2, \cdots, x_k) = (x_3, x_4, \cdots, x_1, x_2), \cdots, T^k(x_1, x_2, \cdots, x_k)$$
$$= (x_1, x_2, \cdots, x_k).$$

Hence T^k is equivalent to the identical transformation I. Thus the transformations $T, T^2, \cdots, T^k = I$ form a cyclic group of order k. Further let us define k more transformations $R_1, R_2, \cdots, R_k$, where $R_i(x_1, x_2, \cdots, x_i, \cdots, x_k) = (x_1, x_2, \cdots, -x_i, \cdots, x_k)$ for $i = 1, 2, \cdots, k$. The $(k + 1)$ transformations $T, R_1, R_2, \cdots, R_k$ generate a nonAbelian group G of $k2^k$ transformations, given by

$$(16.3.1) \quad T^j, T^j R_1, T^j R_2, T^j R_1 R_2, \cdots, T^j R_1 R_2 \cdots R_k \quad j = 1, 2, \cdots, k.$$

The effect of these transformations on a general point $(x_1, x_2, \cdots, x_k)$ is to give rise to the $k2^k$ points with the coordinates

$$(16.3.2) \quad \begin{aligned} &(\pm x_1, \pm x_2, \cdots, \pm x_k), \\ &(\pm x_2, \pm x_3, \cdots, \pm x_1), \cdots, (\pm x_k, \pm x_1, \cdots, \pm x_{k-1}), \end{aligned}$$

and such a set will be denoted by $G(x_1, x_2, \cdots, x_k)$. If each distinct point of $G(x_1, x_2, \cdots, x_k)$ is replicated t times, the distinct runs of $G(x_1, x_2, \cdots, x_k)$ will be denoted by $(1/t)G(x_1, x_2, \cdots, x_k)$.

We now consider three-dimensional second-order rotatable designs constructed from point sets $G(x, y, z)$. The set of points $G(x, y, z)$ satisfies conditions A, B, and C of rotatability (see Section 16.1), and condition D will be satisfied if and only if the function $K(x, y, z)$, defined by

$$(16.3.3) \qquad K(x, y, z) = \tfrac{1}{3}(x^4 + y^4 + z^4 - 3y^2 z^2 - 3z^2 x^2 - 3x^2 y^2),$$

has zero value. With a little algebra, we can show that the point set $G(x, y, z)$, along with n_0 central points, is a rotatable design with the scaled condition $\lambda_2 = 1$ if

$$(16.3.4) \qquad z = \left\{ \frac{n}{8(s + t + 1)} \right\}^{1/2}, \qquad y = t^{1/2} z, \qquad x = s^{1/2} z,$$

where $n = 24 + n_0$, s is any specified positive real number, and t is any positive real number given by

$$(16.3.5) \qquad t = \tfrac{1}{2}\{3(s + 1) \pm \sqrt{5(s^2 + 6s + 1)}\}.$$

For this class of designs

$$(16.3.6) \qquad \lambda_4 = \frac{8(st + s + t)z^4}{n}.$$

This result can be stated as follows:

Theorem 16.3.1. The point set $G(x, y, z)$, along with n_0 central points, generates a second-order rotatable design in three dimensions if the generator (x, y, z) is given by (16.3.4).

If $K(x, y, z)$ is not zero, then this quantity summed over all points of the point set $G(x, y, z)$ is said to be the excess of the set $G(x, y, z)$, written as $Ex[G(x, y, z)]$. Clearly

$$(16.3.7) \quad Ex[G(x, y, z)] = 8(x^4 + y^4 + z^4 - 3y^2z^2 - 3z^2x^2 - 3x^2y^2).$$

Two different point sets with nonzero excesses can be combined to form a rotatable design:

Theorem 16.3.2. The points of $\frac{1}{3}G(a, a, a) \cup \frac{1}{4}G(c, 0, 0)$, along with n_0 central points, form a second-order rotatable design in three dimensions in $14 + n_0$ points if

$$(16.3.8) \qquad\qquad 2c^4 - 16a^4 = 0.$$

Proof. Clearly

$$(16.3.9) \qquad Ex[\tfrac{1}{3}G(a, a, a) \cup \tfrac{1}{4}G(c, 0, 0)] = -16a^4 + 2c^4,$$

and the set of points satisfies condition D of rotatability if this quantity is zero; hence the required result.

We add that the design points will be obtained by using the scaling condition $\lambda_2 = 1$, in which case $8a^2 + 2c^2 = n = 14 + n_0$, and hence $a^2 = n(2 - \sqrt{2})/8$ and $c^2 = n(2 - \sqrt{2})/2\sqrt{2}$.

This method of suitably combining different point sets with nonzero excesses to give zero total excess for constructing rotatable designs can be further extended; we refer the reader to the papers of Bose and Draper (1959) and Draper (1960a).

Draper (1960b,c) and Herzberg (1964) used the transform-group point sets for the construction of third-order rotatable designs. The four designs given by the point sets

$$\tfrac{1}{2}G(a, b, 0) \cup \tfrac{1}{2}G(c, d, 0) \cup \tfrac{1}{3}G(e, e, e);$$

$$\tfrac{1}{3}G(a, a, a) \cup \tfrac{1}{4}G(b, 0, 0) \cup \tfrac{1}{4}G(c, 0, 0) \cup \tfrac{1}{2}G(d, d, 0);$$

$$\tfrac{1}{3}G(a, a, a) \cup \tfrac{1}{4}G(b, 0, 0) \cup \tfrac{1}{4}G(c, 0, 0) \cup G(d, e, e);$$

$$\tfrac{1}{3}G(a, a, a) \cup \tfrac{1}{4}G(b, 0, 0) \cup \tfrac{1}{4}G(b, 0, 0) \cup \tfrac{1}{4}G(c, 0, 0) \cup G(d, e, e);$$

are third-order rotatable designs.

In addition to the excess function (16.3.7) defined on the point set $G(x, y, z)$, Draper (1960b) defined four more functions:

(16.3.10) $Ax[G(x, y, z)] = 8(x^2 + y^2 + z^2);$

(16.3.11) $Gx[G(x, y, z)] = 8(x^4y^2 + y^4z^2 + z^4x^2 - x^2y^4 - y^2z^4 - z^2x^4);$

(16.3.12) $Hx[G(x, y, z)] = 8(x^6 + y^6 + z^6 - 45x^2y^2z^2);$

(16.3.13) $Ix[G(x, y, z)] = 4(x^4y^2 + y^4z^2 + z^4x^2 + x^2y^4$
$$+ y^2z^4 + z^2x^4 - 18x^2y^2z^2).$$

The excess functions (16.3.7), (16.3.10), (16.3.11), (16.3.12), and (16.3.13) operate linearly on point sets, and given point sets represent a third-order rotatable design if all the five excess functions on those point sets have zero values. With the help of these excess functions various third-order rotatable designs were constructed by Draper (1960b,c) and Herzberg (1964).

Draper (1960b) showed that, if D_1, D_2, D_3, D_4, D_5, and D_6 represent the point sets

$$\{\tfrac{1}{3}G(a_1, a_1, a_1) \cup \tfrac{1}{4}G(b_1, 0, 0) \cup \tfrac{1}{4}G(c_1, 0, 0)\};$$
$$\{\tfrac{1}{3}G(a_2, a_2, a_2) \cup \tfrac{1}{3}G(b_2, b_2, b_2) \cup \tfrac{1}{4}G(c_2, 0, 0)\};$$
$$\{\tfrac{1}{2}G(a_3, a_3, 0) \cup \tfrac{1}{4}G(b_3, 0, 0) \cup \tfrac{1}{4}G(c_3, 0, 0)\};$$
$$\{\tfrac{1}{2}G(a_4, a_4, 0) \cup \tfrac{1}{3}G(b_4, b_4, b_4) \cup \tfrac{1}{4}G(c_4, 0, 0)\};$$
$$\{G(a_5, b_5, b_5) \cup \tfrac{1}{3}G(c_5, c_5, c_5)\};$$

and

$$\{G(a_6, b_6, b_6) \cup \tfrac{1}{4}G(c_6, 0, 0)\}$$

respectively, then of the 15 possible pairs, $D_1 \cup D_4$, $D_1 \cup D_6$, $D_2 \cup D_3$, $D_2 \cup D_4$, $D_2 \cup D_6$, $D_3 \cup D_4$, $D_3 \cup D_5$, and $D_3 \cup D_6$ all provide third-order rotatable designs for suitable values of the coordinates, whereas $D_1 \cup D_2$, $D_1 \cup D_3$, $D_1 \cup D_5$, and $D_2 \cup D_5$ do not give third-order rotatable designs. It is not known to this author whether the three combinations $D_4 \cup D_5$, $D_4 \cup D_6$, and $D_5 \cup D_6$ represent third-order rotatable designs or not. One can also possibly see whether three or more of the six point sets $D_1, D_2, \cdots, D_6$ represent third-order rotatable designs.

16.4. SECOND-ORDER ROTATABLE DESIGNS IN k DIMENSIONS FROM SECOND-ORDER ROTATABLE DESIGNS IN $k - 1$ DIMENSIONS

Draper (1960a) presented a method of constructing a second-order rotatable design in k dimensions from a second-order rotatable design in $k - 1$ dimensions. Let a second-order rotatable design exist in n' points with $k - 1$ factors and let the n' combinations be

(16.4.1) $[x_{i1}, x_{i2}, \cdots, x_{i(k-1)}]$ $i = 1, 2, \cdots, n'.$

Let

(16.4.2) $\sum_{i=1}^{n'} x_{i\alpha}^2 = A \neq n'$ for every $\alpha = 1, 2, \cdots, k - 1$

and

(16.4.3) $\sum_{i=1}^{n'} x_{ij}^4 = 3 \sum_{i=1}^{n'} x_{i\alpha}^2 x_{i\beta}^2 = 3C$ $\alpha \neq \beta; \alpha, \beta = 1, 2, \cdots, k - 1.$

Then the $2n' + 4$ points given by

(16.4.4)
$$
\begin{array}{llll}
[x_{i1}, x_{i2}, \cdots, x_{i(k-1)} & \pm a], & i = 1, 2, \cdots, n', \\
(0, \quad 0, \quad \cdots, & 0, & \pm b_1), \\
(0, \quad 0, \quad \cdots, & 0, & \pm b_2)
\end{array}
$$

along with n_0 central points constitute a second-order rotatable design in k dimensions if

(16.4.5)
$$ a^2 = \frac{C}{A} $$

and

(16.4.6) $b_i^2 = (A^2 - n'C) + (-1)^i \dfrac{\{2C(3A^2 - n'C) - (A^2 - n'C)^2\}^{1/2}}{2A},$

$$ i = 1, 2. $$

Since b_i^2 $(i = 1, 2)$ must be real and nonnegative, we must have

(16.4.7)
$$ 1 \leqq \frac{(A^2 - n'C)^2}{C(3A^2 - n'C)} = \phi \leqq 2. $$

We now illustrate this construction. Consider the second-order rotatable design in three dimensions with 14 points:

(16.4.8)
$$
\begin{array}{l}
(\pm a, \pm a, \pm a) \\
(\pm 1.682a, 0, 0) \\
(0, \pm 1.682a, 0) \\
(0, 0, \pm 1.682a)
\end{array}
$$

Here

(16.4.9) $n' = 14, \qquad A = 4(2 + \sqrt{2})a^2, \qquad C = 8a^4.$

We note that ϕ of (16.4.7) is 1.55, and we can construct from this arrangement a second-order four-dimensional rotatable design with 32 noncentral points:

$$
\begin{array}{llll}
(\pm a, & \pm a, & \pm a, & \pm 0.765a) \\
(\pm 1.682a, 0, & 0, & \pm 0.765a) \\
(0, & \pm 1.682a, 0, & \pm 0.765a) \\
(0, & 0, & \pm 1.682a, & \pm 0.765a) \\
(0, & 0, & 0, & \pm 2.049a) \\
(0, & 0, & 0, & \pm 1.122a)
\end{array}
$$

(16.4.10)

where a can be so chosen that $\lambda_2 = 1$.

For an alternative method of this embedding, we refer to Herzberg (1967b).

16.5. USE OF BIB DESIGNS AND SYMMETRICAL UNEQUAL-BLOCK ARRANGEMENTS IN THE CONSTRUCTION OF SECOND-ORDER ROTATABLE DESIGNS

Let v, b, r, k, and λ be the parameters of a BIB design. Let M be the incidence matrix of the design. From each column of M we can generate 2^k design points by considering all possible combinations $\pm a$ for the nonzero entries in that column. This procedure is called multiplication of the symbol combinations of that set (column) with the 2^v- factorial combinations (associate combinations) of levels $+1$ and -1. In this manner we can generate $2^k b$ design points, and these points, along with at least one central point, constitute a second-order rotatable design of v dimensions if $r = 3\lambda$ [cf. Box and Behnken (1960)].

As an illustration let us consider

$$
(16.5.1) \qquad M = \begin{bmatrix}
1 & 0 & 1 & 0 & 1 & 0 \\
1 & 0 & 0 & 1 & 0 & 1 \\
0 & 1 & 1 & 0 & 0 & 1 \\
0 & 1 & 0 & 1 & 1 & 0
\end{bmatrix},
$$

which is the incidence matrix of a BIB design with the parameters

$$
(16.5.2) \qquad v = 4, \qquad b = 6, \qquad r = 3, \qquad k = 2, \qquad \lambda = 1.
$$

From this incidence matrix we generate the 24 noncentral design points of a second-order four-dimensional rotatable design, shown in Table 16.5.1.

We denote the noncentral points obtained in the above method by $(v, b, r, k, \lambda) \times 2^k$. Das and Narasimham (1962) showed that second-order rotatable designs can be constructed from BIB designs even when $r \neq 3\lambda$.

Table 16.5.1

F_1	F_2	F_3	F_4	F_1	F_2	F_3	F_4
a	a	0	0	0	a	0	a
a	$-a$	0	0	0	a	0	$-a$
$-a$	a	0	0	0	$-a$	0	a
$-a$	$-a$	0	0	0	$-a$	0	$-a$
0	0	a	a	a	0	0	a
0	0	a	$-a$	a	0	0	$-a$
0	0	$-a$	a	$-a$	0	0	a
0	0	$-a$	$-a$	$-a$	0	0	$-a$
a	0	a	0	0	a	a	0
a	0	$-a$	0	0	a	$-a$	0
$-a$	0	a	0	0	$-a$	a	0
$-a$	0	$-a$	0	0	$-a$	$-a$	0

They showed that the points $(v, b, r, k, \lambda) \times 2^k \cup (1/2^{v-1})G(c, 0, 0, \cdots, 0)$ constitute a second-order rotatable design for a suitable choice of a and c when $r < 3\lambda$, and the points $(v, b, r, k, \lambda) \times 2^k \cup (1/v)G(d, d, \cdots, d)$ constitute a second-order rotatable design for a suitable choice of a and d when $r > 3\lambda$. Das and Narasimham further observed that the multiplication of symbol combinations can be done with a fraction of the factorial combinations based on a suitable identity relation that does not include interactions of four or fewer factors.

As an illustration, we construct the second-order rotatable design from a BIB design with the parameters

(16.5.3) $v = 6, \quad b = 10, \quad r = 5, \quad k = 3, \quad \lambda = 2.$

Since $r < 3\lambda$, the 80 points $(6, 10, 5, 3, 2) \times 2^3$ and the 12 points $\frac{1}{32}G(c, 0, 0, 0, 0, 0)$ provide the necessary 92 noncentral points; c and a will be so chosen that $c^2/a^2 = 2$, and a will be chosen so that $\lambda_2 = 1$.

As another illustration, let us consider a BIB design with the parameters

(16.5.4) $v = 5, \quad b = 10, \quad r = 4, \quad k = 2, \quad \lambda = 1.$

Since $r > 3\lambda$, the 40 points $(5, 10, 4, 2, 1) \times 2^2$ and the (c, c, c, c, c) symbol combination multiplied with a one-half replicate of 2^5 factorial combinations with 16 points provide a second-order rotatable design with 56 noncentral points; c and a will be so chosen that $c^2/a^2 = 1/2\sqrt{2}$, and a will be so chosen that $\lambda_2 = 1$. The 16 points corresponding to the multiplication of the (c, c, c, c, c) symbol combination are

(16.5.5)

$$
\begin{array}{llllll}
(-c, & -c, & -c, & -c, & -c) & \qquad (-c, & -c, & c, & c, & -c) \\
(\ c, & c, & -c, & -c, & -c) & \qquad (-c, & -c, & c, & -c, & c) \\
(\ c, & -c, & c, & -c, & -c) & \qquad (-c, & -c, & -c, & c, & c) \\
(\ c, & -c, & -c, & c, & -c) & \qquad (\ c, & c, & c, & c, & -c) \\
(\ c, & -c, & -c, & -c, & c) & \qquad (\ c, & c, & c, & -c, & c) \\
(-c, & c, & c, & -c, & -c) & \qquad (\ c, & c, & -c, & c, & c) \\
(-c, & c, & -c, & c, & -c) & \qquad (\ c, & -c, & c, & c, & c) \\
(-c, & c, & -c, & -c, & c) & \qquad (-c, & c, & c, & c, & c)
\end{array}
$$

The designs constructible by Das and Narasimham's method are of five-level factors. Raghavarao (1963) found that the point set $(v, b, r, k, \lambda) \times 2^k \cup [2^{k-1}(3\lambda - r)/2^{v-1}]G(a, 0, 0, \cdots, 0)$ constitutes a second-order rotatable design for a suitable choice of a in three-level factors when $r < 3\lambda$.†

Raghavarao (1963) used the SUB arrangements with two unequal block sizes of Chapter 11 in the construction of second-order rotatable designs. Let $v, b, r, k_1, k_2, b_1, b_2$, and λ be the parameters of a SUB arrangement (the parameters $u, n_1, n_2, p_{11}{}^1$ do not play any role in the present method of construction and hence are omitted). Let k_1 be the minimum of the two block sizes k_1 and k_2. If a $1/2^{k_2-k_1}$ replicate of 2^{k_2} factorial combinations exists without including any interaction of four or fewer factors in the identity relation, then by multiplying the unknown combinations arising from sets of size k_1 with 2^{k_1} factorial combinations and the unknown combinations arising from sets of size k_2 with a $1/2^{k_2-k_1}$ replicate of 2^{k_2} factorial combinations, we obtain a second-order rotatable design if $r = 3\lambda$. If $r \neq 3\lambda$, the modified methods discussed in this section can be used to get new designs.

From a SUB arrangement with the parameters

(16.5.6)
$$
\begin{array}{llll}
v = 15, & b = 16, & r = 6, & k_1 = 5, \\
k_2 = 6, & b_1 = 6, & b_2 = 10, & \lambda = 2,
\end{array}
$$

by using the above method, we construct a second-order rotatable design in $6 \times 2^5 + 10 \times 2^5 = 512$ noncentral points.

A list of second-order rotatable designs constructible by these methods was given by Das and Narasimham (1962) and Nigam and Das (1966).

---

† Each point of $(1/2^{v-1})G(a, 0, 0, \cdots, 0)$ is replicated $2^{k-1}(3\lambda - r)$ times in the expression $[2^{k-1}(3\lambda - r)/2^{v-1}] G(a, 0, 0, \cdots, 0)$.

16.6. CONSTRUCTION OF THIRD-ORDER ROTATABLE DESIGNS THROUGH INCOMPLETE BLOCK DESIGNS

Third-order rotatable designs are of two types: (a) sequential and (b) non-sequential. In sequential designs we first fit a second-order response surface by experimenting with the treatment combinations in one block; if the fit is inadequate, we continue the experiment on further points in another block so as to fit a third-order response curve. In nonsequential designs the experiment will be conducted in one occasion so as to fit a third-order response surface. In the case of sequential plans to estimate the polynomial coefficients clearly the levels of the ith factor must satisfy

$$(16.6.1) \qquad \frac{\sum_1 x_i{}^2}{\sum_2 x_i{}^2} = \frac{n_1 + n_{10}}{n_2 + n_{20}},$$

where n_1 and n_2 are the numbers of points excluding central points in the two blocks, respectively; n_{10} and n_{20} denote the central points that may have to be added to the two blocks; and $\sum_1$ and $\sum_2$ denote the summation over the points in the first and second blocks, respectively. It may be noted that condition (16.6.1) will be made to be satisfied by suitably choosing n_{10} and n_{20}.

If an arbitrary BIB design is taken and the $b2^k$ points $(v, b, r, k, \lambda) \times 2^k$ are formed as indicated in the preceding section, the plan obtained may not satisfy the condition $C_1(\text{ii})$ (see Section 16.1). However, if the design is a doubly balanced incomplete block design as defined in Chapter 7, the arrangement satisfies condition $C_1(\text{ii})$, but conditions $D_1(\text{i})$ and $D_1(\text{ii})$ may not be satisfied. By adding additional points, we can meet conditions $D_1(\text{i})$ and $D_1(\text{ii})$. Das and Narasimham (1962) showed that the $2^k b + 2^v + 2v + 2v(v - 1)$ points

$$(v, b, r, k, \lambda) \times 2^k \cup \frac{1}{v} G(c, c, \cdots, c) \cup \frac{1}{2^{v-1}} G(d, 0, \cdots, 0)$$

$$\cup \{(\pm e, \pm e, 0, \cdots, 0), (\pm e, 0, \pm e, \cdots, 0), \cdots, (\pm e, 0, 0, \cdots, \pm e),$$

$$\cdots, (0, 0, 0, \cdots, \pm e, \pm e)\}$$

form a third-order nonsequential rotatable design if a, c, d, e are chosen to satisfy conditions D, $D_1(\text{i})$, $D_1(\text{ii})$, and $\lambda_2 = 1$, where (v, b, r, k, λ) is a doubly balanced incomplete block design with suitable δ. It may be added that fractional replicates may be taken of the associate combinations while forming the multiplication with symbol combinations. It will also be necessary to include in the same design more than one set of the same type for getting positive solutions for all the levels.

As an illustration, let us consider the doubly balanced design with the parameters

(16.6.2) $v = 4 = b,$ $r = 3 = k,$ $\lambda = 2,$ $\delta = 1.$

From this design we can construct a nonsequential third-order rotatable design with 72 noncentral design points. Of these 32 will be of the form $(4, 4, 3, 3, 2) \times 2^3$; 8 points $(\pm c, 0, 0, 0), (0, \pm c, 0, 0), (0, 0, \pm c, 0), (0, 0, 0, \pm c)$; 8 points $(\pm d, 0, 0, 0), (0, \pm d, 0, 0), (0, 0, \pm d, 0), (0, 0, 0, \pm d)$; and 24 points $(\pm e, \pm e, 0, 0), (\pm e, 0, \pm e, 0), (\pm e, 0, 0, \pm e), (0, \pm e, \pm e, 0), (0, \pm e, 0, \pm e), (0, 0, \pm e, \pm e)$. Here a, c, d, e will be found such that $a^2/e^2 = 0.793701$, $c^2/e^2 = 2.577472$, and $d^2/e^2 = 0.957168$, and e will be so selected that $\lambda_2 = 1$.

Sequential third-order rotatable designs can also be constructed from the same design points. We have seen earlier that the distinct points of $G(c, c, \cdots, c)$ and $G(d, 0, \cdots, 0)$ will form noncentral points of a second-order rotatable design for a proper choice of c and d. These points constitute the noncentral points of the first block. In the second block we choose from a doubly balanced design $(v, b, r, k, \lambda) \times 2^k$ points and the distinct points of $G(e, 0, 0, 0, \cdots, 0)$ to be the noncentral points. The central points for both the blocks will be selected to satisfy (16.6.1). It will also be necessary to include in the same design more than one set of the same type for getting positive solutions for all the levels.

As an example, let us construct a sequential third-order rotatable design for six factors. The first block consists of 76 noncentral points $(\pm c, \pm c, \cdots, \pm c)$, $(\pm d, 0, \cdots, 0), (0, \pm d, \cdots, 0), (0, 0, \cdots, \pm d)$ and the second block consists of 184 points $(6, 20, 10, 3, 4) \times 2^3$ and $(\pm e, 0, \cdots, 0), (0, \pm e, \cdots, 0), \cdots, (0, 0, \cdots, \pm e), (\pm f, 0, \cdots, 0), (0, \pm f, \cdots, 0), \cdots, (0, 0, \cdots, \pm f)$, where $d^2/c^2 = 8$, $a^2/c^2 = 2.519842$, $e^2/c^2 = 5.039684$, and $f^2/c^2 = 5.039684$; c will be selected to satisfy $\lambda_2 = 1$, and the central points for the two blocks will be selected from (16.6.1).

Since each of the point sets, D_1, D_2, D_3, D_4, D_5, and D_6 described at the end of Section 16.3 are second-order rotatable designs, the third-order designs of Draper described there are sequential.

16.7. GENERALIZATIONS OF ROTATABLE DESIGNS

Herzberg (1966, 1967a) generalized the concept of rotatable designs to cylindrically rotatable designs that are rotatable for all factors except 1. Das and Dey (1967) further generalized this concept to group-divisible rotatable designs. According to them group-divisible rotatable designs are those for which the v factors of the design can be split into two groups, and the design is rotatable for each group of factors when the levels of the factors in the other group are held constant.

REFERENCES

1. Bose, R. C., and Draper, N. R. (1959). Second-order rotatable designs in three dimensions. *Ann. Math. Stat.*, **30**, 1097–1107.
2. Box, G. E. P., and Behnken, D. W. (1960). Some new three level designs for the study of quantitative variables. *Technometrics*, **2**, 455–475.
3. Box, G. E. P., and Hunter, J. S. (1957). Multifactor experimental designs for exploring response surfaces. *Ann. Math. Stat.*, **28**, 195–241.
4. Das, M. N., and Dey, A. (1967). Group-divisible rotatable designs. *Ann. Inst. Stat. Math.*, **19**, 331–347.
5. Das, M. N., and Narasimham, V. L. (1962). Construction of rotatable designs through B.I.B. designs. *Ann. Math. Stat.*, **33**, 1421–1439.
6. Draper, N. R. (1960a). Second order rotatable designs in four or more dimensions. *Ann. Math. Stat.*, **31**, 23–33.
7. Draper, N. R. (1960b). Third order rotatable designs in three dimensions. *Ann. Math. Stat.*, **31**, 865–874.
8. Draper, N. R. (1960c). Third order rotatable designs in four dimensions. *Ann. Math. Stat.*, **31**, 875–877.
9. Gardiner, D. A., Grandage, A. H. F., and Hader, R. J. (1959). Third order rotatable designs for exploring response surfaces. *Ann. Math. Stat.*, **30**, 1082–1096.
10. Herzberg, A. M. (1964). Two third order rotatable designs in four dimensions. *Ann. Math. Stat.*, **35**, 445–446.
11. Herzberg, A. M. (1966). Cyclindrically rotatable designs. *Ann. Math. Stat.*, **37**, 242–274.
12. Herzberg, A. M. (1967a). Cyclindrically rotatable designs of types 1, 2 and 3. *Ann. Math. Stat.*, **38**, 167–176.
13. Herzberg, A. M. (1967b). A method for the construction of second order rotatable designs in K dimensions. *Ann. Math. Stat.*, **38**, 177–180.
14. Nigam, A. K., and Das, M. N. (1966). On a method of construction of rotatable designs with smaller number of points controlling the number of levels. *Calcutta Stat. Assn. Bull.*, **15**, 147–157.
15. Raghavarao, D. (1963). Construction of second order rotatable designs through incomplete block designs. *J. Indian Stat. Assn.*, **1**, 221–225.

BIBLIOGRAPHY

Bose, R. C., and Carter, R. L. Complex representation in the construction of rotatable designs. *Ann. Math. Stat.*, **30**, 771–780 (1959).

Box, G. E. P., and Behnken, D. W. Simplex sum designs. A class of second order rotatable designs derivable from those of the first order. *Ann. Math. Stat.*, **31**, 838–864 (1960).

Box, G. E. P., and Draper, N. R. The choice of a second order rotatable design. *Biometrika*, **50**, 335–352 (1963).

Das, M. N. Construction of rotatable designs from factorial designs. *J. Indian Soc. Agric. Stat.*, **13**, 169–194 (1961).

Das, M. N. Construction of second order rotatable designs through B.I.B. designs with unequal block sizes. *Calcutta Stat. Assoc. Bull.*, **12**, 31–46 (1963).

Debaun, R. M. Response surface designs for three factors at three levels. *Technometrics*, **1**, 1–8 (1959).

Dey, A., and Nigam, A. K. Group divisible rotatable designs—some further considerations. *Ann. Inst. Stat. Math.*, **20**, 477–481 (1968).

Draper, N. R. Third order rotatable designs in three dimensions: some specific designs. *Ann. Math. Stat.*, **32**, 910–913 (1961).

Tyagi, B. N. A note on the construction of a class of second order rotatable designs. *J. Indian Stat. Assn.*, **2**, 52–54 (1964).

CHAPTER 17

Weighing Designs

17.1. INTRODUCTION

Yates (1935) was the first to observe that, when measurements are to be made of several quantities, by making the measurements on sets of objects rather than on individuals, increased accuracy of measurements can be achieved. Without loss of generality, we consider the measurements on weighing the objects, devoting this chapter to the study of weighing designs in detail.

Let a chemist be interested in weighing five objects whose true weights are w_1, w_2, w_3, w_4, and w_5 on a chemical balance that needs zero correction. We further assume that each recorded weight has variance σ^2. To find the weights of the five objects we need in all six weighings, and the standard technique assigns the weight to each object as the difference between the readings of the scale when carrying that object and when empty. Thus the standard error for an estimated weight is $\sqrt{2\sigma^2}$.

However, we can slightly reduce the standard error by taking an initial reading and weighing in sets of four, so that each object is weighed four times altogether—three times with any other object and once without it. Calling the readings from the scale $y_1, y_2, \cdots, y_6$, we then have the following equations to determine $w_1, w_2, \cdots, w_5$ and the bias w_0:

$$
\begin{aligned}
w_0 &= y_1, \\
w_0 + w_1 + w_2 + w_3 + w_4 &= y_2, \\
w_0 + w_1 + w_2 + w_3 \quad\;\; + w_5 &= y_3, \\
w_0 + w_1 + w_2 \quad\quad + w_4 + w_5 &= y_4, \\
w_0 + w_1 \quad\quad + w_3 + w_4 + w_5 &= y_5, \\
w_0 + \quad\; w_2 + w_3 + w_4 + w_5 &= y_6.
\end{aligned}
$$

(17.1.1)

305

From this observational setup, $w_1, w_2, \cdots, w_5$ can be estimated by the least squares technique, and

$$(17.1.2) \qquad \hat{w}_1 = \frac{-y_1 + y_2 + y_3 + y_4 + y_5 - 3y_6}{4},$$

where the circumflex stands for the estimated value. Similar expressions hold for $\hat{w}_2$, $\hat{w}_3$, $\hat{w}_4$, and $\hat{w}_5$. Here the standard error of each $\hat{w}_i$ ($i = 1, 2, \cdots, 5$) can be easily verified to be $(\frac{7}{8}\sigma^2)^{1/2}$. Thus by this technique the standard error of the estimate has been reduced.

A further improvement can be made by modifying the previous technique to include the objects in the other pan, which were not weighed in the previous technique. Calling the readings in this case $z_1, z_2, \cdots, z_6$, we have

$$(17.1.3) \qquad \begin{aligned} w_0 - w_1 - w_2 - w_3 - w_4 - w_5 &= z_1, \\ w_0 + w_1 + w_2 + w_3 + w_4 - w_5 &= z_2, \\ w_0 + w_1 + w_2 + w_3 - w_4 + w_5 &= z_3, \\ w_0 + w_1 + w_2 - w_3 + w_4 + w_5 &= z_4, \\ w_0 + w_1 - w_2 + w_3 + w_4 + w_5 &= z_5, \\ w_0 - w_1 + w_2 + w_3 + w_4 + w_5 &= z_6. \end{aligned}$$

The least-squares estimate for w_1 for the setup (17.1.3) is

$$(17.1.4) \qquad \hat{w}_1 = \frac{-z_1 + z_2 + z_3 + z_4 + z_5 - 3z_6}{8}.$$

Similar expressions hold for $\hat{w}_2$, $\hat{w}_3$, $\hat{w}_4$, and $\hat{w}_5$. The standard error of each estimate $\hat{w}_i$ ($i = 1, 2, \cdots, 5$) is $(\frac{7}{32}\sigma^2)^{1/2}$, which is a considerable improvement over the previous two techniques.

Thus we observe that we can attain greater efficiency in estimating the weights by weighing the objects in sets than by weighing them separately. We shall later obtain lower bounds for the variance of estimated weights and discuss the plans that yield the minimum variances for the estimated weights.

Weighing designs are of two kinds: (a) chemical-balance designs and (b) spring-balance designs; we study important problems connected with them in subchapters A and B, respectively. The concept of singular weighing designs was introduced by Raghavarao (1964) and opens up a new field of research; we study these aspects in subchapter C.

For an expository article on the work done on weighing designs till 1950 we refer to Banerjee (1950–1951). Chakrabarti (1962) and Federer (1955) also covered weighing designs in their books at considerable length.

A. Chemical-Balance Weighing Designs

17.2. THE MODEL

Let us consider the problem of weighing p objects in n weighings on a chemical balance. Let

$$
\begin{aligned}
x_{ij} = 1 \quad & \text{if the } j\text{th object is kept in the left} \\
& \text{pan in the } i\text{th weighing,} \\
(17.2.1) \qquad = -1 \quad & \text{if the } j\text{th object is kept in the right} \\
& \text{pan in the } i\text{th weighing,} \\
= 0 \quad & \text{if the } j\text{th object is not weighed in the} \\
& i\text{th weighing,}
\end{aligned}
$$

for $i = 1, 2, \cdots, n; j = 1, 2, \cdots, p$. The $n \times p$ matrix $X = (x_{ij})$ is called the weighing-design matrix. Let $w_1, w_2, \cdots, w_p$ be the true weights of the p objects and $\mathbf{w}$ be the column vector of these true weights. Let $y_1, y_2, \cdots, y_n$ be the readings of the scales in the n weighings and $\mathbf{y}$ be a column vector of the y readings. Then the readings can be represented by the matrix equation

$$
(17.2.2) \qquad\qquad \mathbf{y} = X\mathbf{w} + \mathbf{e},
$$

where $\mathbf{e}$ is a column vector of $e_1, e_2, \cdots, e_n$, which are the errors between observed and expected readings. We assume that $\mathbf{e}$ is a random variable distributed with mean 0 and dispersion matrix $\sigma^2 I_n$. If the balance is not corrected for bias, we can assume one of the true weights, say the first, to be representing the bias, and we choose the first column of X to be $E_{n,1}$. In our further discussion we thus omit the consideration of bias and assume that the balance is corrected for bias.

The least-squares estimate of the true weights is given by $\hat{\mathbf{w}}$, where

$$
(17.2.3) \qquad\qquad (X'X)\hat{\mathbf{w}} = X'\mathbf{y}.
$$

If $X'X$ is nonsingular—that is, X is of rank p—we call it a nonsingular weighing design; if $X'X$ is singular, we call it a singular weighing design. For a nonsingular weighing design we have

$$
(17.2.4) \qquad\qquad \hat{\mathbf{w}} = (X'X)^{-1}X'\mathbf{y}
$$

and

$$
(17.2.5) \qquad\qquad \operatorname{var}(\hat{\mathbf{w}}) = \sigma^2(X'X)^{-1}.
$$

In subchapters A and B, we limit our discussion to nonsingular weighing designs only.

17.3. VARIANCE LIMIT OF ESTIMATED WEIGHTS

We prove the following result giving the lower bound for the variance of the estimated weights:†

Theorem 17.3.1. For any weighing design X the variance of $\hat{w}_i$ for a particular i $(1 \leq i \leq p)$ cannot be less than σ^2/n.

Proof. Let x_i be the ith column of X and c_j be the jth column of $X(X'X)^{-1}$. Then from (17.2.4) and (17.2.5) it follows that

$$(17.3.1) \qquad\qquad \hat{w}_i = c_i'y$$

and

$$(17.3.2) \qquad\qquad \text{var}(\hat{w}_i) = \sigma^2(c_i'c_i).$$

Since $(X'X)^{-1}(X'X) = I_p$, we have $c_i'x_j = \delta_{ij}$, where δ_{ij} is the Kronecker delta. Applying the Schwarz inequality, we have

$$(17.3.3) \qquad\qquad (x_i'x_i)(c_i'c_i) \geq (x_i'c_i)^2 = 1.$$

Hence

$$(17.3.4) \qquad\qquad \text{var}(\hat{w}_i) \geq \frac{\sigma^2}{x_i'x_i} \geq \frac{\sigma^2}{n},$$

because x_i is a column vector with elements 1, -1, or 0 only.

We now investigate the necessary and sufficient conditions under which minimum variance is attained for the estimated weights.

Theorem 17.3.2. For a design matrix X the variances of each of the estimated weights are minimum if and only if $X'X = nI_p$.

Proof. To prove the necessity part we observe that the equality holds in (17.3.3) if and only if $c_i = kx_i$, for a constant k, which implies that $(X'X)^{-1}$ is a diagonal matrix such that $x_i'x_i = n$ and hence $X'X = nI_p$. The sufficiency part of the proof is obvious.

We consider a weighing design X to be optimal if it estimates each of the weights with minimum variance.

---

† This result was originally proved by Hotelling (1944), but here we give the proof of Moriguti (1954).

17.4. HADAMARD MATRICES AND OPTIMUM DESIGNS

Definition 17.4.1. A Hadamard matrix H_n of order n is an nth order square matrix with elements $+1$ and -1 such that $H'_n H_n = n I_n$.

It is obvious that if a Hadamard matrix H_n exists, then choosing any p of its columns we can form an optimum weighing design to weigh p objects in n weighings. Thus Hadamard matrices play a very useful role in the consideration of optimum weighing designs. In the preceding chapters we have seen various interrelationships of Hadamard matrices with other designs. We devote this section to the study of Hadamard matrices and related topics.

Theorem 17.4.1. A necessary condition for the existence of a Hadamard matrix H_n is that either $n = 2$ or $n \equiv 0 \pmod 4$.

Proof. Clearly

$$H_2 = \begin{bmatrix} 1 & 1 \\ 1 & -1 \end{bmatrix}$$

is a Hadamard matrix of order 2. For $n > 2$ the necessary condition follows by using the condition that any three of its columns be pairwise orthogonal.

It is to be remarked that it is not known whether the condition of Theorem 17.4.1 is sufficient for the existence of H_n though the matrices were constructed for a large number of $n \equiv 0 \pmod 4$.

Theorem 17.4.2. If H_m and H_n are two Hadamard matrices of orders m and n, respectively, the Kronecker product $H_m \times H_n$ is a Hadamard matrix of order mn.

Proof. Clearly $H_m \times H_n$ is an mnth order square matrix with elements $+1$ and -1, and

$$(17.4.1) \quad (H_m \times H_n)'(H_m \times H_n) = (H'_m H_m) \times (H'_n H_n) = mn I_{mn},$$

and hence $H_m \times H_n$ is a Hadamard matrix of order mn.

Theorem 17.4.3. Let $n - 1 = p^h \equiv 3 \pmod 4$, where p is a prime and h is a positive integer, and let $\alpha_0 = 0, \alpha_1, \alpha_2, \cdots, \alpha_{n-2}$ be elements of $GF(p^h)$. Let $K = (k_{ij})$ be an $(n-1) \times (n-1)$ matrix with elements

$$(17.4.2) \quad \begin{aligned} k_{ii} &= -1 & i &= 0, 1, \cdots, n-2, \\ k_{ij} &= \psi(\alpha_j - \alpha_i) & i &\neq j;\ i, j = 0, 1, \cdots, n-2, \end{aligned}$$

where $\psi(a)$ is the Legendre symbol, which has the value $+1$ if a is a quadratic residue of p^h, and -1 otherwise.† Then

(17.4.3)
$$H_n = \begin{bmatrix} +1 & E_{1,\,n-1} \\ E_{n-1,\,1} & K \end{bmatrix}$$

is a Hadamard matrix of order n.

Proof. If h_{ij} is the (i,j)th element of H_n, the scalar product of the zeroth and the ith columns

(17.4.4)
$$= h_{00} h_{0i} + h_{i0} h_{ii} + \sum_{\substack{j=0 \\ j \neq i}}^{n-2} \psi(\alpha_i - \alpha_j)$$

$$= 1 - 1 + 0 = 0.$$

The scalar product of the i_1th and the i_2th columns $(i_1 \neq i_2)$

(17.4.5)
$$= h_{0i_1} h_{0i_2} + h_{i_1 i_1} h_{i_1 i_2} + h_{i_2 i_2} h_{i_2 i_1} + \sum_{\substack{j=0 \\ j \neq i_1,\, i_2}}^{n-2} \psi(\alpha_{i_1} - \alpha_j)\psi(\alpha_{i_2} - \alpha_j)$$

$$= 1 - \psi(i_1 - i_2) - \psi(i_2 - i_1) - 1$$

$$= 0,$$

since $p^h \equiv 3 \pmod 4$ and $\psi(-a) = -\psi(a)$. Thus H_n is a Hadamard matrix.

For studying another method of constructing Hadamard matrices we need the following concept:

Definition 17.4.2. An S_n matrix is an nth order square matrix with elements 1, -1, and 0 such that the diagonal elements are filled by zero and the off-diagonal elements by 1 or -1, satisfying $S'_n S_n = (n-1)I_n$.

We show later that these matrices provide best weighing designs under some conditions. These matrices have been found to be useful by Belevitch (1950) in conference telefony. Shrikhande and Singh (1962) showed the relationship of these matrices to the pseudocyclic association scheme of Chapter 8. In this subsection we use them in the construction of Hadamard matrices. First we study the construction of these S_n-matrices.

---

† Though the classical Legendre symbol (a/p) is defined for a prime p as given in Appendix A, we call $\psi(a)$ the Legendre symbol, as it is a natural generalization and satisfies all the properties of the Legendre symbol. In fact both (a/p) and $\psi(a)$ agree with the standard notation for a group character of a multiplicative Abelian group. Hall (1967) called $\psi(a)$ simply a character.

Theorem 17.4.4. Let $n = p^h + 1$, where p is an odd prime and h is a positive integer such that $p^h \equiv 1 \pmod 4$. Let $\alpha_0 = 0, \alpha_1, \alpha_2, \cdots, \alpha_{n-2}$ be the elements of $GF(p^h)$. Let $M = (m_{ij})$ be an $(n-1) \times (n-1)$ matrix with the elements

(17.4.6)
$$\begin{aligned} m_{ii} &= 0 & i &= 0, 1, \cdots, n-2; \\ m_{ij} &= \psi(\alpha_j - \alpha_i) & i &\neq j; \ i, j = 0, 1, \cdots, n-2. \end{aligned}$$

Then the matrix

(17.4.7)
$$S_n = \begin{bmatrix} 0 & E_{1, n-1} \\ E_{n-1, 1} & M \end{bmatrix}$$

is the required S_n-matrix.

The proof of this result follows by verification that $S_n' S_n = (n-1)I_n$ along lines similar to those used in Theorem 17.4.3.

It is to be noted that S_n-matrices do not exist for all $n \equiv 2 \pmod 4$:

Theorem 17.4.5. A necessary condition for the existence of an S_n-matrix when $n \equiv 2 \pmod 4$ is that $(n-1, -1)_p = +1$ for all primes p, where $(a, b)_p$ is the Hilbert norm residue symbol.

Proof. Since S_n is a nonsingular matrix, $S_n' S_n \sim I_n$. As $S_n' S_n = (n-1)I_n$, we have

(17.4.8) $\quad C_p(S_n' S_n) = (-1, -1)_p(n-1, -1)_p^{n(n+1)/2} = C_p(I_n) = (-1, -1)_p.$

Since $n \equiv 2 \pmod 4$, the above condition implies that

(17.4.9)
$$(n-1, -1)_p = +1$$

for all primes, which is the required result.

We can easily deduce the following corollary of the above theorem:

Corollary 17.4.5.1. The necessary condition for the existence of an S_n-matrix will be satisfied only if $n - 1 = \prod_{i=1}^m p_i^{h_i}$, where p_i is an odd prime and h_i is a positive integer such that $p_i^{h_i} \equiv 1 \pmod 4$ for $i = 1, 2, \cdots, m$.

The above theorem shows that S_n-matrices are nonexistent for $n = 22, 34, 58, 78$, etc.

We now indicate the method of constructing Hadamard matrices from S_n-matrices by the following theorem:

Theorem 17.4.6 [Paley (1933)]. If S_n exists, H_{2n} can be constructed from it by

(17.4.10)
$$H_{2n} = S_n \times \begin{bmatrix} 1 & 1 \\ 1 & -1 \end{bmatrix} + I_n \times \begin{bmatrix} 1 & -1 \\ -1 & -1 \end{bmatrix},$$

where $\times$ is the Kronecker-product symbol.

The proof is again by verification that the columns of H_{2n} are orthogonal and is left as an exercise.

One may wonder whether the converse of the above theorem exists and S_n can always be constructed from H_{2n} when $n \equiv 2$ (mod 4). The answer is in the negative because H_{44} exists, whereas S_{22} does not.

Hadamard matrices up to $n = 100$, excepting $n = 92$, were constructed by Plackett and Burman (1946), who presented cyclical solutions. The Hadamard matrix for $n = 92$ was constructed by Baumert, Golomb, and Hall (1962) on an electronic computer. In Table 17.4.1 we list the theorems that can be used in constructing the Hadamard matrices of order $n \leq 100$.

In this section we have given methods covering the methods of constructing H_n-matrices for $n \leq 100$. For further details we refer to Hall (1967) and Dembowski (1968). If we consider a Hadamard matrix H_n to be skew if $H_n = I_n + L_n$, where $L'_n = -L_n$, and n to be an H-number if H_n exists, each of the following conditions implies that n is an H-number:

1. n is a product of H-numbers.
2. $n = 2^h, h > 0$.
3. $n = p^h + 1 \equiv 0$ (mod 4), where p is a prime.
4. $n = h(p^m + 1)$, where h is an H-number and p is a prime.
5. $n = m(m - 1)$, where m is a product of the numbers of the forms 2 and 3.
6. $n = m(m + 3)$, where m and $m + 4$ are products of the numbers of the forms 2 and 3.
7. $n = h_1 h_2 p^m(p^m + 1)$, where h_1 and h_2 are both H-numbers and p is a prime.

Table 17.4.1. Methods of constructing Hadamard matrices of order $n \leq 100$

n	Reference Theorem	n	Reference Theorem
2	17.4.1	52	17.4.6
4	17.4.2	56	17.4.2
8	17.4.2	60	17.4.3
12	17.4.6	64	17.4.2
16	17.4.2	68	17.4.3
20	17.4.6	72	17.4.2
24	17.4.2	76	17.4.6
28	17.4.3	80	17.4.2
32	17.4.2	84	17.4.6
36	17.4.6	88	17.4.2
40	17.4.2	92	Electronic computer
44	17.4.3	96	17.4.2
48	17.4.2	100	17.4.6

8. $n = h_1 h_2 s(s + 3)$, where h_1 and h_2 are both H-numbers and $s, s + 4$ are both of the form $p^h + 1$, where p is a prime.

9. $n = (q + 1)^2$, where both q and $q + 2$ are odd prime powers.

10. $n = (m - 1)^3 + 1$, with m as in condition 5.

11. $n = (m - 1)^2$, where $m - 2$ is a prime power $\equiv 1 \pmod 4$ and $m + 1$ is a product of the numbers of the forms 2 and 3.

12. $n = 2m(m + 1)$, where m is a product of the numbers of the form 2 and 3, and $m + 1$ is a prime.

13. $n = 4h(4m - 1)$; $4h$ is an H-number, and there are BIB designs with the parameters $v = 4m - 1 = b$, $r = 2m = k$, $\lambda = m$, and $v^* = 4m - 1 = b^*$, $r^* = k^*$, $\lambda^* = h - m + k^*$, with circulant incidence matrices.

14. $n = qm$, where $m = 1 + p^\alpha + p^{2\alpha} + \cdots + p^{h\alpha} \equiv 3 \pmod 4$, p is a prime, α is a positive integer such that $p^\alpha \equiv 1 \pmod 4$, h is a positive integer ≥ 2, and $q = m + 1 - 4p^{(h-1)\alpha}$ is of the form $2^s \prod_{i=1}^r (p_i^{\alpha_i} + 1)$, where $s \geq 0$, $p_1, p_2, \cdots, p_r$ are primes and $\alpha_1, \alpha_2, \cdots, \alpha_r$ are positive integers such that $p_i^{\alpha_i} \equiv 3 \pmod 4$ $(i = 1, 2, \cdots, r)$.

15. $n = 92, 116, 156, 172,$ or 232.

Condition 1 is our Theorem 17.4.2, condition 2 follows from Theorem 17.4.2 by taking the Kronecker product of H_2 with itself h times, and 3 is given in Theorem 17.4.3. A particular case of condition 4 is given in Theorem 17.4.6. Results 4 through 8 are due to Williamson (1944, 1947), and result 9 was given by Brauer (1953) for the case when q and $q + 2$ are prime. The general case of 9 follows from the BIB designs given by Stanton and Sprott (1958). Result 10 is due to Goldberg (1966), and 11 to Ehlich (1965). Results 12 and 13 were given by Wallis (1969a,b), and 14 was obtained by Spence (1967). Hadamard matrices H_{116} and H_{232} have been constructed by Baumert (1966), H_{156} was given by Baumert and Hall (1965a), and H_{172} was included in by Williamson (1944).

Hadamard matrices of orders 188, 236, 260, 268, 292, 356, and 376 remain unresolved for orders $n \leq 400$. The generalization of Hadamard matrices was considered by Cohn (1965) and Shrikhande (1964).

We now give some more relations between BIB designs and H_n, S_n-matrices.

Theorem 17.4.7. If S_n is a matrix as constructed in Theorem 17.4.4, then

$$(17.4.11) \quad N = \begin{bmatrix} E_{1, n-1} & 0_{1, n-1} \\ \tfrac{1}{2}(E_{n-1, n-1} - M - I_{n-1}) & \tfrac{1}{2}(E_{n-1, n-1} - M + I_{n-1}) \end{bmatrix}$$

is the incidence matrix of a BIB design with the parameters

$$(17.4.12) \quad v = n, \quad b = 2(n - 1), \quad r = n - 1, \quad k = \frac{n}{2}, \quad \lambda = \frac{n}{2} - 1.$$

Proof. From the properties of $\psi(a)$ it follows that

(17.4.13) $$MM' = (n-1)I_{n-1} - E_{n-1,\,n-1}$$

and

(17.4.14) $$E_{1,\,n-1}\,M = (ME_{n-1,\,1})' = 0_{1,\,n-1}.$$

We can easily verify that

(17.4.15) $$NN' = \frac{n}{2}I_n + \left(\frac{n}{2}-1\right)E_{n,\,n},$$

and hence N is the incidence matrix of a BIB design with the parameters given in (17.4.12).

Theorem 17.4.8. The existence of a Hadamard matrix H_n is equivalent to the existence of a symmetrical BIB design with the parameters

(17.4.16) $$v = b = n - 1, \qquad r = k = \frac{n}{2} - 1, \qquad \lambda = \frac{n}{4} - 1.$$

Proof. A Hadamard matrix H_n will remain a Hadamard matrix if any of its rows (or columns) are multiplied by -1. Hence, without loss of generality, we can assume H_n to be of the form

(17.4.17) $$H_n = \begin{bmatrix} 1 & E_{1,\,n-1} \\ E_{n-1,\,1} & D \end{bmatrix}.$$

Let D_1 be the matrix obtained from D by replacing -1 by 0. Then we can easily verify that D_1 is the incidence matrix of a BIB design with the parameters given in (17.4.16).

Conversely, if N is the incidence matrix of a BIB design with the parameters given in (17.4.16), then by changing 0 into -1 in N, we can get a matrix N^* and

(17.4.18) $$\begin{bmatrix} 1 & E_{1,\,n-1} \\ E_{n-1,\,1} & N^* \end{bmatrix}$$

can easily be verified to be a Hadamard matrix of order n. Thus the result is established.

We state the following result of Szekeres (1969) without proof:

Theorem 17.4.9. If $4t$ is an H-number, then a symmetrical BIB design with the parameters

(17.4.19) $$v = b = (4t)^2, \qquad r = k = 2t(4t-1), \qquad \lambda = 2t(2t-1)$$

exists.

17.5. EFFICIENCY CRITERIA

We have seen in the preceding sections that Hadamard matrices provide optimum chemical-balance weighing designs. But since they do not exist for all n and p, there is a need to select a best weighing design for a given situation. For this purpose we motivate three different definitions of efficiency.

Since the purpose of weighing designs is to estimate each of the individual weights, with whatever accuracy we estimate the individual weights, it is reasonable to expect the design to give minimal average variance for all the estimated weights. This led Kishen (1945) and Kiefer (1959) to define the following:

Definition 17.5.1. Of the class of all $n \times p$ weighing designs, design X is A-optimal if it has the least value for the trace of $(X'X)^{-1}$.

If $(X'X)^{-1} = (c_{ij})$, then the efficiency of the weighing design can be measured [Kishen (1945)] by the factor

$$(17.5.1) \qquad\qquad \frac{p}{n \sum_{i=1}^{p} c_{ii}}.$$

Instead of thinking of minimizing the average variance of all estimated weights, one may consider as a best weighing design the one that gives a minimum generalized variance of the estimated weights. Since $\sigma^2 (X'X)^{-1}$ is the dispersion matrix of the estimated weights, one may be interested in minimizing $|(X'X)^{-1}|$ or consequently maximizing $|X'X|$. This led Mood (1946) and Kiefer (1959) to define the following:

Definition 17.5.2. Of the class of all $n \times p$ weighing designs, design X is D-optimal if it has the maximum value of $|X'X|$.

If $l'w$ is a linear function of true weights, subject to the condition $l'l = 1$, then we can easily verify that the maximum value for the variance of $l'\hat{w}$ for all choices of l satisfying $l'l = 1$ is $\sigma^2 \lambda_{max}$, where λ_{max} is the maximum characteristic root of $(X'X)^{-1}$. This led Ehrenfeld (1955) and Kiefer (1959) to make the following definition:

Definition 17.5.3. Of the class of all $n \times p$ weighing designs, design X is E-optimal if it has the least value for λ_{max}, which is the maximum characteristic root of $(X'X)^{-1}$.

We can easily verify that the best weighing designs introduced earlier remain optimal in view of each of the three above definitions. In the next subsection we derive best weighing designs for the above three definitions when n is odd or $n \equiv 2 \pmod 4$.

17.6. BEST WEIGHING DESIGNS WHEN n IS ODD
OR $n \equiv 2$ (MOD 4)

Since the weights of p objects can be determined with at least p weighings, we derive our results for $n = p$. In conformity with the usual assumptions in the design of experiments, we assume the following:

1. The variances of the estimated weights are equal.
2. The estimated weights are equally correlated.

Subject to the above conditions, a weighing design X satisfies

(17.6.1) $$X'X = (c - d)I_n + dE_{n,n}.$$

Since $|X|$ is real and not equal to zero, we necessarily have

(17.6.2) $$c > d$$

and

(17.6.3) $$c + d(n - 1) > 0.$$

Relation (17.6.3) holds good when d is nonnegative. It also holds good for $c = n$ and $d = -1$. However, $d \leq -2$ leads to a contradiction, since then $0 < c + d(n - 1) \leq c - 2(n - 1)$; that is, $c > 2(n - 1) \geq n$, which is impossible as $c \leq n$ from the structure of $X'X$. Thus in this subsection we consider only the values of c and d that satisfy

(17.6.4) $$c > d \geq 0 \quad \text{or} \quad c = n, d = -1.$$

We now prove the following lemma, which will be useful for our later discussion:

Lemma 17.6.1

(a) Let $c = n$. Then d cannot be even (including zero) when n is odd and d cannot be odd (including -1) when n is even.

(b) Let $c = n - 1$. Then d cannot be even (including zero) when n is odd and d cannot be odd when n is even.

Proof. Let $\mathbf{x_i}$ and $\mathbf{x_j}$ be any two columns of X.

(a) When $c = n$, $\mathbf{x_i'x_j}$ will have n terms, each term being either $+1$ or -1. Since $\mathbf{x_i'x_j} = d$, amongst the n terms $(n - |d|)$ terms sum to zero. Hence n and $|d|$ should either be both odd or both even, and the statement follows.

(b) When $c = n - 1$, $\mathbf{x_i'x_j}$ will have n terms, each term being either $+1$, -1, or 0. Since $\mathbf{x_i'x_j} = d$, amongst the n terms $(n - d)$ terms sum to zero. If n is odd and d is even, $(n - d)$ will be odd and the $(n - d)$ terms cannot sum to zero unless there is a single zero term. Both $\mathbf{x_i}$ and $\mathbf{x_j}$ will contribute a single zero term to $x_i' x_j$ when and only when the zeros of x_i and x_j are in the same row. This is also the case for any two columns of X. Hence, if n is odd and d is even, we get a row of zeros in X, and in this case $|X|$ will be equal to zero, contrary to our assumption. Therefore d cannot be even when n is odd. Similarly it can be proved that d cannot be odd when n is even.

Theorem 17.6.1. Of the class of all weighing designs for odd n, design X, satisfying

$$(17.6.5) \qquad X'X = (n-1)I_n + E_{n,n},$$

is A-, D-, and E-optimal.

Proof. When X is a weighing design such that $X'X = (c-d)I_n + dE_{n,n}$, the variance of each estimated weight is

$$(17.6.6) \qquad \frac{[c + d(n-2)]\sigma^2}{(c-d)[c + d(n-1)]}.$$

Therefore the trace of $(X'X)^{-1}$ is

$$(17.6.7) \qquad \frac{n[c + d(n-2)]}{(c-d)[c + d(n-1)]},$$

which should be minimized in order for the design to be A-optimal. This amounts to maximizing

$$(17.6.8) \qquad \frac{(c-d)[c + d(n-1)]}{n[c + d(n-2)]} = f(c, d), \text{ say.}$$

We see that

$$(17.6.9)$$

$$f(n, 1) - f(c, d) = \frac{(n-1)d(2n - 2c + 2d - 1) + (2n - 2c - 1)(c - d)}{2n[c + d(n-2)]} > 0,$$

when $c < n$. If $c = n$, then

$$(17.6.10) \qquad f(n, 1) - f(n, d) = \frac{d(d-1)(n-2) + n(d^2 - 1)}{2n[n + d(n-2)]},$$

which is again greater than zero for all values of d, excepting 0 and 1; $d = 1$ is trivial, and for $d = 0$ (17.6.10) is less than zero. But Lemma 17.6.1 proves that d cannot be zero, since n is odd. Hence the best weighing design in this case has efficiency $f(n, 1)$, and this is attained when X satisfies (17.6.5). Thus it is A-optimal.

The maximum characteristic root of $(X'X)^{-1}$ is the minimum characteristic root of $(X'X)$, and we maximize this to make the design E-optimal. The minimum characteristic root for $X'X$ as given in (17.6.1) is $c - d$ unless $c = n$, $d = -1$, in which case 1 is the minimum characteristic root. The maximum value is attained for this minimum characteristic root when $c = n$, $d = 1$, in which case the design matrix X satisfies (17.6.5). Thus it is E-optimal.

If we let $f_1(c, d)$ be the value of $|X'X|$, as given in (17.6.1), we can see that $f_1(c, d)$ is a monotonic increasing function in c for a fixed d and also a monotonic decreasing function in d for a fixed c. Now

$$(17.6.11) \qquad f_1(n, 1) - f_1(n - 1, 0) = n(n - 1)^{n-1} > 0.$$

Again, after a little algebra, we can show that

$$(17.6.12) \qquad f_1(n, 1) > f_1(n, -1).$$

From (17.6.11) and (17.6.12), noting that $d \neq 0$ since n is odd, it follows that $|X'X|$ is maximum when X satisfies (17.6.5). Thus it is D-optimal.

Hence the proof is complete.

Definition 17.6.1. A P_n-matrix is an nth order square matrix with elements $+1$ and -1 such that $P_n'P_n = (n - 1)I_n + E_{n, n}$.

Clearly n is odd if P_n exists. Furthermore, in view of Theorem 17.6.1, P_n-matrices are A-, D-, and E-optimal weighing designs. As shown by the following theorem, P_n does not exist for all n:

Theorem 17.6.2. A necessary condition for the existence of P_n is that

$$(17.6.13) \qquad n = \frac{\alpha^2 + 1}{2},$$

where α is an odd integer.

Proof. We have

$$(17.6.14) \qquad |P_n'P_n| = |P_n|^2 = (n - 1)^{n-1}(2n - 1).$$

Since P_n is a matrix with integral elements, the right-hand side of (17.6.14) must be a perfect square. Since n is odd, this implies that $2n - 1$ must be a perfect square, say α^2. Then $n = (\alpha^2 + 1)/2$. Since n is an integer, α is an odd integer, and this completes the proof.

The construction of P_n-matrices is given by the following theorem:

Theorem 17.6.3. If $n = (\alpha^2 + 1)/2$, where α is an odd integer, and if there exists a symmetrical BIB design with the parameters

$$(17.6.15) \qquad v = b = n, \qquad r = k = \frac{n \pm \alpha}{2}, \qquad \lambda = \frac{n \pm 2\alpha + 1}{4},$$

then, by changing the zeros into -1 in the incidence matrix of the BIB design, we get a P_n-matrix.

Proof. Let the column vectors of the incidence matrix after the zeros are changed to -1 be $\mathbf{p_1}, \mathbf{p_2}, \cdots, \mathbf{p_n}$. The negative contribution to $\mathbf{p'_i p_j} = 2(r - \lambda) = (n - 1)/2$ $(i \neq j; i, j = 1, 2, \cdots, n)$. Therefore the positive contribution to $\mathbf{p'_i p_j} = (n + 1)/2$. Hence $\mathbf{p'_i p_j} = 1$ $(i \neq j; i, j = 1, 2, \cdots, n)$. Also $\mathbf{p'_i p_i} = n(i = 1, 2, \cdots, n)$. Thus $P'_n P_n = (n - 1)I_n + E_{n, n}$.

P_n-Matrices exist for a limited number of n, and the matrices for $n = 5, 13$, and 25 were given by Raghavarao (1959).

The following two theorems can be proved along lines similar to those used in Theorem 17.6.1:

Theorem 17.6.4. When $n \equiv 2 \,(\mathrm{mod}\ 4)$ and $n \neq 2$, the S_n-matrices of Definition 17.4.2 are A- and E-optimal.

Theorem 17.6.5. When $n \equiv 2 \,(\mathrm{mod}\ 4)$ and $n \neq 2$, the matrix X satisfying

$$(17.6.16) \qquad\qquad X'X = (n - 2)I_n + 2E_{n, n}$$

is D-optimal.

We can easily show that a necessary condition for the existence of the D-optimal X-matrix satisfying (17.6.16) is

$$(17.6.17) \qquad\qquad n = \frac{4 + (3\beta^2 + 4)^{1/2}}{3},$$

where β is an integer; this necessary condition will be satisfied in a few cases like $n = 6$ and $n = 66$. Such a design for $n = 6$ was given by Raghavarao (1960).

All the results of this section are due to Raghavarao (1959, 1960). In cases where best weighing designs are not available as given by Theorems 17.6.1, 17.6.4 and 17.6.5, Bhaskararao (1966) considered the second best weighing designs.

B.   Spring-Balance Weighing Designs

17.7. VARIANCE LIMITS OF ESTIMATED WEIGHTS AND OPTIMUM WEIGHING DESIGNS

In the case of spring-balance weighing designs the design matrix $X = (x_{ij})$ will have two elements 1 and 0 only, x_{ij} being equal to 1 when the jth object is weighed in the ith weighing and otherwise 0. If bias is present, it can be assumed to be one object, and its value is estimated by taking a column of the elements 1 in X corresponding to the bias. Moriguti (1954) showed along lines similar to those used in Theorem 17.3.1 that the minimum attainable variance for each of the estimated weights for a biased spring-balance weighing design is $4\sigma^2/n$. If we assume that the balance is corrected for bias, we prove the following:

Theorem 17.7.1. Of the class of all weighing designs for weighing $4t - 1$ objects in $4t - 1$ weighings, the incidence matrix of a symmetrical BIB design with the parameters

$$(17.7.1) \qquad v = 4t - 1 = b, \qquad r = 2t - 1 = k, \qquad \lambda = t - 1$$

is D-optimal as a spring-balance weighing design.

Proof. Consider the Hadamard matrix H_{4t}, whose first row and first column contain the element $+1$. In such an H_{4t}-matrix subtract the first row from each of the other $4t - 1$ rows and multiply each of the 2nd, 3rd, $\cdots$, 4th rows by $-\frac{1}{2}$ to get the matrix

$$(17.7.2) \qquad \begin{bmatrix} 1 & E_{1,\,4t-1} \\ 0_{4t-1,\,1} & L \end{bmatrix}.$$

Then L can easily be seen to be the incidence matrix of the BIB design with the parameters given in (17.7.1). Furthermore,

$$(17.7.3) \qquad |H_{4t}| = (-1)2^{4t-1}|L|,$$

and since H_{4t} has the maximum determinant value, $|L|$ also has the maximum determinant value and L is D-optimal.

Any spring-balance weighing design weighing n objects in n weighings with the properties that (a) the variances of the estimated weights are equal and (b) the estimated weights are equally correlated is the incidence matrix of a symmetrical BIB design with the parameters

$$(17.7.4) \qquad v = b = n, \qquad r = k, \qquad \lambda.$$

Applying the three known optimality criteria, we can easily prove the following:

Theorem 17.7.2. For a given n the incidence matrix of the symmetrical BIB design with the parameters given in (17.7.4) is A-optimal if $(r^2 - \lambda)/r^2(r - \lambda)$ is minimum; E-optimal if $r - \lambda$ is maximum; D-optimal if $(r - \lambda)^{v-1}r^2$ is maximum.

When $n = 13$, there are only three symmetrical BIB designs with $v = 13$ and the parameters

(17.7.5) $\qquad\qquad v = 13 = b, \qquad r = 4 = k, \qquad \lambda = 1;$

(17.7.6) $\qquad\qquad v = 13 = b, \qquad r = 9 = k, \qquad \lambda = 6;$

and

(17.7.7) $\qquad\qquad v = 13 = b, \qquad r = 12 = k, \qquad \lambda = 11.$

Applying Theorem 17.7.2, we can easily see that as a spring-balance weighing design the incidence matrix of (17.7.6) is D-optimal, that of (17.7.7) is A-optimal, and those of (17.7.5) and (17.7.6) are E-optimal.

The assumption of constant variance in the model is slightly unrealistic, as one can expect the variance to be proportional to the total weight on the balance. The spring-balance weighing-design problem was solved by Raghavarao, Sodhi, and Singh (1968) with the latter assumption.

C.   Singular Weighing Designs

17.8. NEED FOR CONSIDERING SINGULAR WEIGHING DESIGNS

We have already seen that a weighing design X is called singular if its rank is less than p. A singular weighing design may occur, for example, in any of the following cases:

1. *Bad designing.* As there are no tables of best weighing designs, an experimenter who desires to use a weighing design of a particular order has to construct one for himself before starting the weighing operations. In such circumstances bad designing may result in singular weighing designs.

2. *Laboratory observations.* Many scientists are of the opinion that they can achieve greater precision in their readings by repeating their experiments. The process of repeating weighing operations may also lead to singular weighing designs.

3. *Accidental.* Though optimum or best weighing designs have been selected by the experimenter, accidentally some objects may fall down and break in taking the weighings. In that case, if the experimenter continues his weighing operations, putting $x_{ij} = 0$ for the broken objects, he may finally obtain a singular weighing design. For example, let an experimenter plan to weigh four objects in four weighings with the optimum design

(17.8.1)
$$\begin{bmatrix} 1 & 1 & 1 & 1 \\ 1 & 1 & -1 & -1 \\ 1 & -1 & -1 & 1 \\ 1 & -1 & 1 & -1 \end{bmatrix}.$$

After two weighings the last two objects fall down and break. If he continues the weighing operations assuming $x_{ij} = 0$ for $i, j = 3, 4$, he finally gets the singular weighing design

(17.8.2)
$$\begin{bmatrix} 1 & 1 & 1 & 1 \\ 1 & 1 & -1 & -1 \\ 1 & -1 & 0 & 0 \\ 1 & -1 & 0 & 0 \end{bmatrix}.$$

Had the purpose of the experiment been to determine the total weight of the last two objects, then, it is interesting to note, the variance of its estimate would have been the same with both the optimum design (17.8.1) and the singular design (17.8.2), being equal to $\sigma^2/2$.

4. *Minimum variance for the estimated total weight of all the objects.* Banerjee (1966a) observed that the singular weighing design

(17.8.3)
$$\begin{bmatrix} 1 & 1 & 1 \\ 1 & 1 & 1 \\ 1 & -1 & 0 \\ 1 & -1 & 0 \end{bmatrix}$$

gives a variance of $\sigma^2/2$ to the estimated total weight, which is less than $3\sigma^2/4$, the variance of the estimated total weight with the optimum weighing design. Thus it seems that under certain circumstances singular weighing designs are more useful than even the optimum weighing designs.

The results of this subchapter are due to Raghavarao (1961, 1964) and Banerjee (1966a).

17.9. ESTIMABLE PARAMETRIC FUNCTIONS OF THE WEIGHTS FOR SINGULAR WEIGHING DESIGNS

Let the singular weighing design X be of rank r. In this case we can find r independent columns in X. Let us rearrange the columns of X so that the first r columns of X are independent. Let us also renumber the objects accordingly so that the matrix equation of the readings can be written as

(17.9.1) $\mathbf{y} = [X_1 \, X_2]\mathbf{w} + \mathbf{e}$

where X_1 is an $n \times r$ matrix and X_2 is an $n \times (p - r)$ matrix. It can easily be verified that a solution of the normal equations estimating $\hat{\mathbf{w}}$ is

(17.9.2)
$$\hat{\mathbf{w}} = \begin{bmatrix} (X_1' X_1)^{-1} X_1' \mathbf{y} \\ \mathbf{0} \end{bmatrix},$$

where $\hat{\mathbf{w}}$ is the estimate of $\mathbf{w}$. If $\mathbf{l}'\mathbf{w}$ is any linear function of the weights, we can easily see that it is estimable if and only if

(17.9.3) $\text{rank}\,(X') = \text{rank}\,(X' \mid \mathbf{l}).$

In weighing designs we are interested in finding unbiased estimates of the individual weights. Hence we try to find a necessary and sufficient condition for the estimability of the linear function of the weights $\boldsymbol{\rho}_i'\mathbf{w}$, where $\boldsymbol{\rho}_i$ $(i = 1, 2, \cdots, p)$ is the ith column vector of I_p. Let $\mathbf{x}_1, \mathbf{x}_2, \cdots, \mathbf{x}_p$ be the column vectors of X. Then we have the following:

Theorem 17.9.1.† A necessary and sufficient condition for the linear function of the weights $\boldsymbol{\rho}_i'\,\mathbf{w}$ to be estimable is that

(17.9.4) $\text{rank}\,[\mathbf{x}_1, \mathbf{x}_2, \cdots, \mathbf{x}_{i-1}, \mathbf{x}_{i+1}, \cdots, \mathbf{x}_p] = r - 1.$

The above theorem can be easily proved from the necessary and sufficient condition (17.9.3) for the estimability of the linear function of the weights $\boldsymbol{\rho}_i'\mathbf{w}$ and is left as an exercise.

An immediate consequence of the above theorem is the following corollary:

Corollary 17.9.1.1. The linear function of the weights $\boldsymbol{\rho}_{r+j}'\mathbf{w}$ is not estimable $(j = 1, 2, \cdots, p - r)$.

Let $\mathbf{l}'\mathbf{w}$ be an estimable linear function of the weights for a singular weighing design. With the view of increasing the precision of the estimate, one may like to take m more weighings by a design matrix Z. The additional weighings sometimes increase the precision of the estimate of $\mathbf{l}'\mathbf{w}$ and sometimes keep it unchanged. In fact we state the relevant result in the following theorem without proof:

Theorem 17.9.2. If $\mathbf{l}'\mathbf{w}$ is an estimable linear function of weights for a singular weighing design X, the additional weighings by a design matrix Z do not increase the precision of the estimate of $\mathbf{l}'\mathbf{w}$ if the rows of Z are independent of the rows of X. This condition is not necessary for retaining the equal precision under the design matrices X and $\begin{pmatrix} X \\ Z \end{pmatrix}$.

In fact a still stronger result holds. Subject to the condition of the above theorem, the best linear unbiased estimator of $\mathbf{l}'\mathbf{w}$ will be same under the

† This result is similar to the result of Problem 1.5 of Scheffe (1959).

design matrices X and $\begin{pmatrix} X \\ Z \end{pmatrix}$. A particular case of Theorem 17.9.2 was considered by Banerjee (1966b).

In case the experimenter wants to find unbiased estimates of all the weights, he can do so by taking at least $p - r$ additional weighings. The methods of taking these additional weighings and some related topics form the contents of the subsequent sections.

17.10. A SIMPLE WAY OF TAKING $p - r$ ADDITIONAL WEIGHINGS FOR A SINGULAR WEIGHING DESIGN

A simple and obvious way of taking $p - r$ additional weighings, to obtain unbiased estimates of all the weights, is to use the additional design matrix

$$(17.10.1) \qquad [0_{(p-r),r} \quad A],$$

where A is a nonsingular square matrix of order $p - r$ with elements 1, -1, and 0 only.

Let $p - r$ additional weighings be taken, on the same balance on which the first n weighings were made, with the design matrix (17.10.1). Let $z_{(p-r),1}$ be the column vector of the results recorded in these additional weighings and ω the column vector of the errors in these results. These $p - r$ weighings can be represented by the matrix equation

$$(17.10.2) \qquad z = [0_{(p-r),r} \quad A] w + \omega.$$

The least-squares estimate $\hat{w}$ of w can be obtained from (17.9.1) and (17.10.2) and is

$$(17.10.3) \qquad \begin{bmatrix} X_1' X_1 & X_1' X_2 \\ X_2' X_1 & X_2' X_2 + A'A \end{bmatrix} \hat{w} = \begin{bmatrix} X_1' & 0_{r,(p-r)} \\ X_2' & A' \end{bmatrix} \begin{bmatrix} y \\ z \end{bmatrix};$$

that is

$$(17.10.4) \qquad S_1 \hat{w} = \begin{bmatrix} X_1' y \\ X_2' y + A'z \end{bmatrix}.$$

By applying Theorem A.4.4, we get

$$(17.10.5) \qquad |S_1| = |X_1' X_1| |A'A|,$$

and since $X_1' X_1$ and $A'A$ are nonsingular by hypothesis, S_1 is nonsingular, and we have

$$(17.10.6) \qquad \hat{w} = \begin{bmatrix} (X_1' X_1)^{-1} X_1' y - J(A'A)^{-1} A'z \\ (A'A)^{-1} A'z \end{bmatrix},$$

where

$$(17.10.7) \qquad J = (X_1' X_1)^{-1} X_1' X_2.$$

From (17.10.6) it is evident that the weights of the last $p - r$ objects are estimable from only the last $p - r$ weighings. The dispersion matrix of the above estimates is

$$(17.10.8) \qquad \sigma^2 \begin{bmatrix} (X_1' X_1)^{-1} + J(A'A)^{-1}J' & -J(A'A)^{-1} \\ -(A'A)^{-1}J' & (A'A)^{-1} \end{bmatrix}.$$

The above results can be summarized as follows:

Theorem 17.10.1. Given a singular weighing design, unbiased estimates of all the weights can be determined by taking $p - r$ additional weighings with the design matrix given in (17.10.1), where A is nonsingular. The estimates thus obtained are given by (17.10.6) with the dispersion matrix (17.10.8).

Let

$$(17.10.9) \qquad J' = [\xi_1 \ \xi_2 \ \cdots \ \xi_r],$$

where each ξ_i is a $(p - r)$th order column vector. Then

$$(17.10.10) \quad J(A'A)^{-1}J' = \begin{bmatrix} \xi_1'(A'A)^{-1}\xi_1 & \xi_1'(A'A)^{-1}\xi_2 & \cdots & \xi_1'(A'A)^{-1}\xi_r \\ \xi_2'(A'A)^{-1}\xi_1 & \xi_2'(A'A)^{-1}\xi_2 & \cdots & \xi_2'(A'A)^{-1}\xi_r \\ \cdot & \cdot & \cdots & \cdot \\ \cdot & \cdot & & \cdot \\ \xi_r'(A'A)^{-1}\xi_1 & \xi_r'(A'A)^{-1}\xi_2 & \cdots & \xi_r'(A'A)^{-1}\xi_r \end{bmatrix}.$$

Since $(A'A)^{-1}$ is positive definite, the diagonal elements of $J(A'A)^{-1}J'$ are nonnegative. In the dispersion matrix (17.10.8), as the variances of the new estimates cannot exceed those of the old estimates, it follows that, if the i_1th, i_2th $\cdots$, i_λth $(1 \leq i_j \leq r; j = 1, 2, \cdots, \lambda)$ objects have estimable weights in the first set of n weighings, then $\xi_{i_1} = \xi_{i_2} = \cdots = \xi_{i_\lambda} = 0_{(p-r),1}$.

Conversely, let $\xi_{i_1} = \xi_{i_2} = \cdots = \xi_{i_\lambda} = 0_{(p-r),1}$ $(1 \leq i_j \leq r; j = 1, 2, \cdots, \lambda)$. Then from (17.10.6) it follows that the estimates of the weights of the i_1th, i_2th, $\cdots$, i_λth objects do not make use of the $p - r$ additional weighings and hence are estimable from the first set of n weighings only. Thus we have the following:

Theorem 17.10.2. The weights of the i_1th, i_2th, $\cdots$, i_λth objects are estimable from the first set of n weighings if and only if $\xi_{i_1} = \xi_{i_2} = \cdots = \xi_{i_\lambda} = 0_{(p-r), 1}$ $(1 \leq i_j \leq r; j = 1, 2, \cdots, \lambda)$.

An interesting result is contained in the following:

Corollary 17.10.2.1. In a real singular weighing design (that is, a design in which each object is weighed at least once) of rank r, at most $r - 1$ objects have estimable weights.

Proof. If the weights of the first r objects are estimable from the above theorem, we have

$$(17.10.11) \qquad \xi_1 = \xi_2 = \cdots = \xi_r = 0_{(p-r), 1}.$$

Thus

$$(17.10.12) \qquad X_2 = 0_{n,(p-r)},$$

contrary to our assumption.

The combined setup, with the design matrix

$$\begin{bmatrix} X_1 & X_2 \\ 0_{(p-r),r} & A \end{bmatrix}$$

is D-optimal if the value of $|S_1|$ is maximum. Thus for a chosen X_1, $|S_1|$ can be maximized by maximizing $|A'A|$. But the choice of X_1 is not unique. Hence we obtain maximum precision for the estimates (17.10.6) for Mood's efficiency definition by choosing the r independent columns of X_1 that give the maximum value for $|X_1'X_1|$ and choosing A to be D-optimal for weighing $p - r$ objects in $p - r$ weighings. Our ideas will become clear with the help of the following:

Illustration 17.10.1. Let us consider the singular weighing design

$$(17.10.13) \qquad \begin{bmatrix} 1 & 1 & 1 & 0 \\ 1 & 1 & 1 & 0 \\ 1 & -1 & 0 & 1 \\ 1 & -1 & 0 & 1 \end{bmatrix}$$

of rank 2. Let x_1, x_2, x_3, x_4 be the four column vectors of the above design matrix. Any two of the four columns are independent, but $|X_1'X_1|$ will be maximum only when X_1 consists of x_1 and x_2. Furthermore, we know that the best weighing design of order 2 for Mood's efficiency definition is the Hadamard matrix H_2. Hence we take the additional weighings with the design matrix

$$(17.10.14) \qquad \begin{bmatrix} 0 & 0 & 1 & 1 \\ 0 & 0 & 1 & -1 \end{bmatrix}.$$

From (17.10.6) we obtain the estimates of the weights as

$$
(17.10.15) \qquad 4\hat{\mathbf{w}} = \begin{bmatrix} y_1 + y_2 + y_3 + y_4 - 2z_1 \\ y_1 + y_2 - y_3 - y_4 - 2z_2 \\ 2z_1 + 2z_2 \\ 2z_1 - 2z_2 \end{bmatrix},
$$

with the dispersion matrix

$$
(17.10.16) \qquad \sigma^2 \begin{bmatrix} \frac{1}{2} & 0 & -\frac{1}{4} & -\frac{1}{4} \\ 0 & \frac{1}{2} & -\frac{1}{4} & \frac{1}{4} \\ -\frac{1}{4} & -\frac{1}{4} & \frac{1}{2} & 0 \\ -\frac{1}{4} & \frac{1}{4} & 0 & \frac{1}{2} \end{bmatrix}.
$$

17.11. BEST WAY OF TAKING THE ADDITIONAL WEIGHING WHEN $r = p - 1$

In this section we consider singular weighing designs of rank $p - 1$ and try to find the optimum way—in view of D-optimum—of choosing the additional weighing, so that the best unbiased estimates of all the weights will be determined. Let an additional weighing be taken, on the same balance on which the first n weighings were made, according to the row vector

$$
(17.11.1) \qquad [\mathbf{A_1} \quad A_2],
$$

where $\mathbf{A_1}$ is a $1 \times p - 1$ row vector and A_2 is a scalar. Let z_1 be the result recorded in this additional weighing and ω_1 be the error in this result. The result of the additional weighing can be written as

$$
(17.11.2) \qquad z_1 = [\mathbf{A_1} \quad A_2]\mathbf{w} + \omega_1.
$$

From (17.9.1) and (17.11.2) we obtain the normal equations

$$
(17.11.3) \qquad \begin{bmatrix} X_1' X_1 + \mathbf{A_1'} \mathbf{A_1} & X_1' X_2 + \mathbf{A_1'} A_2 \\ X_2' X_1 + A_2 \mathbf{A_1} & X_2' X_2 + A_2 A_2 \end{bmatrix} \mathbf{L} = \begin{bmatrix} X_1' & \mathbf{A_1'} \\ X_2' & A_2 \end{bmatrix} \begin{bmatrix} \mathbf{y} \\ z \end{bmatrix};
$$

that is,

$$
(17.11.4) \qquad S_2 \hat{\mathbf{w}} = \begin{bmatrix} X_1' \mathbf{y} + \mathbf{A_1'} z_1 \\ X_2' \mathbf{y} + A_2 z_1 \end{bmatrix}.
$$

We now prove the following:

Lemma 17.11.1

(17.11.5) $|S_2| = |X_1' X_1|(A_2 - \mathbf{A_1}\mathbf{J})^2,$

where $\mathbf{J}$ is the column vector as defined in (17.10.8).

Proof. Applying Theorems A.4.4 and A.4.14, we get

$$(17.11.6) \quad |S_2| = |X_1'X_1 + \mathbf{A_1'}\mathbf{A_1}| \, |X_2'X_2 + A_2 A_2 - (X_2'X_1 + A_2\mathbf{A_1})$$
$$\times (X_1'X_1 + \mathbf{A_1'}\mathbf{A_1})^{-1}(X_1'X_2 + \mathbf{A_1'}A_2)|$$
$$= |X_1'X_1 + \mathbf{A_1'}\mathbf{A_1}| \, |X_2'X_2 + A_2^2 - (X_2'X_1 + A_2\mathbf{A_1})$$
$$\times \{(X_1'X_1)^{-1} - (X_1'X_1)^{-1}\mathbf{A_1'}[1 + \mathbf{A_1}(X_1'X_1)^{-1}\mathbf{A_1'}]^{-1}$$
$$\times \mathbf{A_1}(X_1'X_1)^{-1}\}(X_1'X_2 + \mathbf{A_1'}A_2)|$$
$$= |X_1'X_1 + \mathbf{A_1'}\mathbf{A_1}|(A_2 - \mathbf{A_1}\mathbf{J})^2[1 + \mathbf{A_1}(X_1'X_1)^{-1}\mathbf{A_1'}]^{-1}$$
$$= |X_1'X_1|(A_2 - \mathbf{A_1}\mathbf{J})^2,$$

since

(17.11.7) $|X_1'X_1 + \mathbf{A_1'}\mathbf{A_1}| = |X_1'X_1|\{1 + \mathbf{A_1}(X_1'X_1)^{-1}\mathbf{A_1'}\}.$

The lemma is thus established.

In order to make S_2 nonsingular, we choose $\mathbf{A_1}$ and A_2 such that $A_2 - \mathbf{A_1}\mathbf{J} \neq 0$. The value of the determinant of S_2 can be maximized for a given X_1, by choosing $\mathbf{A_1'}$ to be the column vector obtained from $\mathbf{J}$, where the nonnull elements are replaced by $+1$ or -1, depending on whether the element is positive or negative, and finally taking $A_2 = -1$. It is to be noted that $|S_2|$ can also be maximized for a given X_1 by choosing $\mathbf{A_1'}$ to have elements -1 or $+1$ according as the corresponding element of $\mathbf{J}$ is positive or negative and then selecting $A_2 = +1$.

Let the column vectors of $[X_1 \quad \mathbf{X_2}]$ be reshuffled so that another set of $p - 1$ independent column vectors occupies the first $p - 1$ positions. Let the newly obtained matrix be $[Y_1 \quad \mathbf{Y_2}]$ and let

(17.11.8) $\mathbf{L} = (Y_1'Y_1)^{-1}Y_1'\mathbf{Y_2}.$

Let $[\mathbf{H_1} \quad H_2]$ be a row vector, where $\mathbf{H_1'}$ is the column vector obtained from $\mathbf{L}$, where the nonnull elements are replaced by $+1$ or -1, depending on whether the element is positive or negative, and $H_2 = -1$. We now prove the following:

Lemma 17.11.2

(17.11.9) $|X_1'X_1|(A_2 - \mathbf{A_1}\mathbf{J})^2 = |Y_1'Y_1|(H_2 - \mathbf{H_1}\mathbf{L})^2.$

Proof. Let $\mathbf{x}_1, \mathbf{x}_2, \cdots, \mathbf{x}_p$ be the columns of $[X_1 \quad X_2]$. Without loss of generality, we assume that Y_1 consists of the columns $\mathbf{x}_1, \mathbf{x}_2, \cdots, \mathbf{x}_{t-1}$, $\mathbf{x}_p, \mathbf{x}_{t+1}, \cdots, \mathbf{x}_{p-1}$. Let $\mathbf{J}'$ be the row vector $(a_1, a_2, \cdots, a_{p-1})$. Then

$$(17.11.10) \qquad \mathbf{x}_p = a_1 \mathbf{x}_1 + a_2 \mathbf{x}_2 + \cdots + a_{p-1} \mathbf{x}_{p-1}.$$

We can easily see that $\mathbf{L}'$ is the row vector $-a_t^{-1}(a_1, \cdots, a_{t-1}, -1, a_{t+1}, \cdots, a_{p-1})$. Now evaluation of the determinants gives

$$(17.11.11) \qquad |Y_1'Y_1| = a_t^2 |X_1'X_1|.$$

We have

$$
\begin{aligned}
|X_1'X_1|(A_2 - \mathbf{A_1 J})^2 &= |X_1'X_1| \left\{ \sum_{i=1}^{p-1} |a_i| + 1 \right\}^2 \\
&= a_t^{-2} |Y_1'Y_1| \left\{ \sum_{i=1}^{p-1} |a_i| + 1 \right\}^2 \\
(17.11.12) \qquad &= |Y_1'Y_1| \left\{ \sum_{i=1}^{p-1} |a_i| a_t^{-1} + a_t^{-1} \right\}^2 \\
&= |Y_1'Y_1| \left\{ \sum_{\substack{i=1 \\ i \neq t}}^{p-1} |a_i| a_t^{-1} + 1 + a_t^{-1} \right\}^2 \\
&= |Y_1'Y_1| (H_2 - \mathbf{H_1 L})^2,
\end{aligned}
$$

and hence the lemma.

From the above lemma it is obvious that whatever X_1 is selected, by following the technique indicated in the paragraph just below Lemma 17.11.1, in selecting the additional row vector, $|S_2|$ can be uniquely maximized.

By selecting S_2 to be nonsingular, the solution of the normal equations (17.11.4) can be seen to be

$$(17.11.13)$$
$$(A_2 - \mathbf{A_1 J})\hat{\mathbf{w}} = \begin{bmatrix} (A_2 - \mathbf{A_1 J})(X_1'X_1)^{-1}X_1'\mathbf{y} + \mathbf{J A_1}(X_1'X_1)^{-1}X_1'\mathbf{y} - \mathbf{J}z_1 \\ z_1 - \mathbf{A_1}(X_1'X_1)^{-1}X_1'\mathbf{y} \end{bmatrix}.$$

The dispersion matrix of the estimates can be seen to be

$$(17.11.14)$$
$$\sigma^2 \begin{bmatrix} (X_1'X_1)^{-1} + \mathbf{J}F^{-1}\mathbf{J}' & -[\mathbf{J} + (X_1'X_1)^{-1} \\ + \mathbf{J}F^{-1}(A_2 - \mathbf{A_1 J})D\mathbf{A_1}(X_1'X_1)^{-1} & \times \mathbf{A_1}'D(A_2 - \mathbf{A_1 J})]F^{-1} \\ + (X_1'X_1)^{-1}\mathbf{A_1}'D(A_2 - \mathbf{A_1 J})F^{-1}\mathbf{J}' & \\ -F^{-1}[\mathbf{J}' + (A_2 - \mathbf{A_1 J})'D\mathbf{A_1}(X_1'X_1)^{-1}] & F^{-1} \end{bmatrix}.$$

where

(17.11.15) $D = \{1 + A_1(X_1'X_1)^{-1}A_1'\}^{-1}$, $F = (A_2 - A_1J)^2 D$.

These results are summarized as follows:

Theorem 17.11.1. Given a singular weighing design of rank $p - 1$, the objects will be efficiently estimated, for Mood's efficiency definition, by taking an additional weighing as the row vector $[A_1 \quad A_2]$ determined as in the paragraph below Lemma 17.11.1. The estimates of the weights are obtained from (17.11.13) with the dispersion matrix (17.11.14).

Illustration 17.11.1. Let us consider the singular weighing design

(17.11.16)
$$\begin{bmatrix} 1 & 1 & 1 \\ 1 & 1 & 1 \\ 1 & -1 & 0 \\ 1 & -1 & 0 \end{bmatrix}.$$

This singular weighing design has rank 2. Since the first two column vectors of (17.11.16) are independent, we can choose them for our X_1. Then we easily see that J, as defined by (17.10.8), is the column vector $(\frac{1}{2}, \frac{1}{2})'$. In order to maximize $|S_2|$, the additional weighing is to be selected according to the row vector

(17.11.17) $(1, \quad 1, \quad -1)$.

The estimated weights are given by

(17.11.18) $\hat{w} = \frac{1}{8} \begin{bmatrix} 2z_1 + y_1 + y_2 + 2y_3 + 2y_4 \\ 2z_1 + y_1 + y_2 - 2y_3 - 2y_4 \\ 2y_1 + 2y_2 - 4z_1 \end{bmatrix}$,

with the dispersion matrix

(17.11.19) $\sigma^2 \begin{bmatrix} \frac{7}{32} & -\frac{1}{32} & -\frac{1}{16} \\ -\frac{1}{32} & \frac{7}{32} & -\frac{1}{16} \\ -\frac{1}{16} & -\frac{1}{16} & \frac{3}{8} \end{bmatrix}.$

17.12. SOME RESULTS IN THE GENERAL CASE

In Section 17.10 we considered the problem of taking $p - r$ additional weighings, making use of only the last $p - r$ objects in the additional weighings. There is no reason why we should restrict our additional weighings to the last $p - r$ objects only. We now consider the general way of taking the $p - r$ additional weighings $(r < p - 1)$, making use of all the p objects in the additional weighings, to get unbiased estimates of the weights of all the objects. These results, however, are not so elegant as the results of the preceding subsection.

Let $p - r$ additional weighings be taken on the same balance on which the first n weighings were made, according to the design matrix

$$(17.12.1) \qquad\qquad [B_1 \quad B_2],$$

where B_1 is a $(p - r) \times r$ matrix and B_2 is a $(p - r) \times (p - r)$ matrix. Let $\mathbf{z}$ be the column vector of the results recorded in these additional weighings and $\boldsymbol{\omega}$ the column vector of errors in these results. The results of the additional weighings can be written as

$$(17.12.2) \qquad\qquad \mathbf{z} = [B_1 \quad B_2]\mathbf{w} + \boldsymbol{\omega}.$$

From (17.9.1) and (17.12.2) we obtain the normal equations

$$(17.12.3) \qquad \begin{bmatrix} X_1'X_1 + B_1'B_1 & X_1'X_2 + B_1'B_2 \\ X_2'X_1 + B_2'B_1 & X_2'X_2 + B_2'B_2 \end{bmatrix} \hat{\mathbf{w}} = \begin{bmatrix} X_1' & B_1' \\ X_2' & B_2' \end{bmatrix}\begin{bmatrix} \mathbf{y} \\ \mathbf{z} \end{bmatrix};$$

that is,

$$(17.12.4) \qquad\qquad S_3\hat{\mathbf{w}} = \begin{bmatrix} X_1'\mathbf{y} + B_1'\mathbf{z} \\ X_2'\mathbf{y} + B_2'\mathbf{z} \end{bmatrix}.$$

Analogous to Lemma 17.11.1, we can easily prove the following:

Lemma 17.12.1

$$(17.12.5) \qquad\qquad |S_3| = |X_1'X_1||B_2 - B_1J|^2,$$

where J is as defined in (17.10.8).

If B_1 and B_2 are selected so as to make S_3 nonsingular, it can be shown that the solution of the normal equations (17.12.4) is

(17.12.6)

$$\hat{\mathbf{w}} = \begin{bmatrix} (X_1'X_1)^{-1}X_1'\mathbf{y} + J(B_2 - B_1J)^{-1}B_1(X_1'X_1)^{-1}X_1'\mathbf{y} - J(B_2 - B_1J)^{-1}\mathbf{z} \\ (B_2 - B_1J)^{-1}\mathbf{z} - (B_2 - B_1J)^{-1}B(X_1'X_1)^{-1}X_1'\mathbf{y} \end{bmatrix}.$$

with the dispersion matrix

(17.12.7)

$$\sigma^2 \begin{bmatrix} \begin{array}{l} (X_1'X_1)^{-1} + JF^{-1}J' + JF^{-1}(B_2 - B_1J)' \\ \quad \times DB_1(X_1'X_1)^{-1} + (X_1'X_1)^{-1} \\ \quad \times B_1'D(B_2 - B_1J)F^{-1}J' \\ -F^{-1}\{J' + (B_2 - B_1J)'DB_1(X_1'X_1)^{-1}\} \end{array} & \begin{array}{l} -\{J + (X_1'X_1)^{-1}B_1'D \\ \quad \times (B_2 - B_1J)\}F^{-1} \\ \\ F^{-1} \end{array} \end{bmatrix},$$

where

(17.12.8)
$$D = \{I_{p-r} + B_1(X_1'X_1)^{-1}B_1'\}^{-1};$$
$$F = (B_2 - B_1J)'D(B_2 - B_1J).$$

Theorem 17.12.1. Given a singular weighing design, unbiased estimates of all the weights can be determined by taking $p - r$ additional weighings with the design matrix (17.12.1) such that $|B_2 - B_1J| \neq 0$. The estimates thus obtained are given by (17.12.6) with the dispersion matrix (17.12.7).

For a given choice of X_1 the estimated weights (17.12.6) will be efficiently determined, for Mood's efficiency definition, if the determinant of S_3 is maximum, which in turn means that $|B_2 - B_1J|^2$ is maximum. It is very difficult to give rules of determining B_1 and B_2 that maximize $|B_2 - B_1J|^2$. However, we can set an upper bound for this determinant value, given by the following:

Lemma 17.12.2. For the choice of elements 1, -1, and 0 for B_1 and B_2,

(17.12.9)
$$|B_2 - B_1J|^2 \leqq (p - r)^{p-r} \prod_{i=1}^{p-r} \left\{ 1 + \sum_{k=1}^{r} |j_{ki}| \right\}^2,$$

where j_{ki} is the (k, i)th element of J.

Proof. Since B_1 is a matrix with elements 1, -1, and 0, every element in the ith column of B_1J lies in the closed interval $[-\sum_{k=1}^{r}|j_{ki}|, \sum_{k=1}^{r}|j_{ki}|]$. Furthermore, since B_2 is also a matrix with elements 1, -1, and 0 only, every element in the ith column of $(B_2 - B_1J)$ lies in the closed interval $[-1, -\sum_{k=1}^{r}|j_{ki}|, +1 + \sum_{k=1}^{r}|j_{ki}|]$ for $i = 1, 2, \cdots, p - r$. Now

$$|B_2 - B_1J|^2 = |(B_2 - B_1J)'(B_2 - B_1J)|$$

(17.12.10)
$$\leqq (p - r)^{p-r} \prod_{i=1}^{p-r} \left\{ 1 + \sum_{k=1}^{r} |j_{ki}| \right\}^2,$$

since the maximum value of the (i, i)th element of $(B_2 - B_1J)'(B_2 - B_1J)$ is $\{1 + \sum_{k=1}^{r}|j_{ki}|\}^2 (p - r)$. The proof is complete.

It can be easily seen that $|B_2 - B_1J|^2 = (p - r)^{p-r} \prod_{i=1}^{p-r}\{1 + \sum_{k=1}^{r}|j_{ki}|\}^2$, when a Hadamard matrix of order $p - r$ exists and every row of J contains not more than one element. In such cases the method of constructing B_1 and B_2 such that

(17.12.11)
$$|B_2 - B_1J|^2 = (p - r)^{p-r} \prod_{i=1}^{p-r} \left\{ 1 + \sum_{k=1}^{r} |j_{ki}| \right\}^2$$

is as follows:

Let H_{p-r} be a Hadamard matrix of order $p - r$ and let $\mathbf{h}_1, \mathbf{h}_2, \cdots, \mathbf{h}_{p-r}$ be the column vectors of H_{p-r}. Let us choose $B_2 = H_{p-r}$ and select the row vectors $\mathbf{b}'_1, \mathbf{b}'_2, \cdots, \mathbf{b}'_{p-r}$ such that

$$(17.12.12) \qquad \begin{bmatrix} \mathbf{b}'_1 \, \mathbf{J}_i \\ \mathbf{b}'_2 \, \mathbf{J}_i \\ \vdots \\ \mathbf{b}'_{p-r} \, \mathbf{J}_i \end{bmatrix} = - \sum_{k=1}^{r} |j_{ki}| \, \mathbf{h}_i \qquad i = 1, 2, \cdots, p - r,$$

where $\mathbf{J}_i$ is the ith column vector of J. Then

$$(17.12.13) \qquad B_1 = \begin{bmatrix} \mathbf{b}'_1 \\ \mathbf{b}'_2 \\ \vdots \\ \mathbf{b}'_{p-r} \end{bmatrix}$$

is our required B_1. Choosing the $\mathbf{b}_i$ vectors to satisfy (17.12.12) is possible since every row of J contains at most one element.

The above considerations of selecting B_1 and B_2 are now illustrated as follows:

Illustration 17.12.1. Let us consider the singular weighing design

$$(17.12.14) \qquad X = \begin{bmatrix} 1 & 1 & 1 & 1 & 1 & 1 & 1 & 1 \\ -1 & 1 & -1 & 1 & -1 & 1 & 1 & -1 \\ 1 & 1 & 1 & -1 & -1 & 1 & -1 & -1 \\ -1 & 1 & -1 & -1 & 1 & 1 & -1 & 1 \\ -1 & 1 & 1 & 1 & 1 & -1 & -1 & -1 \\ 1 & 1 & -1 & 1 & -1 & -1 & -1 & 1 \end{bmatrix}.$$

The rank of X is 6. Let the first six columns constitute X_1. Then J, as defined by (17.10.8), is

$$(17.12.15) \qquad J' = \begin{bmatrix} 0 & -1 & 0 & 1 & 0 & 1 \\ 1 & 0 & -1 & 0 & 1 & 0 \end{bmatrix}.$$

This J contains not more than one element in each row. Furthermore, $p - r = 2$. Hence we choose

$$(17.12.16) \qquad B_2 = \begin{bmatrix} 1 & 1 \\ 1 & -1 \end{bmatrix}.$$

When we choose $\mathbf{b}_1' = (-1, 1, 1, -1, -1, -1)$ and $\mathbf{b}_2' = (1, 1, -1, -1, 1, -1)$, we get

$$(17.12.17) \qquad \begin{bmatrix} \mathbf{b}_1' \, \mathbf{J}_1 \\ \mathbf{b}_2' \, \mathbf{J}_1 \end{bmatrix} = -3 \begin{bmatrix} 1 \\ 1 \end{bmatrix} \quad \text{and} \quad \begin{bmatrix} \mathbf{b}_1' \, \mathbf{J}_2 \\ \mathbf{b}_2' \, \mathbf{J}_2 \end{bmatrix} = -3 \begin{bmatrix} 1 \\ -1 \end{bmatrix},$$

satisfying the requirements (17.12.12). Thus we have

$$(17.12.18) \qquad B_1 = \begin{bmatrix} -1 & 1 & 1 & -1 & -1 & -1 \\ 1 & 1 & -1 & -1 & 1 & -1 \end{bmatrix}.$$

We can easily see that

$$(17.12.19) \qquad \begin{bmatrix} X \\ B_1 & B_2 \end{bmatrix}$$

is a Hadamard matrix of order 8 and hence has a maximum value for its determinant.

We close this chapter with the remark that the concept of singular weighing designs introduced by Raghavarao (1964) opens a new branch with a large scope for further research. The method of selecting the additional weighings so as to make the resulting design A- and E-optimal is still an open problem for research in this area.

REFERENCES

1. Banerjee, K. S. (1950–1951). Weighing designs. *Calcutta Stat. Assn. Bull.*, **3**, 64–76.
2. Banerjee, K. S. (1966a). Singularity in Hotelling's weighing designs and a generalized inverse. *Ann. Math. Stat.*, **37**, 1021–1032.
3. Banerjee, K. S. (1966b). On non-randomized fractional weighing designs. *Ann. Math. Stat.*, **37**, 1836–1841.
4. Baumert, L. D. (1966). Hadamard matrices of orders 116 and 232. *Bull. Am. Math. Soc.*, **72**, 237.
5. Baumert, L. D., Golomb, S. W., and Hall, M., Jr. (1962). Discovery of a Hadamard matrix of order 92. *Bull. Am. Math. Soc.*, **68**, 237–238.
6. Baumert, L. D., and Hall, M. (1965). A new construction for Hadamard matrices. *Bull. Am. Math. Soc.*, **71**, 169–170.
7. Belevitch, V. (1950). Theory of $2n$-terminal net works with application to conference telephony. *Elect. Commun.*, **27**, 231–244.
8. Bhaskararao, M. (1966). Weighing designs when n is odd. *Ann. Math. Stat.*, **37**, 1371–1381.
9. Brauer, A. (1953). On a new class of Hadamard determinants. *Math. Z.*, **58**, 219–225.

10. Chakrabarti, M. C. (1962). *Mathematics of Design and Analysis of Experiments.* Asia Publishing House, Bombay.
11. Cohn, J. H. E. (1965). Hadamard matrices and some generalizations. *Am. Math. Monthly*, **72**, 515–518.
12. Dembowski, P. (1968). *Finite Geometries.* Springer Verlag.
13. Ehlich, H. (1965). Neue Hadamard-Matrizen. *Arch. Math.*, **16**, 34–36.
14. Ehrenfeld, S. (1955). On the efficiencies of experimental designs. *Ann. Math. Stat.*, **26**, 247–255.
15. Federer, W. T. (1955). *Experimental Design: Theory and Application.* Macmillan, New York.
16. Goldberg, K. (1966). Hadamard matrices of order cube plus one. *Proc. Am. Math. Soc.*, **17**, 744–746.
17. Hall, M., Jr. (1967). *Combinatorial Theory.* Blaisdell Publishing Company.
18. Hotelling, H. (1944). Some improvements in weighing and other experimental techniques. *Ann. Math. Stat.*, **15**, 297–305.
19. Kiefer, J. (1959). Optimum experimental designs. *J. Roy. Stat. Soc.*, **B21**, 272–304.
20. Kishen, K. (1945). On the design of experiments for weighing and making other types of measurements. *Ann. Math. Stat.*, **16**, 294–300.
21. Mood, A. M. (1946). On Hotelling's weighing problem. *Ann. Math. Stat.*, **17**, 432–446.
22. Moriguti, S. (1954). Optimality of orthogonal designs. *Rep. Stat. Appl. Res.*, **3**, 1–24.
23. Paley, R. E. A. C. (1933). On orthogonal matrices. *J. Math. Phys. Mass. Inst. Technol.*, **12**, 311–320.
24. Plackett, R. L., and Burman, J. P. (1946). Designs of optimum multifactorial experiments. *Biometrika*, **33**, 305–325.
25. Raghavarao, D. (1959). Some optimum weighing designs. *Ann. Math. Stat.*, **30**, 295–303.
26. Raghavarao, D. (1960). Some aspects of weighing designs. *Ann. Math. Stat.*, **31**, 878–884.
27. Raghavarao, D. (1961). Some contributions to the design and analysis of experiments. Unpublished Ph.D. thesis submitted to Bombay University.
28. Raghavarao, D. (1964). Singular weighing designs. *Ann. Math. Stat.*, **35**, 673–680.
29. Raghavarao, D., Sodhi, J. S., and Singh, R. (1968). A new assumption in spring balance weighing designs leading to an application in sampling. Submitted to *Calcutta Stat. Assn. Bull.*
30. Scheffe, H. (1959). *The Analysis of Variance.* Wiley, New York.
31. Shrikhande, S. S. (1964). Generalized Hadamard matrices and orthogonal arrays of strength two. *Can. J. Math.*, **16**, 736–740.
32. Shrikhande, S. S., and Singh, N. K. (1962). On a method of constructing symmetrical balanced incomplete block designs. *Sankhyā*, **A24**, 25–32.
33. Spence, E. (1967). A new class of Hadamard matrices. *Glasgow Math. J.*, **8**, 59–62.

34. Stanton, R. G., and Sprott, D. A. (1958). A family of difference sets. *Can. J. Math.*, **10**, 73–77.
35. Szekeres, G. (1969). A new class of symmetric block designs. *J. Comb. Theory*, **6**, 219–221.
36. Wallis, J. (1969a). A class of Hadamard matrices. *J. Comb. Theory*, **6**, 40–44.
37. Wallis, J. (1969b). A note on a class of Hadamard matrices. *J. Comb. Theory*, **6**, 222–233.
38. Williamson, J. (1944). Hadamard's determinant theorem and the sum of four squares. *Duke Math. J.*, **11**, 65–82.
39. Williamson, J. (1947). Note on Hadamard's determinant theorem. *Bull. Am. Math. Soc.*, **53**, 608–613.
40. Yates, F. (1935). Complex experiments. *J. Roy. Stat. Soc.*, **B2**, 181–223.

BIBLIOGRAPHY

Banerjee, K. S. Weighing designs and balanced incomplete blocks. *Ann. Math. Stat.*, **19**, 394–399 (1948).

Banerjee, K. S. On certain aspects of spring balance designs. *Sankhyā*, **9**, 367–376 (1948).

Banerjee, K. S. On variance factor in weighing designs. *Calcutta Stat. Assn. Bull.*, **2**, 38–43 (1949).

Banerjee, K. S. Some contributions to Hotelling's weighing designs. *Sankhyā*, **10**, 371–382 (1950).

Banerjee, K. S. Some observations on the practical aspects of weighing designs. *Biometrika*, **38**, 248–251 (1951).

Banerjee, K. S. Weighing designs and partially balanced incomplete block designs. *Calcutta Stat. Assn. Bull.*, **4**, 36–38 (1951–1953).

Baumert, L. D., and Hall, M. Hadamard matrices of the Williamson type. *Math. Comp.*, **19**, 442–447 (1965).

Dade, E. C., and Goldberg, K. The construction of Hadamard matrices. *Michigan Math. J.*, **6**, 247–250 (1959).

Dey, A. A note on weighing designs. *Ann. Inst. Stat. Math.*, **21**, 343–346 (1969).

Kempthorne, O. The factorial approach to the weighing problem. *Ann. Math. Stat.*, **19**, 238–245 (1948).

Menon, P. K. Certain Hadamard designs. *Proc. Am. Math. Soc.*, **13**, 524–531 (1962).

Rao, C. R. On most efficient designs in weighing. *Sankhyā*, **7**, 440–441 (1945–1946).

Shrikhande, S. S., and Bhagwandas. A note on the embedding of Hadamard matrices. *Essays in Probability and Statistics.* University of North Carolina Press, Chapel Hill, pp. 673–688 (1970).

Zacks, S. Randomized fractional weighing designs. *Ann. Math. Stat.*, **37**, 1382–1395 (1966).

APPENDIX A

Mathematics for Statisticians

A.1. MATHEMATICAL SYSTEMS

We assume that our readers are familiar with sets and various algebraic operations on sets. Between two sets S and T, if in some manner there is established a law of correspondence that assigns to each element of S a definite element of T, then such a correspondence is called a mapping or map or function f of S into T. We denote the image of an element $x \in S$ by xf. The set S is called the domain of the map. The set of all images contained in the map is called the range of the map. If the range of the map is the whole of set T, the map is said to be onto or surjective. A map from S onto T is said to be a $1 - 1$ map or bijective if elements of S that are distinct have distinct images. A binary operation "$*$" on the elements of S is a map from $S \times S$ into S.

We now introduce some special abstract systems of great use.

Definition A.1.1. A group $\mathscr{G}$ is defined to be an abstract system $\mathscr{G} = \{G; *\}$, having only one closed binary operation satisfying the following postulates:

1. The binary operation $*$ is associative.
2. An identity element $e \in G$ exists for the binary operation $*$.
3. Each $g \in G$ has an inverse $g^{-1} \in G$ for the binary operation $*$.

Furthermore, the group $\mathscr{G}$ is called commutative, or Abelian, if

4. The binary operation $*$ is commutative.

We can easily see that (a) the rational numbers with addition as the binary operation, (b) the nonzero rational numbers with multiplication as the binary operation, (c) the integers with addition as the binary operation, and (d) the n complex nth roots of unity with multiplication as the operation are some examples of groups.

A generalization of groups leads us to rings, defined as below:

Definition A.1.2. Let $\mathscr{R}$ be a system $\{R, +, \cdot\}$ consisting of a nonempty set R on which two binary operations called addition $(+)$ and multiplication $(\cdot)$ are defined. Then $\mathscr{R}$ is called a ring if

1. $\{R, +\}$ is a commutative group.
2. Multiplication $(\cdot)$ is associative.
3. Multiplication $(\cdot)$ is distributive over addition $(+)$.

The multiplication symbol $\cdot$ can be replaced by the juxtaposition of the elements to indicate multiplication. It is to be noted that a ring need not have a multiplicative identity u such that $au = ua = a$ for every $a \in R$. A ring having a multiplicative identity will be called a ring with unity element. A commutative ring is one for which multiplication is a commutative operation. It can easily be seen that the set of integers form a commutative ring.

A further specialization of rings leads to the concept of fields as defined by the following:

Definition A.1.3. A field $\mathscr{F}$ is defined to be a system $\mathscr{F} = \{F; +, \cdot\}$ having two binary operations and satisfying the following postulates:

1. The system $\{F, +\}$ is a commutative group whose identity will be denoted by 0.
2. The system $\{F_0, \cdot\}$ is a commutative group whose identity will be denoted by 1, where $F_0 = \{x \in F \mid x \neq 0\}$.
3. The operation multiplication $(\cdot)$ is distributive over the operation addition $(+)$.

Rational numbers, real numbers, complex numbers, etc., can be seen to be some examples of fields. A field can consist of either infinite or finite number of elements. In the former case it will be called an infinite field, and in the latter case it will be called a finite field. In all modern algebra textbooks it is proved that the number of elements in a finite field is a prime or a prime power. Finite fields are also called Galois fields. We devote Section A.2 to the study of the construction of finite fields.

A.2. GALOIS FIELDS

Before studying Galois fields, we introduce the number-theory concept of congruences.

Definition A.2.1. The quantity a is said to be congruent to b to modulus n if $a - b$ is divisible by n. We denote this relation by $a \equiv b \pmod{n}$.

Congruences can be treated as equations for addition, subtraction, and multiplication. If $a \equiv b \pmod{n}$, then $ac \equiv bc \pmod{n}$ and $a \pm c \equiv b \pm c \pmod{n}$. Furthermore, if $d \equiv e \pmod{n}$, then $ad \equiv be \pmod{n}$ and $a \pm d \equiv b \pm e \pmod{n}$. However, division cannot be treated as it is in equations. If $ac \equiv bc \pmod{n}$, we cannot conclude that $a \equiv b \pmod{n}$. But $ac \equiv bc \pmod{n}$ implies that $a \equiv b \pmod{n/d}$, where d is the greatest common divisor of n and c.

If $x \equiv a \pmod{n}$, then a is called a residue of x to modulus n. If $0 \leq a \leq n - 1$, then a is the least nonzero residue of x to modulus n. Clearly two integers a and b will have the same least nonzero residue to modulus n if and only if $a \equiv b \pmod{n}$. A class of residues to modulus n is the class of all integers congruent to a given residue $\pmod{n}$, and every member of the class is called a representative of the class. There are in all n classes, represented by $(0), (1), (2), \cdots, (n - 1)$, and the representative $0, 1, 2, \cdots, n - 1$ of these classes are called a complete system of incongruent residues to modulus n.

If i and j are any two members of a complete system of incongruent residues to modulus n, then addition and multiplication between i and j are defined by

(A.2.1) $$i + j = (i + j)(\bmod p),$$

(A.2.2) $$i \cdot j = ij(\bmod p),$$

where the right-hand-side elements are members of complete system of residues $\pmod{n}$. It is easy to see that a complete system of residues $\pmod{n}$ forms a commutative ring with unity element. Let s be any nonzero element of these residues. Then s will possess a multiplicative inverse among the complete system of residues $\pmod{n}$ if and only if n is a prime, p. Thus if p is a prime, a complete system of residues $\pmod{p}$ forms a Galois (or finite) field and is denoted by $\mathrm{GF}(p)$. In general, a Galois field of p^m elements is obtained as follows: Let $P(x)$ be any given polynomial in x of degree m with coefficients belonging to $\mathrm{GF}(p)$ and let $F(x)$ be any polynomial in x with integral coefficients. Then $F(x)$ may be expressed as

(A.2.3) $$F(x) = f(x) + p \cdot q(x) + P(x) \cdot Q(x),$$

where

(A.2.4) $$f(x) = a_0 + a_1 x + a_2 x^2 + \cdots + a_{m-1} x^{m-1}$$

and the coefficients a_i ($i = 0, 1, \cdots, m - 1$) belong to $\mathrm{GF}(p)$. This relationship may be written as

(A.2.5) $$F(x) \equiv f(x) \bmod \{p, P(x)\},$$

and we say $f(x)$ is the residue of $F(x)$ modulus p and $P(x)$. The functions $F(x)$ that satisfy (A.2.5) when $f(x)$, p, and $P(x)$ are kept fixed form a class.

If p and $P(x)$ are kept fixed but $f(x)$ is varied, p^m classes may be formed, since each coefficient in $f(x)$ may take the p values of GF(p). It may be readily verified that the classes defined by $f(x)$ form a commutative ring, which will be a field if and only if $P(x)$ is irreducible over GF(p).

The finite field formed by the p^m classes of residues is called a Galois field of order p^m and is denoted by GF(p^m). The function $P(x)$ is said to be a minimum function for generating the elements of GF(p^m). The nonzero elements may be represented either as polynomials of degree at most $m - 1$, as in the preceding paragraph, or as powers of a primitive root x such that $x^{p^m - 1} = 1$, but $x^d \neq 1$ for d dividing $p^m - 1$.

To obtain a minimum function we divide $x^{p^m - 1} - 1$ by the least common multiple of all factors like $x^d - 1$, where d is a divisor of $p^m - 1$. Then we get the cyclotomic equation, that is, the equation that has for its roots all primitive roots of the equation $x^{p^m - 1} = 1$. The order of this equation will be $\phi(p^m - 1)$, where $\phi(k)$ denotes the number of positive integers less than k and relatively prime to it. In this equation, by replacing each coefficient by its least nonzero residue to modulus p, we get the cyclotomic polynomial of order $\phi(p^m - 1)$. Let $P(x)$ be an irreducible factor of this polynomial. Then $P(x)$ is a minimum function, which is in general not unique.

As an example, let us construct a minimum function for generating the elements of GF(2^3). The cyclotomic polynomial is $(x^7 - 1)/(x - 1) = x^6 + x^5 + x^4 + x^3 + x^2 + x + 1$, the factors of which are $(x^3 + x^2 + 1)$ and $(x^3 + x + 1)$. Hence $P(x)$ can be taken as either $x^3 + x^2 + 1$ or $x^3 + x + 1$. We choose $P(x) = x^3 + x^2 + 1$ to generate the elements of GF(2^3). If x is a primitive root, the elements are

$$\alpha_0 = 0, \qquad \alpha_1 = x, \qquad \alpha_2 = x^2, \qquad \alpha_3 = x^3 = x^2 + 1,$$

(A.2.6) $\qquad \alpha_4 = x^4 = x^2 + x + 1, \qquad \alpha_5 = x^5 = x + 1,$

$$\alpha_6 = x^2 + x, \qquad \alpha_7 = x^7 = 1.$$

We now list a few minimum functions that are needed in the construction of designs presented in the text:

GF	Minimum Function
2^2	$x^2 + x + 1$
2^3	$x^3 + x^2 + 1$
2^4	$x^4 + x^3 + 1$
3^2	$x^2 + x + 2$
3^3	$x^3 + 2x + 1$
5^2	$x^2 + 2x + 3$

Let $\alpha_0 = 0, \alpha_1, \alpha_2, \cdots, \alpha_{p^m-1}$ be the elements of $\mathrm{GF}(p^m)$. The addition table of the elements of $\mathrm{GF}(p^m)$ is a p^mth order square array whose (i, j)th cell is filled by $\alpha_i + \alpha_j$ for $i, j = 0, 1, 2, \cdots, p^m - 1$.

A.3. VECTOR ALGEBRA

Vector spaces are somewhat more complicated than the abstract systems introduced in Section A.1. In this case there are two sets of elements involved. One set V consists of abstract elements called vectors, denoted by boldface greek or roman letters, the other set F is a field whose elements are called scalars. In addition to the two field operations $+$ and $\cdot$, two operations involving vectors are postulated, the addition of vectors, denoted by $\boldsymbol{\alpha} \oplus \boldsymbol{\beta}$, and the scalar multiplication of a scalar and a vector, denoted by $a \odot \boldsymbol{\alpha}$. Vector spaces are defined as follows:

Definition A.3.1. A system $\mathscr{V} = \{V, F; +, \cdot, \oplus, \odot\}$ is called a vector space over a field $\mathscr{F}$ if and only if

1. $\{F, +, \cdot\}$ is a field $\mathscr{F}$ whose additive and multiplicative identity elements are denoted by 0 and 1.
2. $\{V, \oplus\}$ is a commutative group whose identity element is denoted by $\boldsymbol{0}$.
3. For all $a, b \in F$ and all $\boldsymbol{\alpha}, \boldsymbol{\beta} \in V$, $a \odot \boldsymbol{\alpha} \in V$, and
    (i) $(a + b) \odot (\boldsymbol{\alpha}) = (a \odot \boldsymbol{\alpha}) \oplus (b \odot \boldsymbol{\alpha})$,
    (ii) $a \odot (\boldsymbol{\alpha} \oplus \boldsymbol{\beta}) = (a \odot \boldsymbol{\alpha}) \oplus (a \odot \boldsymbol{\alpha})$,
    (iii) $ab \odot (\boldsymbol{\alpha}) = a \odot (b \odot \boldsymbol{\alpha})$,
    (iv) $1 \odot \boldsymbol{\alpha} = \boldsymbol{\alpha}$.

As an example, it can be verified that for a fixed positive integer n the set of all n-tuples of real numbers forms a vector space if the addition of vectors and the multiplication of a scalar and a vector are defined by

(A.3.1)
$$(x_1, x_2, \cdots, x_n) + (y_1, y_2, \cdots, y_n) = (x_1 + y_1, x_2 + y_2, \cdots, x_n + y_n)$$

and

(A.3.2)
$$a \odot (x_1, x_2, \cdots, x_n) = (ax_1, ax_2, \cdots, ax_n).$$

As a further example, we can see that for a fixed n the set of all polynomials in x with real coefficients and of degree not exceeding n, together with the zero polynomial, forms a vector space where the vector sum and the scalar multiplication are defined as the usual polynomial sum and product by real numbers.

We drop the notation $\oplus$ and $\odot$ and use the symbol $+$ and juxtaposition, as the operations will be evident from the types of letters involved in an expression.

We now define the following:

Definition A.3.2. Given a vector space $\mathscr{V}$ over $\mathscr{F}$, a subset S of vectors is said to form a subspace $\mathscr{S}$ of $\mathscr{V}$ if and only if the subsystem

$$\mathscr{S} = \{S, F; +, \cdot, \oplus, \odot\}$$

is a vector space.

It is easy to verify that a nonvoid subset S of V is a subspace if and only if S is closed under the two operations of vector addition and scalar multiplication, as defined for $\mathscr{V}$.

Given a vector space $\mathscr{V}$, let A be a set of vectors. Then the set $[A]$ of all linear combinations of vectors of A is the collection of all finite sums of the form

(A.3.3) $a_1\alpha_1 + a_2\alpha_2 + \cdots + a_m\alpha_m,$

where $a_i \in \mathscr{F}$, $\alpha_i \in A$, and $m = 1, 2, 3, \cdots$. The set $[A]$ can easily be verified to be a subspace of $\mathscr{V}$. Vector spaces are also called linear spaces because of the wide use of linear combinations of vectors in the study of vector spaces. The space $[A]$ is said to be spanned by the vectors of A.

We now introduce the concepts of linear independence of vectors and the basis of a vector space.

Definition A.3.3. A set $\{\alpha_1, \alpha_2, \cdots, \alpha_k\}$ of vectors is said to be linearly independent if and only if the equation

(A.3.4) $a_1\alpha_1 + a_2\alpha_2 + \cdots + a_k\alpha_k = 0$

implies that $a_1 = a_2 = \cdots = a_k = 0$. Otherwise the vectors are said to be linearly dependent.

Definition A.3.4. A maximal, linearly independent set of vectors spanning a vector space $\mathscr{V}$ is called a basis for $\mathscr{V}$. The number of vectors in a basis is called its dimensionality.

The following two theorems can be established from the definitions:

Theorem A.3.1. Every vector of $\mathscr{V}$ has a unique representation as a linear combination of the vectors of a given basis of $\mathscr{V}$.

Theorem A.3.2. Every basis for a finite dimensional vector space $\mathscr{V}$ has the same number of vectors.

We introduce the following definition:

Definition A.3.5. Let $\mathcal{V} = \{V, F; +, \cdot, \oplus, \odot\}$ and $\mathcal{W} = \{W, F; +, \cdot, +, \cdot\}$ be vector spaces over a field $\mathcal{F}$. A mapping H of V into W is called a homomorphism, provided that for all $\alpha, \beta \in V$ and all $a \in F$

(A.3.5) $(\alpha \oplus \beta)H = \alpha H + \beta H$

and

(A.3.6) $(a \odot \alpha)H = a \cdot \alpha H.$

If every vector of W is in the range of H, H is said to be a homomorphism of $\mathcal{V}$ onto $\mathcal{W}$. A $1:1$ homomorphism of $\mathcal{V}$ onto $\mathcal{W}$ is called an isomorphism. If such a mapping exists, $\mathcal{V}$ and $\mathcal{W}$ are said to be isomorphic.

If $\alpha_1, \alpha_2, \cdots, \alpha_n$ is a basis for $\mathcal{V}$ and if $\xi = a_1 \alpha_1 + a_2 \alpha_2 + \cdots + a_n \alpha_n \in V$ is mapped on the n-tuple $(a_1, a_2, \cdots, a_n)$, we can easily show that the mapping is an isomorphism. Thus we have the following theorem:

Theorem A.3.3. Any n-dimensional vector space $\mathcal{V}$ over $\mathcal{F}$ is isomorphic to the space ε_n of all n-tuples of elements of $\mathcal{F}$.

In view of this theorem, it is sufficient to study the vector spaces formed by ordered n-tuples. In fact many books define vectors only as n-tuples instead of giving the general exposition as above. We now restrict ourselves to the study of n-tuples as vectors.

Definition A.3.6. If $\mathbf{x} = (x_1, x_2, \cdots, x_n)$ and $\mathbf{y} = (y_1, y_2, \cdots, y_n)$ are two vectors, then their scalar product (or inner product), denoted by $\mathbf{x} \cdot \mathbf{y}$ is

(A.3.7) $$\mathbf{x} \cdot \mathbf{y} = \sum_{i=1}^{n} x_i y_i.$$

Definition A.3.7. The norm (or length) of a vector $\mathbf{x} = (x_1, x_2, \cdots, x_4)$, denoted by $\| \mathbf{x} \|$, is

(A.3.8) $$\| \mathbf{x} \| = \sqrt{\mathbf{x} \cdot \mathbf{x}}.$$

Definition A.3.8. A vector $\mathbf{x}$ is said to be normalized if it has unit length.

Definition A.3.9. Two vectors $\mathbf{x}$ and $\mathbf{y}$ are called orthogonal if $\mathbf{x} \cdot \mathbf{y} = 0$.

Definition A.3.10. A basis $\mathbf{x_1}, \mathbf{x_2}, \cdots, \mathbf{x_n}$ of a vector space $\mathcal{V}$ is called orthonormal if every pair of vectors in the basis are orthogonal and every vector is normalized.

The following result is well known as the Gram–Schmidt process:

Theorem A.3.4. Given an arbitrary basis $\{\alpha_1, \alpha_2, \cdots, \alpha_n\}$ for $\mathcal{V}$, there exists an orthonormal basis $\{\gamma_1, \gamma_2, \cdots, \gamma_n\}$ for $\mathcal{V}$ such that γ_i is a linear combination of $\alpha_1, \alpha_2, \cdots, \alpha_i$.

Proof. Let $\beta_1 = \alpha_1$. Construct $\beta_2 = \alpha_2 - b_{21}\beta_1$ such that β_2 is orthogonal to β_1. Thus $b_{21} = (\beta_1 \cdot \alpha_2)/(\beta_1 \cdot \beta_1)$. $\beta_2 \neq \theta$, as otherwise α_1 and α_2 will be dependent. Again construct $\beta_3 = \alpha_3 - b_{31}\beta_1 - b_{32}\beta_2$ such that β_3 is orthogonal to β_2 and β_1 separately. Thus $b_{31} = (\alpha_3 \cdot \beta_1)/(\beta_1 \cdot \beta_1)$ and $b_{32} = (\alpha_3 \cdot \beta_2)/(\beta_2 \cdot \beta_2)$. $\beta_3 \neq \theta$, because $\alpha_1, \alpha_2, \alpha_3$ are independent. Following similarly, we can construct a basis $\{\beta_1, \beta_2, \cdots, \beta_n\}$ for $\mathscr{V}$, where the vectors are pairwise orthogonal. Defining $\gamma_i = \beta_i/\|\beta_i\|$, we can see that $\{\gamma_1, \gamma_2, \cdots, \gamma_n\}$ is the required orthonormal basis for $\mathscr{V}$.

We introduce the following definitions:

Definition A.3.11. Let V_r be an r-dimensional subspace of an n-dimensional space V_n and let $\mathbf{x} \in V_n$. Then $\mathbf{x}$ is said to be orthogonal to V_r if and only if $\mathbf{x}$ is orthogonal to every vector in V_r.

Definition A.3.12. If V_s is an s-dimensional subspace of the r-dimensional subspace V_r of the n-dimensional space V_n, then the totality of vectors in V_r that are orthogonal to V_s is called the orthocomplement of V_s in V_r.

A.4. MATRIX ALGEBRA

We introduce the following definitions:

Definition A.4.1. A rectangular array

$$(A.4.1) \qquad A = A_{m,n} = \begin{bmatrix} a_{11} & a_{12} & \cdots & a_{1n} \\ a_{21} & a_{22} & \cdots & a_{2n} \\ \cdot & \cdot & \cdots & \cdot \\ a_{m1} & a_{m2} & \cdots & a_{mn} \end{bmatrix}$$

of m rows and n columns of elements belonging to a field $\mathscr{F}$ is called a matrix over $\mathscr{F}$ of order $m \times n$.

The matrix can be enclosed in either brackets or parentheses. It can be written in the abbreviated form $A = A_{m,n} = (a_{ij})$, denoting its general element.

Definition A.4.2. Two $m \times n$ matrices $A_{m,n} = (a_{ij})$ and $B_{m,n} = (b_{ij})$ are said to be equal if $a_{ij} = b_{ij}$ for every i and j.

Definition A.4.3. The transpose of $A_{m,n}$, written A', is the $n \times m$ matrix obtained from A by interchanging rows and columns.

Definition A.4.4. The sum of two $m \times n$ matrices $A_{m,n} = (a_{ij})$ and $B_{m,n} = (b_{ij})$ is the $m \times n$ matrix $A + B = (a_{ij} + b_{ij})$.

Definition A.4.5. The product of an $m \times n$ matrix $A_{m,n} = (a_{ij})$ with a scalar c is the $m \times n$ matrix $cA_{m,n} = (ca_{ij})$.

Definition A.4.6. If $A_{m,n} = (a_{ij})$ and $B_{n,r} = (b_{jk})$ are two given matrices such that the number of columns of A is same as the number of rows of B, then the product of A and B is defined to be the $m \times r$ matrix $C_{m,r} = AB = (c_{ik})$, where

$$(A.4.2) \qquad c_{ik} = \sum_{j=1}^{n} a_{ij} b_{jk}.$$

It is to be remembered here that matrix multiplication need not be commutative.

The following special kinds of matrices need mention here:

1. A zero matrix $0_{m,n}$ is an $m \times n$ matrix with each entry 0. It may be noted that we may have $A_{m,n} B_{n,r} = 0_{m,r}$, whereas $A_{m,n} \neq 0_{m,n}$ and $B_{n,r} \neq 0_{n,r}$. An $m \times n$ matrix with all positive unit elements is denoted by $E_{m,n}$.

2. An $m \times n$ matrix is called a square matrix if $m = n$, and we denote it by $A = A_n = (a_{ij})$. A square matrix $A = (a_{ij})$ is called symmetrical if $a_{ij} = a_{ji}$ and skew symmetrical if $a_{ij} = -a_{ji}$. A square matrix $A = (a_{ij})$ is called upper triangular if $a_{ij} = 0$ for $i > j$ and lower triangular if $a_{ij} = 0$ for $i < j$.

3. A square matrix of order n, written I_n, is called an identity matrix if $I_n = (\delta_{ij})$, where δ_{ij} is the Kronecker delta. We note that $A_{m,n} I_n = A_{m,n}$ and $I_m A_{m,n} = A_{m,n}$. A diagonal matrix is a square matrix whose off-diagonal elements are zeros.

4. An n-tuple vector $\mathbf{x}$ may be regarded as an $n \times 1$ matrix.

We now introduce the concept of partitioned matrices. Suppose the $m \times n$ matrix A is divided into $\alpha\beta$ submatrices A_{ij} by means of horizontal and vertical dividing lines, as indicated below:

$$(A.4.3) \quad A_{m,n} =
\begin{array}{c}
\begin{array}{cccc}
\quad n_1 & n_2 & & n_\beta \\
\text{columns} & \text{columns} & \cdots & \text{columns}
\end{array} \\
\begin{array}{l}
m_1 \text{ rows} \\
m_2 \text{ rows} \\
\\
m_\alpha \text{ rows}
\end{array}
\left[
\begin{array}{c|c|c|c}
A_{11} & A_{12} & \cdots & A_{1\beta} \\
\hline
A_{21} & A_{22} & \cdots & A_{2\beta} \\
\hline
\cdot & \cdot & \cdots & \cdot \\
\hline
A_{\alpha 1} & A_{\alpha 2} & \cdots & A_{\alpha\beta}
\end{array}
\right]
\end{array}$$

so that A_{ij} is $m_i \times n_j$, and $\sum_{i=1}^{\alpha} m_i = m$, $\sum_{j=1}^{\beta} n_j = n$. Suppose further that the $n \times r$ matrix B is likewise partitioned, with the restriction that the partitioning of the n rows of B is the same as that of the n columns of A, so that the rows are partitioned into sets of $n_1, n_2, \cdots, n_\beta$. Let the columns of B be partitioned into sets $r_1, r_2, \cdots, r_\gamma$, such that $\sum_{k=1}^{\gamma} r_k = r$. This partitions B into $\beta\gamma$ submatrices B_{jk}, where B_{jk} is $n_j \times r_k$. Now form the product $C = AB$ and partition its m rows the same way as the m rows of A and its r columns the same way as the r columns of B, so that there are $\alpha\gamma$ submatrices C_{ik}, where C_{ik} is $m_i \times r_k$. We now have

$$(A.4.4) \quad \begin{bmatrix} A_{11} & A_{12} & \cdots & A_{1\beta} \\ A_{21} & A_{22} & \cdots & A_{2\beta} \\ \cdot & \cdot & \cdots & \cdot \\ A_{\alpha 1} & A_{\alpha 2} & \cdots & A_{\alpha\beta} \end{bmatrix} \begin{bmatrix} B_{11} & B_{12} & \cdots & B_{1\gamma} \\ B_{21} & B_{22} & \cdots & B_{2\gamma} \\ \cdot & \cdot & \cdots & \cdot \\ B_{\beta 1} & B_{\beta 1} & \cdots & B_{\beta\gamma} \end{bmatrix} = \begin{bmatrix} C_{11} & C_{12} & \cdots & C_{1\gamma} \\ C_{21} & C_{22} & \cdots & C_{2\gamma} \\ \cdot & \cdot & \cdots & \cdot \\ C_{\alpha 1} & C_{\alpha 2} & \cdots & C_{\alpha\gamma} \end{bmatrix}.$$

It may be verified that the submatrices in the product satisfy the relation

$$(A.4.5) \quad C_{ik} = \sum_{j=1}^{\beta} A_{ij} B_{jk},$$

as in ordinary matrix multiplication.

We shall sometimes have occasion to use the notation $(A \mid B), (A + B \mid C)$ for partitioned matrices.

With respect to every square matrix A, we can associate a quantity called its determinant, written $|A|$, and we assume that the readers are familiar with its definition and elementary properties. We introduce the following definition:

Definition A.4.7. A square matrix A is called singular if $|A| = 0$ and nonsingular if $|A| \neq 0$.

Definition A.4.8. If for a square matrix of order n there exists another square matrix B of the same order n, such that

$$(A.4.6) \quad AB = BA = I_n,$$

then B is called the inverse of A and is written A^{-1}.

From these definitions the following theorem can be easily established:

Theorem A.4.1. The square matrix A_n has an inverse if and only if A is nonsingular. Then A^{-1} is unique, and if $n > 1$, the (i, j)th element of A^{-1} is $A_{ji}/|A|$, where A_{ij} is the cofactor of a_{ij} in $|A|$.

The following theorem gives us a method of calculating the inverse of a partitioned symmetrical matrix, which can be proved by verification:

Theorem A.4.2. If

$$(A.4.7) \qquad A = \begin{bmatrix} P & Q \\ Q' & R \end{bmatrix}$$

is a partitioned symmetrical matrix such that A and P are nonsingular, then

$$(A.4.8) \qquad A^{-1} = \begin{bmatrix} P^{-1}(I + QS^{-1}Q'P^{-1}) & -P^{-1}QS^{-1} \\ -S^{-1}Q'P^{-1} & S^{-1} \end{bmatrix},$$

where

$$(A.4.9) \qquad S = R - Q'P^{-1}Q.$$

The following theorem, whose proof is obvious, gives us the inverse of a product:

Theorem A.4.3. If A_n and B_n are nonsingular matrices, so is their product AB and we have

$$(A.4.10) \qquad (AB)^{-1} = B^{-1}A^{-1}.$$

The determinant of a partitioned square matrix is given by the following theorem:

Theorem A.4.4. If A is a partitioned square matrix

$$(A.4.11) \qquad A = \begin{bmatrix} P & Q \\ R & S \end{bmatrix},$$

then

$$(A.4.12) \qquad |A| = |P| \, |S - RP^{-1}Q|,$$

when P is nonsingular.

Proof. We verify

$$(A.4.13) \qquad \begin{bmatrix} I & 0 \\ -RP^{-1} & S \end{bmatrix} \begin{bmatrix} P & Q \\ R & S \end{bmatrix} \begin{bmatrix} I & -P^{-1}Q \\ 0 & I \end{bmatrix} = \begin{bmatrix} P & 0 \\ 0 & S - RP^{-1}Q \end{bmatrix}.$$

Taking determinants on both sides, we get the required result (A.4.12).

As a particular application of the above theorem, we have the following:

Theorem A.4.5.

$$(A.4.14) \quad |A| = \frac{\left| \begin{array}{ccc|ccc} & & & e_{1,s+1} & \cdots & e_{1,s+t} \\ \multicolumn{3}{c|}{(\alpha - \beta)I_s + \beta E_{s,s}} & \cdot & \cdots & \cdot \\ & & & e_{s,s+1} & \cdots & e_{s,s+t} \\ \hline e_{s+1,1} & \cdots & e_{s+1,s} & e_{s+1,s+1} & \cdots & e_{s+1,s+t} \\ \cdot & \cdots & \cdot & \cdot & \cdots & \cdot \\ e_{s+t,1} & \cdots & e_{s+t,s} & e_{s+t,s+1} & \cdots & e_{s+t,s+t} \end{array} \right|}{}$$

$$= \{\alpha + (s-1)\beta\}^{-t+1}(\alpha - \beta)^{s-t-1}|B_t|,$$

where B_t is of order $t \times t$ and the (j, u)th element of B_t is

$$b_{ju} = \{\alpha + (s-1)\beta\}(\alpha - \beta)e_{s+j,s+u} - \{\alpha + (s-1)\beta\}$$

$$(A.4.15) \qquad \times \sum_{i=1}^{s} e_{i,s+u}e_{s+j,i} + \beta \sum_{i=1}^{s} e_{i,s+u} \sum_{i=1}^{s} e_{s+j,i}.$$

We introduce the following definitions:

Definition A.4.9. The row (or column) rank of a matrix $A_{m,n}$ is the maximum number of independent rows (or column) in A.

Definition A.4.10. The rank (or determinant rank) of a matrix $A_{m,n}$ is the maximum order of nonzero subdeterminants.

The following result is well known:

Theorem A.4.6. The row rank, the column rank, and the rank of an $m \times n$ matrix A are all equal.

We prove the following theorems:

Theorem A.4.7. Rank $(AB) \leq \min \{\text{rank } (A), \text{rank } (B)\}$.

Proof. Since the columns of AB are linear combinations of the columns of A, the number of linearly independent columns in AB cannot exceed the number in A; hence rank $(AB) \leq$ rank (A). Arguing similarly about rows, we get rank $(AB) \leq$ rank (B).

Theorem A.4.9. If A is an $m \times n$ matrix and if P_m and Q_n are nonsingular matrices, then rank $(PAQ) =$ rank (A).

Proof. From Theorem A.4.7 we have rank $(PA) \leq$ rank (A) and rank $(A) \leq$ rank (PA), the latter inequality following because $A = P^{-1}PA$. Thus rank $(A) =$ rank (PA). Similarly we can show that rank $(PA) =$ rank (PAQ).

Definition A.4.11. The system of equations

$$a_{11}x_1 + a_{12}x_2 + \cdots + a_{1n}x_n = l_1,$$

(A.4.16) $\qquad a_{21}x_1 + a_{22}x_2 + \cdots + a_{2n}x_n = l_2,$

$$\cdots$$

$$a_{m1}x_1 + a_{m2}x_2 + \cdots + a_{mn}x_n = l_m,$$

where the a_{ij}'s and the l_i's $(i = 1, 2, \cdots, m; j = 1, 2, \cdots, n)$ are known and the x_j's $(j = 1, 2, \cdots, n)$ are unknown, is called m simultaneous equations in n unknowns.

Equations (A.4.16) can be abbreviated to

(A.4.17) $\qquad\qquad\qquad\qquad A\mathbf{x} = \mathbf{l},$

where

(A.4.18) $\quad A_{m,n} = \begin{bmatrix} a_{11} & a_{12} \cdots a_{1n} \\ a_{21} & a_{22} \cdots a_{2n} \\ \cdot & \cdots \cdot \\ a_{m1} & a_{m2} \cdots a_{mn} \end{bmatrix}, \quad \mathbf{x} = \begin{bmatrix} x_1 \\ x_2 \\ \cdot \\ x_n \end{bmatrix}, \quad \mathbf{l} = \begin{bmatrix} l_1 \\ l_2 \\ \cdot \\ l_m \end{bmatrix}.$

The following is known:

Theorem A.4.9. A system of simultaneous equations will have a solution if and only if

(A.4.19) $\qquad\qquad\qquad \text{rank } (A) = \text{rank } (A \mid \mathbf{l}).$

The proof can be supplied by using the definitions of column rank and the linear dependency of vectors.

Definition A.4.12. The system of simultaneous equations (A.4.17) will be called homogeneous equations if $\mathbf{l} = 0_{m,1}$.

Trivially $\mathbf{x} = 0_{n,1}$ is a solution of a system of homogeneous equations

(A.4.20) $\qquad\qquad\qquad A_{m,n}\mathbf{x}_{n,1} = 0_{m,1},$

which is called a trivial solution. In algebra textbooks it is proved that the following obtains:

Theorem A.4.10. The system of homogeneous equations (A.4.20) will have a nontrivial solution if and only if

(A.4.21) $\qquad\qquad\qquad \text{rank } (A) \leqq n.$

We now introduce the concept of characteristic roots and vectors of a square matrix.

Definition A.4.13. Given a square matrix A, if there exist a column vector $\mathbf{x}$ and a scalar λ such that

(A.4.22) $$A\mathbf{x} = \lambda\mathbf{x},$$

then λ is called a characteristic root of A and $\mathbf{x}$ is called a characteristic vector associated with the root λ.

Clearly we can see that the characteristic roots are solutions of the polynomial equation

(A.4.23) $$|A - \lambda I| = 0.$$

Definition A.4.14. The polynomial $|A - \lambda I|$ is called the characteristic polynomial of the matrix A.

We can see from (A.4.23) that, if $\lambda_1, \lambda_2, \cdots, \lambda_n$ are the n characteristic roots of A, then

(A.4.24) $$\lambda_1\lambda_2 \cdots \lambda_n = |A|$$

and

(A.4.25) $$\lambda_1 + \lambda_2 + \cdots + \lambda_n = \text{sum of the diagonal elements of } A.$$

The sum of the diagonal elements of A is called the *trace* of A and is denoted by tr (A). We prove the following theorem:

Theorem A.4.11

(A.4.26) $$\text{tr } (AB) = \text{tr } (BA),$$

where A and B are any $m \times n$ and $n \times m$ matrices, respectively.

Proof. Clearly

(A.4.27) $$\begin{bmatrix} \lambda I_m - AB & A \\ 0_{n,m} & \lambda I_n \end{bmatrix}\begin{bmatrix} I_m & 0_{m,n} \\ B & I_n \end{bmatrix} = \begin{bmatrix} I_m & 0 \\ B & I_n \end{bmatrix}\begin{bmatrix} \lambda I_m & A \\ 0_{n,m} & \lambda I_n - BA \end{bmatrix}$$

and thus

(A.4.28) $$|\lambda I_m - AB| \lambda^n = |\lambda I_n - BA| \lambda^m.$$

Thus the nonzero characteristic roots of AB and BA are equal and hence tr $(AB) = $ tr (BA), as required.

We can easily establish that the characteristic roots of a symmetrical matrix A_n are real and there exist orthonormal vectors $\mathbf{x}_1, \mathbf{x}_2, \cdots, \mathbf{x_n}$ corresponding to its n roots $\lambda_1, \lambda_2, \cdots, \lambda_n$. If we define

(A.4.29) $$C = (\mathbf{x_1} \ \mathbf{x_2} \ \cdots \ \mathbf{x_n}),$$

then C is an orthogonal matrix satisfying $CC' = C'C = I_n$ and

(A.4.30) $C'AC = \text{diag } [\lambda_1, \lambda_2, \cdots, \lambda_n].$

We introduce the following definition:

Definition A.4.15. The matrices A and B are said to be congruent if there exists a nonsingular matrix P such that

(A.4.31) $P'AP = B.$

If P is orthogonal, then A and B are said to be orthogonally congruent, and if P is a matrix with rational elements, then A and B are said to be rationally congruent.

Relation (A.4.30) implies the following:

Theorem A.4.12. Any symmetrical matrix is orthogonally congruent to a diagonal matrix.

We now prove the following theorem:

Theorem A.4.13. Any nonsingular symmetrical matrix A_n is congruent to I_n.

Proof. Put $P = C \text{ diag } [1/\sqrt{\lambda_1}, 1/\sqrt{\lambda_2}, \cdots, 1/\sqrt{\lambda_n}]$ in (A.4.30).† This gives

(A.4.32) $P'AP = I_n.$

Thus A is congruent to I_n.

We introduce the following definitions:

Definition A.4.16. The minimal polynomial of the square matrix A_n is the monic scalar polynomial

(A.4.33) $M(x) = x^m + a_1 x^{m-1} + \cdots + a_m$

of least degree such that

(A.4.34) $M(A) = A_n^m + a_1 A_n^{m-1} + \cdots + a_m I_n = 0_{n,n}.$

We can easily verify that the distinct characteristic roots of A are solutions of its minimal polynomial equation.

From (A.4.30) it is clear that

(A.4.35) $A = \sum_{i=1}^{n} \lambda_i \mathbf{x_i} \mathbf{x_i'}$

---

† P is not necessarily a matrix of real elements. However, it will be real if and only if all roots of A are positive.

and the representation on the right-hand side of (A.4.35) is called the spectral decomposition of A.

For a square matrix A we have defined the inverse, when A is nonsingular. However, for a symmetrical singular matrix we define the following:

Definition A.4.17. If A is a symmetrical matrix of order n with nonzero characteristic roots λ_1, λ_2, $\cdots$, λ_r with the corresponding orthonormal vectors $\mathbf{x}_1$, $\mathbf{x}_2$, $\cdots$, $\mathbf{x}_r$, then the pseudoinverse of A, denoted by $A^{-\mathscr{P}}$, is given by

$$(A.4.36) \qquad A^{-\mathscr{P}} = \sum_{i=1}^{r} \lambda_i^{-1} \mathbf{x}_i \mathbf{x}_i'.$$

We easily verify that

$$(A.4.37) \qquad AA^{-\mathscr{P}}A = A.$$

It may be noted that this inverse is one form of the generalized inverse A^- of A given by Rao (1962).

We now prove the following:

Theorem A.4.14. Let A be a nonsingular $n \times n$ matrix, let U and V be $n \times m$ and $m \times n$ matrices, respectively, and let $A + UV$ be nonsingular. Then

$$(A.4.38) \qquad (A + UV)^{-1} = A^{-1} - A^{-1}U\{I_m + VA^{-1}U\}^{-1}VA^{-1}.$$

Proof. Applying Theorem A.4.4. we have

$$(A.4.39) \qquad \begin{vmatrix} A_n & U_{n,m} \\ -V_{m,n} & I_m \end{vmatrix} = |A|\,|I_m + VA^{-1}U|.$$

By interchanging the rows and columns on the left-hand-side determinant of (A.4.39), we get

$$(A.4.40) \qquad \begin{vmatrix} A_n & U_{n,m} \\ -V_{m,n} & I_m \end{vmatrix} = \begin{vmatrix} I_m & -V_{m,n} \\ U_{n,m} & A_n \end{vmatrix} = |A + UV|.$$

Equating (A.4.39) and (A.4.40), we get

$$(A.4.41) \qquad |A + UV| = |A|\,|I_m + UA^{-1}U|.$$

Thus $A + UV$ is nonsingular if and only if A and $I_m + VA^{-1}U$ are nonsingular. We can verify (A.4.38) by multiplying its right-hand side by $A + UV$.

We close this section with the following concept of the Kronecker product of matrices:

Definition A.4.18. If $A = (a_{ij})$ is an $m \times n$ matrix and $B = (b_{ij})$ is a $p \times q$ matrix, the Kronecker product (or direct product) of the matrices A and B, denoted by $A \times B$, is an $mp \times nq$ matrix.

$$(A.4.42) \quad (a_{ij} B) = \left[\begin{array}{c|c|c|c} a_{11}B & a_{12}B & \cdots & a_{1n}B \\ \hline a_{21}B & a_{22}B & \cdots & a_{2n}B \\ \hline \cdot & \cdot & \cdots & \cdot \\ \hline a_{m1}B & a_{m2}B & \cdots & a_{mn}B \end{array} \right].$$

It may be remarked that some people define the direct product $A \times B$ as (Ab_{ij}). Our definition differs from this only by a permutation of rows and columns, and the distinction should be noted to properly keep track of symbols and sets in a design. It may also be observed that the direct product of matrices is not necessarily a commutative operation.

We observe that

$$(A.4.43) \qquad\qquad (A \times B)' = A' \times B'$$

and

$$(A.4.44) \qquad\qquad (A \times B) \cdot (C \times D) = AC \times BD.$$

In (A.4.44) we assume that the orders of the matrices are compatible with the operations involved.

A.5. QUADRATIC FORMS

Definition A.5.1. A quadratic form in n variables $x_1, x_2, \cdots, x_n$ is a function of the form

$$(A.5.1) \qquad\qquad Q = \sum_{i,j=1}^{n} a_{ij} x_i x_j,$$

where the a_{ij} are constants such that $a_{ij} = a_{ji}$.

If we write $\mathbf{x}' = (x_1, x_2, \cdots, x_n)$ and $A_n = (a_{ij})$, then Q can be written as

$$(A.5.2) \qquad\qquad Q = \mathbf{x}'A\mathbf{x}.$$

The matrix A is called the matrix of the quadratic form Q.

Definition A.5.2. The quadratic form $\mathbf{x}'A\mathbf{x}$ is said to be positive semi-definite if

(A.5.3) $$\mathbf{x}'A\mathbf{x} \geqq 0$$

for every $\mathbf{x}$. It is said to be positive definite if the equality in (A.5.3) holds if and only if $\mathbf{x} = 0_{n,1}$.

Theorem A.5.1. If A is the matrix of a positive semidefinite quadratic form, then its characteristic roots are nonnegative. However, if A is the matrix of a positive definite quadratic form, then its characteristic roots are all positive.

We define the following:

Definition A.5.3. Let $\mathbf{a}_1, \mathbf{a}_2, \cdots, \mathbf{a}_n$ be n, m component vectors. Then the gramian of the set, U, is defined by

(A.5.4) $$U = (\mathbf{a}_1\ \mathbf{a}_2\ \cdots\ \mathbf{a}_n)'(\mathbf{a}_1\ \mathbf{a}_2\ \cdots\ \mathbf{a}_n).$$

Clearly U is at least a positive semidefinite matrix. If $\mathbf{a}_1, \mathbf{a}_2, \cdots, \mathbf{a}_n$ are independent, then U will be positive definite.

A.6. THE LEGENDRE SYMBOL, THE HILBERT NORM RESIDUE SYMBOL AND THE HASSE–MINKOWSKI INVARIANT

The Legendre symbol is defined for a prime p by

(A.6.1) $$\begin{aligned}(a/p) &= +1 \quad \text{if } a \text{ is a quadratic residue of } p;\\ &= -1 \quad \text{if } a \text{ is a nonquadratic residue of } p.\end{aligned}$$

A slight generalization of the Legendre symbol is the Hilbert norm residue symbol $(a, b)_p$. If a and b are nonzero rational numbers, we define $(a, b)_p$ to have the value $+1$ or -1, depending on whether the congruence

(A.6.2) $$ax^2 + by^2 \equiv 1 \pmod{p^r}$$

has or has not for every value of r rational solutions x_r and y_r. Here p is any prime, including the conventional prime p_∞. We mention the useful properties of (a/p) and $(a, b)_p$ in the three theorems that follow, without proof. Since it is advisable to calculate the Hasse–Minkowski invariant for odd primes for proving the nonexistence of designs, we list the properties of the Hilbert norm residue symbol only for odd primes. Proofs can be obtained from Jones (1950) and Pall (1945).

Theorem A.6.1. For the Legendre symbol we have

(A.6.3) $(a/p) = (b/p)$ if $a \equiv b \pmod{p}$,

(A.6.4) $(ab/p) = (a/p)(b/p)$,

(A.6.5) $(p/q)(q/p) = (-1)^{(p-1)(q-1)/4}$,

(A.6.6) $(-1/p) = (-1)^{(p-1)/2}$,

(A.6.7) $(2/p) = (-1)^{(p^2-1)/8}$,

where p and q denote odd primes.

Theorem A.6.2. If m and m' are integers not divisible by the odd prime p, then

(A.6.8) $(m, m')_p = +1$,

(A.6.9) $(m, p)_p = (m/p)$.

Moreover, if $m \equiv m' \not\equiv 0 \pmod{p}$, then

(A.6.10) $(m, p)_p = (m', p)_p$.

Theorem A.6.3. For arbitrary nonzero integers m, m', n, n', and s and for every odd prime p,

(A.6.11) $(m, -m)_p = +1$,

(A.6.12) $(m, n)_p = (n, m)_p$,

(A.6.13) $(m, nn')_p = (m, n)_p(m, n')_p$,

(A.6.14) $(mm', m - m')_p = (m, -m')_p$,

(A.6.15) $\prod_{j=1}^{m}(j, j+1)_p = ((m+1)!, -1)_p$,

and

(A.6.16) $(as^2, b)_p = (a, b)_p$.

Now let A and B be two rational, symmetrical, and nonsingular matrices of the same order, which are rationally congruent. This fact will be denoted by $A \sim B$. Let D_r be the leading principal minor determinant of order r

and suppose that $D_r \neq 0$ for all r. Define $D_0 = 1$. Then the Hasse–Minkowski invariant of A is given by

$$(A.6.17) \qquad C_p(A) = (-1, -1)_p \prod_{i=0}^{n-1} (D_{i+1}, -D_i)_p$$

for each prime p. Then the following theorem is well known:

Theorem A.6.4 (Hasse). The necessary and sufficient conditions for two positive definite, rational, and symmetrical matrices A and B of the same order to be rationally congruent are that the square free parts of their determinants be the same and that their Hasse–Minowski invariants be equal for all primes p, including p_∞.

We now state, without proofs, certain theorems regarding the Hasse–Minkowski invariant.

Theorem A.6.5. If $A_1, A_2, \cdots, A_m$ are rational, nonsingular, and symmetrical matrices, and if

$$(A.6.18) \qquad A = \text{diag} [A_1, A_2, \cdots, A_m],$$

then

$$(A.6.19) \quad C_p(A) = (-1, -1)_p^{m-1} \left\{ \prod_{i=1}^{m} C_p(A_i) \right\} \left\{ \prod_{\substack{i, j=1 \\ i<j}}^{m} (|A_i|, |A_j|)_p \right\}.$$

As a particular case, we have the following corollary:

Corollary A.6.5.1. The Hasse–Minkowski invariant of

$$(A.6.20) \qquad A = I_m \times B$$

is

$$(A.6.21) \quad C_p(A) = (-1, -1)_p^{m-1} \{C_p(B)\}^m (|B|, -1)_p^{m(m-1)/2}.$$

Theorem A.6.6. For

$$(A.6.22) \qquad A = dI_m$$

we have

$$(A.6.23) \qquad C_p(A) = (-1, -1)_p (-1, d)_p^{m(m+1)/2}.$$

Theorem A.6.7. If ρ is a nonzero rational number,

$$(A.6.24) \qquad C_p(\rho A) = (-1, \rho)_p^{m(m+1)/2} (\rho, |A|)_p^{m-1} C_p(A).$$

Theorem A.6.8. If

$$(A.6.25) \qquad A = eI_m + fE_{m,m},$$

where e and f are nonzero rational numbers, then

$$(A.6.26) \quad C_p(A) = (-1, -1)_p(-1, e)_p^{m(m-1)/2}(-1, g)_p(m, g)_p(m, e)_p(g, e)_p^{m-1},$$

where

$$(A.6.27) \qquad g = e + mf.$$

Theorem A.6.9. If the $m - 1$ rational column vectors $\mathbf{a}_2, \mathbf{a}_3, \cdots, \mathbf{a_m}$ of dimension m are linearly independent and are orthogonal to $E_{m,1}$, then the gramian of the set, U, satisfies

$$(A.6.28) \qquad C_p(U) = (-1, -1)_p.$$

A.7. FINITE GEOMETRIES

A.7.1. Finite Plane Projective Geometry PG(n, s)

Let p be a prime and h a positive integer and let $s = p^h$. Any ordered set of $n + 1$ elements $(x_0, x_1, \cdots, x_n)$, where the x elements belong to GF(s) and are not all simultaneously zero, is called a point of our PG(n, s). Two sets $(x_0, x_1, \cdots, x_n)$ and $(y_0, y_1, \cdots, y_n)$ will represent the same point if $y_i = \rho x_i$ $(i = 0, 1, \cdots, n)$, where ρ is a nonnull element of GF(s). The elements $x_0, x_1, \cdots, x_n$ are called the coordinates of the point $(x_0, x_1, \cdots, x_n)$.

It can be easily seen that the number of points in PG(n, s) is

$$(A.7.1) \qquad Q_n = \frac{s^{n+1} - 1}{s - 1},$$

All the points that satisfy a set of $n - m$ independent, linear, homogeneous equations

$$(A.7.2) \qquad \begin{aligned} & a_{10} x_0 + a_{11} x_1 + \cdots + a_{1n} x_n = 0, \\ & a_{20} x_0 + a_{21} x_1 + \cdots + a_{2n} x_n = 0, \\ & \qquad\qquad \cdots \\ & a_{(n-m)0} x_0 + a_{(n-m)1} x_1 + \cdots + a_{(n-m)n} x_n = 0 \end{aligned}$$

are said to form an m-dimensional subspace, or an m-flat, of PG(n, s). Any other set of $n - m$ independent, linear, homogeneous equations obtained by linear combinations of the above equations will have the same set of solutions and will thus represent the same m-flat.

In (A.7.2) there are $n + 1$ unknowns and $n - m$ equations. The solutions of (A.7.2) will be

(A.7.3)
$$\begin{bmatrix} x_0 \\ x_1 \\ \cdot \\ x_n \end{bmatrix} = a_0 \xi_0 + a_1 \xi_1 + \cdots + a_m \xi_m ,$$

where $\xi_0, \xi_1, \cdots, \xi_m$ are any $m + 1$ linearly independent column vectors satisfying (A.7.2) and the a terms belong to $GF(s)$ and are not simultaneously zero. Each a can be chosen in s ways, and since not all the a terms are zero, the total number of points satisfying (A.7.2) is $s^{m+1} - 1$. But, as $(x_0, x_1, \cdots, x_n)$ and $(\rho x_0, \rho x_1, \cdots, \rho x_n)$, where ρ is a nonzero element of $GF(s)$, represent the same point, the number of distinct points satisfying (A.7.2)—in other words, the number of points lying on the m-flat (A.7.2)—is

(A.7.4)
$$Q_m = \frac{s^{m+1} - 1}{s - 1}.$$

We easily see that, if two points lie on an m-flat, the line joining them wholly lies on the said m-flat.

We now try to obtain the total number of m-flats in the geometry. Each m-flat is determined by any set of $(m + 1)$ independent points lying on it. Hence the total number of m-flats in $PG(n, s)$, denoted by $\phi(n, m, s)$, equals the number of ways of selecting $(m + 1)$ independent points from the Q_n points of the geometry, divided by the number of ways of selecting $(m + 1)$ independent points on an m-flat. From the totality of points of the geometry, the first point can be chosen in Q_n ways and the second point can be chosen in $Q_n - 1 = Q_n - Q_0$ ways. The third point must be chosen in such a way that it is linearly independent of the first and second points, that is, it is not a point on the 1-flat formed by the first two points and there are Q_1 points on a 1-flat. Therefore the third point can be chosen in $Q_n - Q_1$ ways. In general, when l independent points are already chosen, the $(l + 1)$th point must be chosen in such a way that it is linearly independent of the first l points, that is, it does not lie on an $(l - 1)$-flat formed by them and there are Q_{l-1} points on this flat. Therefore the $(l + 1)$th point can be chosen in $Q_n - Q_{l-1}$ ways. Similarly following, we get the number of ways of selecting $(m + 1)$ independent points to be equal to $Q_n(Q_n - Q_0) \cdots (Q_n - Q_{m-1})$. But the same m-flat can be generated by any one of $Q_m(Q_m - Q_0) \cdots (Q_m - Q_{m-1})$ sets of $(m + 1)$ independent points. Therefore the total number of distinct m-flats in $PG(n, s)$ is

$$\phi(n, m, s) = \frac{Q_n(Q_n - Q_0) \cdots (Q_n - Q_{m-1})}{Q_m(Q_m - Q_0) \cdots (Q_m - Q_{m-1})},$$

(A.7.5)

$$= \frac{(s^{n+1} - 1)(s^n - 1) \cdots (s^{n-m+1} - 1)}{(s^{m+1} - 1)(s^m - 1) \cdots (s - 1)}.$$

It can easily be seen that

(A.7.6) $$\phi(n, m, s) = \phi(n, n - m - 1, s).$$

On the lines of the above argument, we can show that the total number of distinct m-flats through a fixed point A is $\phi(n - 1, m - 1, s)$ and the number of distinct m-flats passing through two fixed points A and B is $\phi(n - 2, m - 2, s)$.

A.7.2. Finite Euclidean Geometry EG(n, s)

An ordered set of n elements $(x_1, x_2, \cdots, x_n)$, where the x elements belong to GF(s), is called a point of the finite Euclidean geometry EG(n, s). Two points $(x_1, x_2, \cdots, x_n)$ and $(y_1, y_2, \cdots, y_n)$ are the same if and only if $x_i = y_i$ for $i = 1, 2, \cdots, n$. It is obvious that the total number of points in EG(n, s) is

(A.7.7) $$E_n = s^n.$$

All the points satisfying a set of $n - m$ consistent and independent linear equations

(A.7.8)
$$
\begin{aligned}
a_{10} &+ a_{11}x_1 &+ \cdots + a_{1n}x_n &= 0, \\
a_{20} &+ a_{21}x_1 &+ \cdots + a_{2n}x_n &= 0, \\
&\quad \cdots \\
a_{(n-m)0} &+ a_{(n-m)1}x_1 + \cdots + a_{(n-m)n}x_n &= 0
\end{aligned}
$$

are said to form an m-flat in EG(n, s). As in the case of PG(n, s), we can see that the number of points on an m-flat of EG(n, s) is

(A.7.9) $$E_m = s^m$$

and that there are

(A.7.10) $$s^{n-m} \phi(n - 1, m - 1, s) = \phi(n, m, s) - \phi(n - 1, m, s)$$

distinct m-flats. The total number of distinct m-flats passing through a fixed point A is $\phi(n - 1, m - 1, s)$ and of those passing through two fixed points, $\phi(n - 2, m - 2, s)$.

A.7.3. Relation between $\text{PG}(n, s)$ **and** $\text{EG}(n, s)$

If the first coordinate of a point in $\text{PG}(n, s)$ is not zero, it can be regarded as $(1, x_1/x_0, \cdots, x_n/x_0)$. Consider all the points of $\text{PG}(n, s)$ satisfying $x_0 = 0$. It is an $n - 1$ flat of $\text{PG}(n, s)$. We call this an $(n - 1)$-flat at infinity, with points on it being known as points at infinity. The remaining points are called finite points of $\text{PG}(n, s)$. For these points, since the first coordinate is different from zero, they can always be reduced to the form $(1, x_1', x_2', \cdots, x_n')$. Therefore we can have a $1:1$ correspondence between the finite points $(1, x_1', x_2', \cdots, x_n')$ of $\text{PG}(n, s)$ and the points $(x_1', x_2', \cdots, x_n')$ of $\text{EG}(n, s)$.

An m-flat of $\text{PG}(n, s)$ may be said to be wholly at infinity if all its points are at infinity. All other flats are said to be finite m-flats. To any finite m-flat of $\text{PG}(n, s)$ given by the equations

$$(A.7.11) \qquad a_{i0} x_0 + a_{i1} x_1 + \cdots + a_{in} x_n = 0 \qquad i = 1, 2, \cdots, n - m,$$

let there correspond an m-flat of $\text{EG}(n, s)$, given by the equations

$$(A.7.12) \qquad a_{i0} + a_{i1} x_1 + \cdots + a_{in} x_n = 0 \qquad i = 1, 2, \cdots, n - m.$$

It is easy to see that the latter set of equations is consistent when the m-flat of $\text{PG}(n, s)$ is finite. Thus there exists a $1:1$ correspondence between the finite m-flats of $\text{PG}(n, s)$ and the m-flats of $\text{EG}(n, s)$. The finite points of the m-flats of $\text{PG}(n, s)$ correspond to the points of the m-flats of $\text{EG}(n, s)$. The geometry of $\text{EG}(n, s)$ is derivable from $\text{PG}(n, s)$ by cutting out all the points at infinity and the m-flats wholly lying at infinity. Conversely, from our $\text{EG}(n, s)$ we can construct $\text{PG}(n, s)$ by considering the points in $\text{EG}(n, s)$ as the finite points of $\text{PG}(n, s)$ and adding the $(n - 1)$-flat at infinity $x_0 = 0$ along with the distinct points lying on that flat.

REFERENCES

1. Jones, B. W. (1950). *The Arithmetic Theory of Quadratic Forms*. Wiley, New York.
2. Pall, G. (1945). The arithmetical invariant of quadratic forms. *Bull. Am. Math. Soc.*, **51**, 185–197.
3. Rao, C. R. (1962). A note on the generalized inverse of a matrix with applications to problems in mathematical statistics. *J. Roy. Stat. Soc.*, **24**, 152–158.

BIBLIOGRAPHY

Bose, R. C. On the construction of balanced incomplete block designs. *Ann. Eugenics*, **9**, 353–399 (1939).

Macduffe, C. C. *The Theory of Matrices*. Chelsea Publishing Company, New York, 1946.

APPENDIX B

Statistics for Mathematicians

B.1. PROBABILITY

Let S be the sample space consisting of all elementary events of a random experiment and let R be the class of subsets of S. Then we have the following:

Definition B.1.1. The class R will be called a ring of $A \in R$, and $B \in R$ implies $A \cup B \in R$, and $A - B \in R$.

As an immediate consequence, we verify that $A \cap B \in R$ in a ring R.

Definition B.1.2. A ring R is said to be a σ-ring if $A_i \in R$ $(i = 1, 2, \cdots)$ implies $\bigcup_{i=1}^{\infty} A_i \in R$. A σ-ring of subsets of S consisting of S is called a σ-algebra or a Borel field.

We call the elements of the Borel field of subsets of S random events, with ϕ being called an impossible event and S the sure event. We define the following:

Definition B.1.3. A set function is a function whose domain is a class of sets. A set function μ is called σ-additive if $\mu(\bigcup_{i=1}^{\infty} A_i) = \sum_{i=1}^{\infty} \mu(A_i)$ whenever $A_i \cap A_j = \phi$. It will be called real if its range is a subset of the real field.

Definition B.1.4. The σ-additive, real set function on a class of random events is said to be a probability if

(B.1.1) $$0 \leq \Pr(A) \leq 1$$

and

(B.1.2) $$\Pr(S) = 1,$$

where A is a random event and $\Pr(A)$ stands for the probability of the random event A.

From the additivity property of the probability function, it is clear that, if we assign the probability to each elementary event subject to (B.1.1) and (B.1.2), the probability of any random event will be uniquely determined when the sample space has at most countable number of points. The natural assignment of probability assigns equal probability to each elementary event.

If A and B are any two random events, we define the conditional probability of the event A given that B has occurred $[\Pr(B) > 0]$ by

(B.1.3) $$\Pr(A|B) = \frac{\Pr(AB)}{\Pr(B)}.$$

We say that the events A and B are independent if

(B.1.4) $$\Pr(A|B) = \Pr(A).$$

B.2. RANDOM VARIABLE AND DISTRIBUTION FUNCTIONS

Let X be a real-valued function defined on the class of random events. Then we have the following:

Definition B.2.1. The function X is called a random variable if the inverse image of every interval I on the real line of the form $[-\infty, x]$ is a random event.

If X is a random variable showing the correspondence of the random event A to the subset T, an element of the Borel field of the real field, then

(B.2.1) $$\Pr(X \in T) = \Pr(A).$$

The function $\Pr(X \in T)$ for every Borel set T on the real line is said to be the probability function of the random variable X. We define the following:

Definition B.2.2. The function $F(x) = \Pr(X \leq x)$ is said to be the distribution function of the random variable X.

In standard books on mathematical statistics it is proved that a distribution function $F(x)$ is a nondecreasing function that is everywhere continuous to the right and is such that $F(-\infty) = 0$ and $F(\infty) = 1$. If $F(x)$ is continuous everywhere such that $F'(x) = f(x)$, then the corresponding random variable X is said to be of the continuous type and $f(x)$ is called the frequency function of the random variable. In this case

(B.2.2) $$\Pr(X \in T) = \int_T f(x)\, dx,$$

where T is a Borel set on the real line.

B.3. PARAMETERS OF THE DISTRIBUTION FUNCTION

Definition B.3.1. If $g(x)$ is a single-value function, then the expected value of the random variable $g(x)$, denoted by $\mathscr{E}[g(x)]$, is given by

$$(B.3.1) \qquad \mathscr{E}[g(x)] = \int g(x)\, dF(x),$$

where the integral on the right-hand side is the Riemann–Stieltjes integral and is absolutely convergent.

In particular, when $g(x) = X^k$, the expected value of X^k is called the kth order moment of X. The first-order moment of the random variable X is called its mean or expected value and is a measure of the central tendency of the random variable X. It is commonly denoted by μ. If $g(x) = (X - \mu)^k$, the expected value of $(X - \mu)^k$ is called the kth order central moment. The second-order central moment is called the variance of X and measures the scatter of the random variable from its mean. It is denoted by either μ_2 or σ^2. The square root of σ^2 is known as the standard deviation of the random variable.

B.4. MULTIDIMENSIONAL RANDOM VARIABLE

Definition B.4.1. The collection of n real, single-value functions $\mathbf{X}' = (X_1, X_2, \cdots, X_n)$ defined on S is called an n-dimensional random variable if the inverse image A of every generalized interval I: $[-\infty, x_1; -\infty, x_2; \cdots; -\infty, x_n]$ in the n-dimensional Euclidean space is a random event.

The distribution function of an n-dimensional random variable $\mathbf{X}' = (X_1, X_2, \cdots, X_n)$ is given by the following:

Definition B.4.2. The function

$$(B.4.1) \qquad F(x_1, x_2, \cdots, x_n) = \Pr\,(X_1 \leqq x_1, X_2 \leqq x_2, \cdots, X_n \leqq x_n)$$

is said to be the distribution function of the n-dimensional random variable $\mathbf{X}$.

The expected value of the single-value function $g(X_1, X_2, \cdots, X_n)$ is denoted by $\mathscr{E}[g(X_1, X_2, \cdots, X_n)]$ and is given by the Riemann–Stieltjes integral

$$(B.4.2) \qquad \iint \cdots \int g(x_1, x_2, \cdots, x_n)\, dF(x_1, x_2, \cdots, x_n).$$

If $g(X_1, X_2, \cdots, X_n) = X_1^{v_1} X_2^{v_2} \cdots X_n^{v_n}$, then $\mathscr{E}[g(X_1, X_2, \cdots, X_n)]$ is known as the moment of order $v_1 + v_2 + \cdots v_n$ and is denoted by $\alpha_{v_1, v_2, \cdots, v_n}$. For the first-order moments, we use the notation

$$(B.4.3) \qquad \mu_i = \mathscr{E}(X_i) = \iint \cdots \int x_i\, dF(x_1, x_2, \cdots, x_n).$$

The vector $\mu' = (\mu_1, \mu_2, \cdots, \mu_n)$ is then said to be the mean of the random vector $\mathbf{X}'$ and is denoted by

(B.4.4) $$\mathscr{E}(\mathbf{X}) = \mu.$$

If

$$g(X_1, X_2, \cdots, X_n) = (X_1 - \mu_1)^{v_1}(X_2 - \mu_2)^{v_2} \cdots (X_n - \mu_n)^{v_n},$$

then

$$\mathscr{E}[g(X_1, X_2, \cdots, X_n)]$$

is said to be the central moment of order $v_1 + v_2 + \cdots + v_n$ and is denoted by $\mu_{v_1, v_2, \cdots, v_n}$. The second-order central moments play a vital role. We introduce the notation

(B.4.5) $$\sigma_{ii} = \mathscr{E}(X_i - \mu_i)^2 \qquad i = 1, 2, \cdots, n$$

and

(B.4.6) $$\sigma_{ij} = \mathscr{E}(X_i - \mu_i)(X_j - \mu_j) \qquad i \neq j; i, j = 1, 2, \cdots, n,$$

where σ_{ii} is called the variance of X_i and σ_{ij} is called the covariance of X_i and X_j. The $n \times n$ matrix $\sum = (\sigma_{ij})$ is said to be the dispersion matrix or variance covariance matrix of the random vector $\mathbf{X}$, and this fact is denoted by

(B.4.7) $$\text{var}(\mathbf{X}) = \sum.$$

It can be seen that $\sum$ is at least positive semidefinite; $|\sum|$ is called the generalized variance of the random vector $\mathbf{X}$.

We now introduce the concept of marginal distribution and that of the independence of random variables.

Definition B.4.3. If we have an n-dimensional random variable $\mathbf{X}' = (X_1, X_2, \cdots, X_n)$, with the distribution function $F(x_1, x_2, \cdots, x_n)$, the marginal distribution of any k of the n random variables is obtained from F by replacing the other $n - k$ variables by $+\infty$.

Definition B.4.4. The random variables $X_1, X_2, \cdots, X_n$ are said to be independent if

(B.4.8) $$F(x_1, x_2, \cdots, x_n) = F_1(x_1)F_2(x_2) \cdots F_n(x_n),$$

where $F_1(x_1), F_2(x_2), \cdots, F_n(x_n)$ are the marginal distribution functions of $X_1, X_2, \cdots, X_n$, respectively.

B.5. NORMAL DISTRIBUTION AND RELATED DISTRIBUTIONS

The most useful of the distribution functions is the normal distribution function, which is associated with a continuous random variable X and whose frequency function is given by

$$(B.5.1) \qquad f(x) = \frac{1}{\sqrt{2\pi}\sigma} e^{-(x-\mu)^2/2\sigma^2}.$$

For this random variable we can easily verify that μ and σ^2 are the mean and variance of the random variable X. The transformed variable

$$(B.5.2) \qquad Z = \frac{X - \mu}{\sigma}$$

is said to be the standardized normal variable and has the frequency function

$$(B.5.3) \qquad \phi(z) = \frac{1}{\sqrt{2\pi}} e^{-z^2/2},$$

with the corresponding distribution function

$$(B.5.4) \qquad \Phi(z) = \int_{-\infty}^{z} \frac{1}{\sqrt{2\pi}} e^{-z^2/2} \, dz.$$

If $X_1, X_2, \cdots, X_n$ are n independent random variables, each distributed normally with mean 0 and variance 1, then the distribution of

$$X_1^2 + X_2^2 + \cdots + X_n^2$$

is known as the χ^2 distribution with n degrees of freedom. The frequency function of such a distribution is

$$(B.5.5) \qquad \frac{1}{2^{n/2}\Gamma(n/2)} e^{-(\chi^2/2)}(\chi^2)^{(n/2)-1}, \qquad \chi^2 > 0.$$

If χ_1^2 and χ_2^2 are two independent χ^2 distributions with m and n degrees of freedom, respectively, then the ratio $(\chi_1^2/m)/(\chi_2^2/n)$ is distributed according to the F-ratio, with m and n degrees of freedom, respectively, whose frequency function is

$$(B.5.6) \qquad \frac{m}{n} \frac{\Gamma[(m+n)/2]}{\Gamma(m/2)\Gamma(n/2)} \frac{(mF/n)^{(m/2)-1}}{(1 + mF/n)^{(m+n)/2}}, \qquad F > 0.$$

The probability functions of normal distribution, χ^2 distribution, and F distribution are provided in all standard textbooks on mathematical statistics.

B.6. THEORY OF LINEAR ESTIMATION AND TESTS OF LINEAR HYPOTHESIS

Let $y_1, y_2, \cdots, y_n$ be n random variables such that the expectation of each of them is a known linear combination of unknown parameters. Let $\mathbf{y}' = (y_1, y_2, \cdots, y_n)$ be the vector of n observations and $\boldsymbol{\beta}' = (\beta_1, \beta_2, \cdots, \beta_p)$ the vector of unknown parameters. Let

$$(B.6.1) \qquad\qquad \mathscr{E}(\mathbf{y}) = X\boldsymbol{\beta},$$

where $X = (x_{ij})$ is an $n \times p$ matrix of known constants. Let var $(\mathbf{y}) = \sigma^2 I_n$. Any linear parametric function $l_1\beta_1 + l_2\beta_2 + \cdots + l_p\beta_p$, denoted by $\mathbf{l}'\boldsymbol{\beta}$, where $\mathbf{l}' = (l_1, l_2, \cdots, l_p)$, is said to be estimable if there exists a vector $\mathbf{a}' = (a_1, a_2, \cdots, a_n)$ such that

$$(B.6.2) \qquad\qquad \mathscr{E}(\mathbf{a}'\,\mathbf{y}) = \mathbf{l}'\,\boldsymbol{\beta}.$$

This implies that $\mathbf{a}'\,X\,\boldsymbol{\beta} = \mathbf{l}'\,\boldsymbol{\beta}$ identically in $\boldsymbol{\beta}$ and hence

$$(B.6.3) \qquad\qquad X'\,\mathbf{a} = \mathbf{l}.$$

A necessary and sufficient condition for the existence of a solution $\mathbf{a}$ for (B.6.3) is

$$(B.6.4) \qquad\qquad \text{rank}\,(X') = \text{rank}\,(X'\mid \mathbf{l}).$$

Hence

Theorem B.6.1. A necessary and sufficient condition for a parametric function $\mathbf{l}'\,\boldsymbol{\beta}$ to be estimable is (B.6.4).

There may be many estimators for $\mathbf{l}'\,\boldsymbol{\beta}$. Of all of them, we select the one with minimum variance, which is called the best linear unbiased estimator. Since var $(\mathbf{a}'\,\mathbf{y}) = \sigma^2\mathbf{a}'\,\mathbf{a}$, we minimize $\mathbf{a}'\,\mathbf{a}$ subject to (B.6.3), giving

$$(B.6.5) \qquad\qquad \mathbf{a} = X\boldsymbol{\lambda},$$

where $\boldsymbol{\lambda}$ is a column vector of Lagrangian multipliers; substituting this in (B.6.3), we get

$$(B.6.6) \qquad\qquad (X'\,X)\boldsymbol{\lambda} = \mathbf{l}.$$

Thus if a solution $\boldsymbol{\lambda}$ exists for (B.6.6), the best linear unbiased estimator for $\mathbf{l}'\,\boldsymbol{\beta}$ is $\boldsymbol{\lambda}'X\mathbf{y}$. We can easily see that the following obtains:

Theorem B.6.2. A necessary and sufficient condition for the existence of a best linear unbiased estimator for a linear parametric function $\mathbf{l}'\,\boldsymbol{\beta}$ is that

$$(B.6.7) \qquad\qquad \text{rank}\,(X'\,X) = \text{rank}\,(X'\,X\mid \mathbf{l}).$$

One can easily show that (B.6.4) and (B.6.7) are equivalent, and hence the estimability of $l' \beta$ implies the existence of a best linear unbiased estimator for $l' \beta$. The best linear unbiased estimator for $l' \beta$ can be verified to be unique.

The value of β that minimizes $(y - X\beta)'(y - X\beta)$ is denoted by $\hat{\beta}$, and the equations giving $\hat{\beta}$ are called normal equations. We can see that the normal equations are

$$(B.6.8) \qquad\qquad X' X\hat{\beta} = X' y.$$

A unique solution for $\hat{\beta}$ may or may not exist, depending on whether rank $(X) = p$ or rank $(X) < p$, respectively. In any case $\hat{\beta} = (X' X)^{-\mathscr{P}} X' y$ can be seen to be a solution of (B.6.8). The following result is well known:

Theorem B.6.3 (Gauss–Markoff Theorem). The best linear unbiased estimator of an estimable linear parametric function $l' \beta$ is $l' \hat{\beta}$, where $\hat{\beta}$ is given from (B.6.8). Furthermore the unbiased estimator of σ^2 is $(y - X\hat{\beta})'(y - X\hat{\beta})/(n - r)$, where r is rank (X).

If $l' \beta$ and $m' \beta$ are two estimable functions and $X' X\lambda = l$ and $X' X\mu = m$, then

$$(B.6.9) \qquad\qquad \text{var} (l' \hat{\beta}) = (l' \lambda)\sigma^2$$

and

$$(B.6.10) \qquad\qquad \text{cov} (l' \hat{\beta}, m' \hat{\beta}) = (l' \mu)\sigma^2 \quad \text{or} \quad (\lambda' m)\sigma^2.$$

We introduce the following definition:

Definition B.6.1. A linear function of the observations $e_1 y_1 + e_2 y_2 + \cdots + e_n y_n$, denoted by $e' y$, where $e' = (e_1, e_2, \cdots, e_n)$, is said to belong to the error space if

$$(B.6.11) \qquad\qquad \mathscr{E}(e' y) = 0.$$

We can easily verify that the functions of the error space form a vector space of $n - r$ dimensions, and the estimable functions belong to a vector space of r dimensions that is orthogonal to the former.

For testing any relations between the unknown parameters we assume that the observations $y_1, y_2, \cdots, y_n$ are normally distributed.† Let the hypothesis to be tested about s estimable functions $l'_i \beta$ $(i = 1, 2, \cdots, s)$ be $H: l'_i \beta = g_i$ $(i = 1, 2, \cdots, s)$. Put

$$(B.6.12) \qquad\qquad z_i = l'_i \beta - g_i \qquad (i = 1, 2, \cdots, s)$$

† Otherwise transformations will be applied.

and

(B.6.13)
$$D = \begin{bmatrix} \mathbf{l}_1' \lambda_1 & \mathbf{l}_1' \lambda_2 & \cdots & \mathbf{l}_1' \lambda_s \\ \mathbf{l}_2' \lambda_1 & \mathbf{l}_2' \lambda_2 & \cdots & \mathbf{l}_2' \lambda_s \\ \cdot & \cdot & \cdots & \cdot \\ \mathbf{l}_s' \lambda_1 & \mathbf{l}_s' \lambda_2 & \cdots & \mathbf{l}_s' \lambda_s \end{bmatrix},$$

where

(B.6.14) $\qquad\qquad X' X \lambda_i = \mathbf{l}_i \qquad i = 1, 2, \cdots, s.$

It is proved in textbooks on the design and analysis of experiments that

(B.6.15)
$$\frac{\mathbf{z}' D^{-1} \mathbf{z}}{\sigma^2}$$

is distributed as a χ^2 distribution with s degrees of freedom and is independently distributed with

(B.6.16)
$$\frac{(\mathbf{y} - X\hat{\boldsymbol{\beta}})'(\mathbf{y} - X\hat{\boldsymbol{\beta}})}{\sigma^2},$$

which is also distributed as a χ^2 distribution, but with $n - r$ degrees of freedom. Under the hypothesis H the quantity

(B.6.17)
$$\frac{\mathbf{z}' D^{-1} \mathbf{z}/s}{(\mathbf{y} - X\hat{\boldsymbol{\beta}})'(\mathbf{y} - X\hat{\boldsymbol{\beta}})/(n - r)}$$

is distributed as an F-ratio with s and $(n - r)$ degrees of freedom. This ratio can be used for testing the significance of the hypothesis H. It may be noted that $\mathbf{z}' D^{-1} \mathbf{z}$ is called the sum of squares due to the hypothesis and $(\mathbf{y} - X\hat{\boldsymbol{\beta}})'(\mathbf{y} - X\hat{\boldsymbol{\beta}})$ is called the error sum of squares.

B.7. BLOCK DESIGNS

Let there be b blocks, the ith block consisting of k_i plots. Let v treatments be applied to these plots, the jth treatment being replicated r_j times. The incidence relation between the treatments and blocks is denoted by the matrix $N = (n_{ij})$, known as the incidence matrix, where n_{ij} denotes the number of times the ith treatment occurs in the jth block. We see that

(B.7.1) $\qquad\qquad N E_{b,1} = \mathbf{r} \quad \text{and} \quad E_{1,v} N = \mathbf{k}',$

where $\mathbf{r}' = (r_1, r_2, \cdots, r_v)$ and $\mathbf{k}' = (k_1, k_2, \cdots, k_b)$. The design is said to be a complete block design if $k_1 = k_2 = \cdots = k_b = v$, and it is said to be an incomplete block design if $k_i < v$ $(i = 1, 2, \cdots, b)$.

If y_{ij} is the yield of the plot in the ith block to which the jth treatment is applied, we assume that

$$(B.7.2) \qquad \mathcal{E}(y_{ij}) = \mu + \beta_i + t_j \qquad \begin{aligned} i &= 1, 2, \cdots, b; \\ j &= 1, 2, \cdots, v. \end{aligned}$$

where μ is the general effect, β_i is the ith block effect ($i = 1, 2, \cdots, b$), and t_j is the jth treatment effect ($j = 1, 2, \cdots, v$). We assume that the y_{ij}'s are normally and independently distributed with constant variance σ^2.

Let y_i be the column vector of the k_i observations in the ith block and let U_i be a $k_i \times v$ matrix such that the jth row has a unit in the column corresponding to the treatment applied to the jth plot of the ith block ($i = 1, 2, \cdots, b$). All the $\sum_{i=1}^{b} k_i = n$ observations can then be written in the form

$$(B.7.3) \quad \mathcal{E}\begin{bmatrix} y_1 \\ y_2 \\ \cdot \\ y_b \end{bmatrix} = \begin{bmatrix} E_{n,1} & \text{diag}\,[E_{k_1,1}, E_{k_2,1}, \cdots, E_{k_b,1}] & \begin{matrix} U_1 \\ U_2 \\ \cdot \\ U_b \end{matrix} \end{bmatrix}\begin{bmatrix} \mu \\ \beta \\ t \end{bmatrix},$$

Where $\beta' = (\beta_1, \beta_2, \cdots, \beta_b)$ and $t' = (t_1, t_2, \cdots, t_v)$. The normal equations can be seen to be

$$(B.7.4) \quad \begin{bmatrix} n & k' & r' \\ k & \text{diag}\,[k_1, k_2, \cdots, k_b] & N' \\ r & N & \text{diag}\,[r_1, r_2, \cdots, r_v] \end{bmatrix}\begin{bmatrix} \hat{\mu} \\ \hat{\beta} \\ \hat{t} \end{bmatrix} = \begin{bmatrix} G \\ B \\ T \end{bmatrix},$$

where the circumflex on a parameter indicates the estimates, G is the grand total, B is the column vector of the block totals, and T is the column vector of the treatment totals. By eliminating $\hat{\beta}$ from second and third equations, we can easily see that $\hat{t}$ is given by the equation

$$(B.7.5) \qquad C\hat{t} = Q,$$

where

$$(B.7.6) \quad C = \text{diag}\,[r_1, r_2, \cdots, r_v] - N\{\text{diag}\,[k_1^{-1}, k_2^{-1}, \cdots, k_b^{-1}]\}N'$$

and

$$(B.7.7) \qquad Q = T - N\{\text{diag}\,[k_1^{-1}, k_2^{-1}, \cdots, k_b^{-1}]\}B.$$

Equation (B.7.5) is known as the equation for estimating the treatment effects, and the matrix defined by (B.7.6) is the well-known C-matrix, extensively used in Chapter 4. Shah (1959) proved that a solution of (B.7.5) is

$$(B.7.8) \qquad \hat{t} = (C + aE_{v,v})^{-1}Q = (c^{ij})Q,$$

where a is a nonzero real number.

One can easily verify that the linear function $l_1 t_1 + l_2 t_2 + \cdots + l_v t_v$ is estimable if and only if $l_1 + l_2 + \cdots + l_v = 0$, in which case the linear function $\mathbf{l}'\mathbf{t}$ is called a contrast. Contrasts of the form $t_i - t_j$ $(i \neq j)$ are called elementary contrasts. The best linear unbiased estimator of $t_i - t_j$ is $\hat{t}_i - \hat{t}_j$, where $\hat{t}_i$ $(i = 1, 2, \cdots, v)$ is given by (B.7.8) and

$$(\text{B.7.9}) \qquad \text{var}\,(\hat{t}_i - \hat{t}_j) = (c^{ii} + c^{jj} - 2c^{ij})\sigma^2.$$

The sum of squares for testing the hypothesis $t_1 = t_2 = \cdots = t_v$ can be seen to be $\hat{\mathbf{t}}'\mathbf{Q}$, with $v - 1$ degrees of freedom against the error sum of squares $\sum_{i,j} y_{ij}^2 - \sum_{i=1}^{b} B_i^2 / k_i - \hat{\mathbf{t}}'\mathbf{Q}$, with $n - b - v + 1$ degrees of freedom.

REFERENCE

Shah, B. V. (1959). A generalization of partially balanced incomplete block designs. *Ann. Math. Stat.*, **30**, 1041–1050.

BIBLIOGRAPHY

Chakrabarti, M. C. *Mathematics of Design and Analysis of Experiments*. Asia Publishing House, Bombay, India, 1962.

Cochran, W. G., and Cox, G. M. *Experimental Designs*. Wiley, New York, 1967.

Cramer, H. *Mathematical Methods of Statistics*. Princeton Mathematical Series, Princeton University Press, 1947.

Fisz, M. *Probability Theory and Mathematical Statistics*. Wiley, New York, 1963.

Kempthorne, O. *The Design and Analysis of Experiments*. Wiley, New York, 1952.

Rao, C. R. *Linear Statistical Inference and Its Applications*. Wiley, New York, 1965.

Scheffe, H. *The Analysis of Variance*. Wiley, New York, 1959.

Tocher, K. D. The design and analysis of block experiments. *J. Roy. Stat. Soc.*, **B14**, 45–91 (1952).

Miscellaneous Exercises

1. Divide the 81 numbers 1, 2, $\cdots$, 81 into nine sets of nine numbers each so that the totals of each set of numbers are equal.

2. Construct complete sets of mutually orthogonal latin squares of orders 7 and 9.

3. Construct $s^2 + s - 2$ mutually orthogonal latin cubes of the first order of size s.

4. For a given size of the experimental material show that a completely randomized design is more efficient if each treatment is replicated the same number of times.

5. Show that unequal-block-sized balanced designs exist.

6. Show that the BIB designs with the parameters

$$v^* = \frac{(k-1)(k-2)}{2}, \qquad b^* = \frac{k(k-1)}{2}, \qquad r^* = k, \qquad k^* = k-2, \qquad \lambda^* = 2,$$

and

$$v' = b' = \frac{k(k-1)}{2} + 1, \qquad r' = k' = k, \qquad \lambda' = 2$$

coexist.

7. Show that the existence of a symmetrical BIB design with the parameters $v = b = 4s^2$, $r = 2s^2 - s = k$, $\lambda = s^2 - s$ implies the existence of a Hadamard matrix H_{4s^2}.

8. Let N be the incidence matrix of an asymmetrical BIB design with the parameters v, b, r, k, and λ, which can be written without loss of generality as

$$N = \begin{bmatrix} E_{1,r} & 0_{1,b-r} \\ N_1 & N_2 \end{bmatrix};$$

then $N^* = (N_1 \quad E_{v-1,b-r} - N_2)$ is the incidence matrix of a BIB design if and only if N is the incidence matrix of a BIB design with the parameters

$$v = 2t + 2, \qquad b = 2r, \qquad r = m(2t+1), \qquad k = t+1, \qquad \lambda = mt.$$

371

Then N^* is the incidence matrix of another BIB design with the parameters

$$v_1 = 2t + 1, \qquad b_1 = 2m(2t + 1), \qquad r_1 = 2mt, \qquad k_1 = t, \qquad \lambda_1 = m(t - 1).$$

9. If $12t + 1$ is a prime or a prime power and x is a primitive root of $GF(12t + 1)$, show that the t initial sets $(0, x^0, x^{4t}, x^{8t})$; $(0, x^2, x^{4t+2}, x^{8t+2})$; $\cdots$; $(0, x^{2t-2}, x^{6t-2}, x^{10t-2})$ form a difference set for a BIB design with the parameters

$$v = 12t + 1, \qquad b = t(12t + 1), \qquad r = 4t, \qquad k = 4, \qquad \lambda = 1.$$

10. If $4t + 1$ is a prime or a prime power and x is a primitive root of $GF(4t + 1)$, show that the $3t + 1$ initial sets

$$(x_1{}^0, x_1{}^{2t}, x_2{}^\alpha, x_2^{\alpha+2t}); (x_1{}^2, x_1^{2t+2}, x_2^{\alpha+2}, x_2^{\alpha+2t+2}); \cdots;$$

$$(x_1^{2t-2}, x_1^{4t-2}, x_2^{\alpha+2t+2}, x_2^{\alpha+4t-2}); (x_2{}^0, x_2{}^{2t}, x_3{}^\alpha, x_3^{\alpha+2t});$$

$$(x_2{}^2, x_2^{2t+2}, x_3^{\alpha+2}, x_3^{\alpha+2t+2}); \cdots;$$

$$(x_2^{2t-2}, x_2^{4t-2}, x_3^{\alpha+2t-2}, x_3^{\alpha+4t-2}); (x_3{}^0, x_3{}^{2t}, x_1{}^\alpha, x_1^{\alpha+2t});$$

$$(x_3{}^2, x_3^{2t+2}, x_1^{\alpha+2}, x_1^{\alpha+2t+2}); \cdots;$$

$$(x_3^{2t-2}, x_3^{4t-2}, x_1^{\alpha+4t-2}, x_1^{\alpha+4t-2});$$

$$(\infty, 0_1, 0_2, 0_3)$$

form a difference set for a BIB design with the parameters

$$v = 12t + 4, \qquad b = (4t + 1)(3t + 1), \qquad r = 4t + 1, \qquad k = 4, \qquad \lambda = 1.$$

11. If $20t + 1$ is a prime or a prime power and x is a primitive root of $GF(20t + 1)$, show that the t initial sets $(x^0, x^{4t}, x^{8t}, x^{12t}, x^{16t})$; $(x^2, x^{4t+2}, x^{8t+2}, x^{12t+2}, x^{16t+2})$; $\cdots$; $(x^{2t-2}, x^{6t-2}, x^{10t-2}, x^{14t-2}, x^{18t-2})$ form a difference set for a BIB design with the parameters

$$v = 20t + 1, \qquad b = t(20t + 1), \qquad r = 5t, \qquad k = 5, \qquad \lambda = 1.$$

12. If $4t + 1$ is a prime or a prime power and x is a primitive root of $GF(4t + 1)$, show that the $5t + 1$ initial sets

$$
\begin{array}{ll}
(x_1^{2i}, x_1^{2i+2t}, x_3^{2i+\alpha}, x_3^{2i+\alpha+2t}, 0_2) & i = 0, 1, \cdots, t - 1, \\
(x_2^{2i}, x_2^{2i+2t}, x_4^{2i+\alpha}, x_4^{2i+\alpha+2t}, 0_3) & i = 0, 1, \cdots, t - 1, \\
(x_3^{2i}, x_3^{2i+2t}, x_5^{2i+\alpha}, x_5^{2i+\alpha+2t}, 0_4) & i = 0, 1, \cdots, t - 1, \\
(x_4^{2i}, x_4^{2i+2t}, x_1^{2i+\alpha}, x_1^{2i+\alpha+2t}, 0_5) & i = 0, 1, \cdots, t - 1, \\
(x_5^{2i}, x_5^{2i+2t}, x_2^{2i+\alpha}, x_2^{2i+\alpha+2t}, 0_1) & i = 0, 1, \cdots, t - 1, \\
\end{array}
$$

$$(0_1, 0_2, 0_3, 0_4, 0_5)$$

form a difference set for a BIB design with the parameters

$$v = 20t + 5, \qquad b = (5t + 1)(20t + 5), \qquad r = 5t + 1, \qquad k = 5, \qquad \lambda = 1.$$

13. If $6t + 1$ is a prime or a prime power and x is a primitive root of GF($6t + 1$), show that the t initial sets $(0, x^i, x^{2t+i}, x^{4t+i})$, $i = 0, 1, \cdots, t - 1$, form a difference set for a BIB design with the parameters

$$v = 6t + 1, \quad b = t(6t + 1), \quad r = 4t, \quad k = 4, \quad \lambda = 2.$$

14. If $4t + 1$ is a prime or a prime power and x is a primitive root of GF($4t + 1$), show that the t initial sets $(x^i, x^{t+i}, x^{2t+i}, x^{3t+i})$, $i = 0, 1, \cdots, t - 1$, form a difference set for a BIB design with the parameters

$$v = 4t + 1, \quad b = t(4t + 1), \quad r = 4t, \quad k = 4, \quad \lambda = 3.$$

15. If $10t + 1$ is a prime or a prime power and x is a primitive root of GF($10t + 1$), show that the t initial sets $(x^i, x^{2t+i}, x^{4t+i}, x^{6t+i})$, $i = 0, 1, \cdots, t - 1$, form a difference set for a BIB design with the parameters

$$v = 10t + 1, \quad b = t(10t + 1), \quad r = 5t, \quad k = 5, \quad \lambda = 2.$$

16. If $5t + 1$ is a prime or a prime power and x is a primitive root of GF($5t + 1$), show that the t initial sets $(x^i, x^{t+i}, x^{2t+i}, x^{3t+i}, x^{4t+i})$, $i = 0, 1, \cdots, t - 1$, form a difference set for a BIB design with the parameters

$$v = 5t + 1, \quad b = t(5t + 1), \quad r = 5t, \quad k = 5, \quad \lambda = 4.$$

17. If $2t + 1$ is a prime or a prime power and x is a primitive root of GF($2t + 1$), show that the $3t + 2$ initial sets

$$(x_1{}^i, x_1^{t+i}, x_2^{q+i}, x_2^{q+t+i}) \qquad i = 0, 1, \cdots, t - 1,$$
$$(x_2{}^i, x_2^{t+i}, x_3^{q+i}, x_3^{q+t+i}) \qquad i = 0, 1, \cdots, t - 1,$$
$$(x_3{}^i, x_3^{t+i}, x_1^{q+i}, x_1^{q+t+i}) \qquad i = 0, 1, \cdots, t - 1,$$
$$(\infty, 0_1, 0_2, 0_3), (\infty, 0_1, 0_2, 0_3)$$

form a difference set for a BIB design with the parameters

$$v = 2(3t + 2), \quad b = (2t + 1)(3t + 2), \quad r = 4t + 2, \quad k = 4, \quad \lambda = 2.$$

18. If $2t + 1$ is a prime or a prime power and x is a primitive root of GF($2t + 1$), show that the $5t + 2$ initial sets

$$(x_1{}^i, x_1^{i+t}, x_3^{i+\alpha}, x_3^{i+\alpha+t}, 0_2) \qquad i = 0, 1, \cdots, t - 1,$$
$$(x_2{}^i, x_2^{i+t}, x_4^{i+\alpha}, x_4^{i+\alpha+t}, 0_3) \qquad i = 0, 1, \cdots, t - 1,$$
$$(x_3{}^i, x_3^{i+t}, x_5^{i+\alpha}, x_5^{i+\alpha+t}, 0_4) \qquad i = 0, 1, \cdots, t - 1,$$
$$(x_4{}^i, x_4^{i+t}, x_1^{i+\alpha}, x_1^{i+\alpha+t}, 0_5) \qquad i = 0, 1, \cdots, t - 1,$$
$$(x_5{}^i, x_5^{i+t}, x_2^{i+\alpha}, x_2^{i+\alpha+t}, 0_1) \qquad i = 0, 1, \cdots, t - 1,$$
$$(0_1, 0_2, 0_3, 0_4, 0_5), (0_1, 0_2, 0_3, 0_4, 0_5),$$

form a difference set for a BIB design with the parameters

$$v = 10t + 5, \quad b = (5t + 2)(2t + 1), \quad r = 5t + 2, \quad k = 5, \quad \lambda = 2.$$

19. If $tk + 1$ is a prime or a prime power and x is a primitive root of $GF(tk + 1)$, show that the t initial sets

$$(x^i, x^{i+t}, x^{i+2t}, \cdots, x^{i+(k-1)t}), \quad i = 0, 1, \cdots, t - 1,$$

form a difference set for a BIB design with the parameters

$$v = tk + 1, \qquad b = t(tk + 1), \qquad r = tk, \qquad k, \qquad \lambda = k - 1.$$

20. If $2t(2\lambda + 1) + 1$ is a prime or a prime power and x is a primitive root of $GF[2t(2\lambda + 1) + 1]$, show that the t initial sets $(x^i, x^{i+2t}, x^{i+4t}, \cdots, x^{i+4\lambda t})$, $i = 0, 1, \cdots, t - 1$, form a difference set for a BIB design with the parameters

$$v = 2t(2\lambda + 1) + 1, \qquad b = t[2t(2\lambda + 1) + 1], \qquad r = t(2\lambda + 1), \qquad k = 2\lambda + 1, \qquad \lambda.$$

21. If $2t(2\lambda - 1) + 1$ is a prime or a prime power and x is a primitive root of $GF[2t(2\lambda - 1) + 1]$, show that the t initial sets $(0, x^i, x^{i+2t}, \cdots, x^{i+(4\lambda-1)t})$, $i = 0$, $1, \cdots, t - 1$, form a difference set for a BIB design with the parameters

$$v = 2t(2\lambda - 1) + 1, \qquad b = t[2t(2\lambda - 1) + 1], \qquad r = 2t\lambda, \qquad k = 2\lambda, \qquad \lambda.$$

22. If $t(\lambda - 1) + 1$ is a prime or a prime power and x is a primitive root of $GF[t(\lambda - 1) + 1]$, show that the t initial sets $(0, x^i, x^{i+2t}, \cdots, x^{i+(\lambda-2)t})$, $i = 0$, $1, \cdots, t - 1$, form a difference set for a BIB design with the parameters

$$v = t(\lambda - 1) + 1, \qquad b = t[t(\lambda - 1) + 1], \qquad r = t\lambda, \qquad k = \lambda, \qquad \lambda.$$

23. If $4t - 1$ is a prime or a prime power and x is a primitive root of $GF(4t - 1)$, show that the initial sets $(0, x^0, x^2, \cdots, x^{4(t-1)})$; $(\infty, x, x^3, \cdots, x^{4t-3})$ form a difference set for a BIB design with the parameters

$$v = 4t, \qquad b = 2(4t - 1), \qquad r = 4t - 1, \qquad k = 2t, \qquad \lambda = 2t - 1.$$

24. If $2k - 1$ is a prime or a prime power and x is a primitive root of $GF(2k - 1)$, show that the four initial sets $(0, x^i, x^{i+2}, \cdots, x^{i+2k-4})$; $(\infty, x^{i+1}, x^{i+3}, \cdots, x^{i+2k-3})$, $i = 0, 1$, form a difference set for a BIB design with the parameters

$$v = 2k, \qquad b = 4(2k - 1), \qquad r = 2(2k - 1), \qquad k, \qquad \lambda = 2(k - 1).$$

25. For a semiregular group-divisible design show that $\lambda_1 < \lambda_2$.

26. Show that a regular group-divisible design is selfdual if $r^2 - v\lambda_2$ and $\lambda_1 - \lambda_2$ are coprime.

27. Show that the partial geometry (r, k, r) is a linked block design.

28. Show that the following series of PBIB designs coexist:

$$A_1 : v = \binom{2n}{2}, \qquad b = (2n - 1)(2n - 3), \qquad r = 2n - 3, \qquad k = n, \qquad \lambda_1 = 0, \qquad \lambda_2 = 1,$$

$$n_1 = 2(2n - 2), \qquad p_{11}{}^1 = 2n - 2;$$

$$A_1^* : v = (2n - 1)(2n - 3), \qquad b = \binom{2n}{2}, \qquad r = n, \qquad k = 2n - 3, \qquad \lambda_1 = 0, \qquad \lambda_2 = 1,$$

$$n_1 = 2(n - 1)^2, \qquad p_{11}{}^1 = (n - 1)^2;$$

$$A_2 : v = \binom{2n-1}{2}, \quad b = (2n-1)(2n-3), \quad r = 2n-3, \quad k = n-1, \quad \lambda_1 = 0,$$

$$\lambda_2 = 1, \quad n_1 = 2(2n-3), \quad p_{11}{}^1 = 2n-3;$$

$$A_2^* : v = (2n-1)(2n-3), \quad b = \binom{2n-1}{2}, \quad r = n-1, \quad k = 2n-3, \quad \lambda_1 = 1,$$

$$\lambda_2 = \lambda_3 = 0, \quad n_1 = 2(n-1)(n-2), \quad n_2 = 2(n-2), \quad p_{11}{}^1 = (n-2)^2,$$

$$p_{12}{}^1 = n-3, \quad p_{22}{}^1 = 0, \quad p_{11}{}^2 = (n-1)(n-3), \quad p_{12}{}^2 = 0, \quad p_{22}{}^2 = 2n-5,$$

$$p_{11}{}^3 = (n-2)^2, \quad p_{12}{}^3 = n-2, \quad p_{22}{}^3 = 0.$$

29. If there exists a two-class association scheme with the parameters $n_1, n_2, p_{jk}{}^i$ $(i, j, k = 1, 2)$ such that either (a) $p_{11}{}^1 = p_{11}{}^2 = \lambda$ or (b) $p_{22}{}^1 = p_{22}{}^2 = \lambda$, show that we can always construct a symmetrical BIB design with the parameters $v, r = n_1$, $\lambda = \lambda$ or $v, r = n_2, \lambda = \lambda$, depending on whether condition (a) or (b) is satisfied.

30. Using the above exercise, prove that the existence of $s - 2$ MOLS of order $2s$ implies the existence of a symmetrical BIB design with the parameters

$$v = 4s^2 = b, \quad r = s(2s-1) = k, \quad \lambda = s(s-1).$$

31. Derive the conditions under which each set of the rectangular design has the same number of symbols from (a) each row and (b) each column of the association scheme.

32. Show that no rectangular design is a linked block design.

33. If N is the incidence matrix of a BIB design, show that $N \times N \times N$, where $\times$ denotes the Kronecker product of matrices, is the incidence matrix of a cubic design. Derive the parameters of the cubic design in terms of the parameters of the BIB design.

34. Show that the necessary conditions for the existence of regular, symmetrical, rectangular designs are that

$$\rho_1{}^{\alpha_1} \rho_2{}^{\alpha_2} \rho_3{}^{\alpha_3} \sim 1$$

and, if the above is satisfied, then

$$\left\{ \prod_{i=1}^{3} (-1, \rho_i)_p^{z_i(\alpha_i + 1)/2} \right\} (\rho_1, \rho_2)_p{}^{z_3} (\rho_1 \rho_2, -1)_p{}^{z_3} (\rho_3, vn^m m^n)_p (\rho_1, vn^m)_p (\rho_2, vm^n)_p = 1$$

for all primes p, where

$$\rho_1 = r - \lambda_2 + (n-1)(\lambda_1 - \lambda_3), \quad \rho_2 = r - \lambda_1 + (m-1)(\lambda_2 - \lambda_3),$$

$$\rho_3 = r - \lambda_1 - \lambda_2 + \lambda_3,$$

$$\alpha_1 = m-1, \quad \alpha_2 = n-1, \quad \alpha_3 = (m-1)(n-1).$$

35. Show that the necessary conditions for the existence of regular, symmetrical, cubic designs in $v = s^3$ symbols are that

$$\rho_1{}^{\alpha_1} \rho_2{}^{\alpha_2} \rho_3{}^{\alpha_3} \sim 1$$

and, if the above is satisfied, then

$$\left\{\prod_{i=1}^{3}(-1, \ \rho_i)_p^{z_i(\alpha_i+1)/2}\right\}(\rho_1, \ \rho_2)_p^{z_1z_2}(\rho_1, \ \rho_3)_p^{z_1z_3}(\rho_2, \ \rho_3)_p^{z_2z_3}(s, \ \rho_1)_p(s, \ \rho_2\rho_3)_p^{s-1}=+1$$

for all primes p, where

$$\rho_1 = r + (2s-3)\lambda_1 + (s-1)(s-3)\lambda_2 - (s-1)^2\lambda_3,$$
$$\rho_2 = r + (s-3)\lambda_1 - (2s-3)\lambda_2 + (s-1)\lambda_3, \ \rho_3 = r - 3\lambda_1 + 3\lambda_2 - \lambda_3,$$
$$\alpha_1 = 3(s-1), \qquad \alpha_2 = 3(s-1)^2, \qquad \alpha_3 = (s-1)^3.$$

36. Show that the necessary conditions for the existence of regular, symmetrical, right-angular designs are that

$$\rho_1\rho_3^{l-1}\rho_4^{l-1} \sim 1$$

and if the above is satisfied, then

$$(-1, \ \rho_1)_p(-1, \ \rho_3\rho_4)_p^{l(l-1)/2}(-1, \ \rho_2)_p^{l(s-1)}(\rho_3\rho_4, \ l(2s)^{l-1})_p^{l-2}(\rho_3^{l-1}, \ \rho_4^{l-1}l(2s)^{l-1})_p$$

$$(l(2s)^{l-1}, \ \rho_4^{l-1})_p = +1$$

for all primes p, where

$$\rho_1 = r - \lambda_1 + s(\lambda_1 - \lambda_2) + s(l-1)(\lambda_3 - \lambda_4),$$
$$\rho_2 = r - \lambda_1, \ \rho_3 = r - \lambda_1 + s(\lambda_1 + \lambda_2 - \lambda_3 - \lambda_4),$$
$$\rho_4 = r - \lambda_1 + s(\lambda_1 - \lambda_2 - \lambda_3 + \lambda_4).$$

37. Show that in any affine resolvable equiblock-sized incomplete block design k^2/v is an integer.

38. Construct all the confounded designs for the types of confounding given on page 261 of O. Kempthorne's *Design and Analysis of Experiments* (Wiley, New York, 1952).

39. If X is a symmetrical, nonsingular spring-balance weighing design, show that the estimates of the weights obtainable with the assumption that the error variance in each weighing is proportional to the total weights of the objects in that weighing are same as the usual estimates obtainable with the equal-error-variance assumption.

40. Let $A = (a_{ij})$ be a $v \times v$ matrix of integers and let $k(k-1) = \lambda(v-1)$ and $AA' = A'A = (k-\lambda)I_v + \lambda E_{v,v}$. Then show that A or $-A$ is the incidence matrix of a symmetrical BIB design.

41. A set of words {READ, ACCEPT, DO, CALL, PAUSE, STOP} is to be transmitted to a computer by representing each word by one of the letters in that word, such that the words will be represented uniquely. Discuss the possibility of a solution in the light of a system of distinct representatives.

42. In a symmetrical BIB design with the parameters $v = b = 4t + 3, r = k = 2t + 1, \lambda = t$ show that a subset of t symbols cannot occur in four sets. Furthermore, if t is even, then a subset of t symbols cannot occur in three sets.

43. Show that if

$$N = \begin{bmatrix} E_{1,r} & 0_{1,b-r} \\ N_1 & N_2 \end{bmatrix}$$

is the incidence matrix of a symmetrical BIB design with the parameters $v = b = 4t + 3$, $r = k = 2t + 1$, $\lambda = t$, then

$$N^* = \begin{bmatrix} E_{1,r} & 0_{1,b-r} \\ N_1 & E_{v-1,b-r} - N_2 \end{bmatrix}$$

is the incidence matrix of a symmetrical BIB design with the same parameters.

44. Show that the method of block section and block intersection can be applied to certain PBIB designs.

45. Show that the parameters of cubic and rectangular association schemes uniquely define the corresponding association schemes.

Index